Pervasive Haptics

Hiroyuki Kajimoto • Satoshi Saga • Masashi Konyo
Editors

Pervasive Haptics

Science, Design, and Application

Springer

Editors
Hiroyuki Kajimoto
The University of Electro-Communications
Tokyo, Japan

Satoshi Saga
University of Tsukuba
Tsukuba, Japan

Masashi Konyo
Tohoku University
Sendai, Japan

ISBN 978-4-431-55771-5 ISBN 978-4-431-55772-2 (eBook)
DOI 10.1007/978-4-431-55772-2

Library of Congress Control Number: 2016936574

Springer Tokyo Heidelberg New York Dordrecht London

Printed on acid-free paper

Springer Japan KK is part of Springer Science+Business Media (www.springer.com)

Preface

We mainly rely on visual and auditory channels to understand information about a surrounding space. Recently, however, a third, haptic channel has drawn the attention of researchers. Whereas visual and auditory information require a process of understanding, haptic information is considered more "intuitive", i.e., the information is believable, or can be directly linked to the haptic action. This feature of haptics leads to diverse application areas such as medicine, rehabilitation, sensory substitution, navigation, education, and gaming.

Haptics is a relatively immature and rapidly growing research field. The brain processes involved are only partially understood, and many sensory illusions have recently been found. Novel tactile displays and sensors based on new principles are proposed on a yearly basis, leading to the expansion of applications.

This book presents the state of the art of diverse haptics-related research. The book is structured in four parts.

Part I comprises three chapters on the fundamentals of haptics including the dimensionality of tactile sensation, brain processes, and force perception. These fundamentals constitute the basis of haptics research, and the chapters provide readers with the background of this field.

Part II comprises eight chapters on the component design of haptic devices, including tactile display using the micro-vibration of shape–memory alloy wires, electrical stimulation, ultrasonic vibration, lateral-force-based design, airborne ultrasonic tactile display, thermal display, and force sensors. The chapters provide readers with state-of-the-art research on the basic principles of haptic devices.

Part III comprises four chapters on the practical application of haptics, including education through touch, tactile scoring in therapy, sensory substitution, and industrial and medical applications. Most chapters include feedback from field studies or workshops, and illustrate the bridge between academic work and industry.

Part IV comprises four chapters on possible application of haptics. This part is presented separately from Part III because the application of haptics has not yet been fully explored, and fundamental considerations are needed to find new applications. Readers, with guidance from the authors, are expected to propose potential applications of haptics.

This book is a resource for not only active haptic researchers but also general readers wishing to gain an understanding of this interdisciplinary area of science, technology, and design.

Finally, the editors would like to thank all the authors for their contributions to this book. It is our great pleasure to work with them in this era of pervasive haptics.

Tokyo, Japan — Hiroyuki Kajimoto
Tsukuba, Japan — Satoshi Saga
Sendai, Japan — Masashi Konyo
July 2015

Contents

Part I Fundamentals of Hapitcs

1 **Psychophysical Dimensions of Material Perception and Methods to Specify Textural Space** 3
Shogo Okamoto, Hikaru Nagano, and Hsin-Ni Ho

2 **The Brain Network for Haptic Object Recogniton** 21
Ryo Kitada

3 **Understanding Force Perception Characteristics of a Human and Its Applications** 39
Yuichi Kurita

Part II Component Design

4 **Tactile Display Using the Micro-vibration of Shape-Memory Alloy Wires and Its Application to Tactile Interaction Systems** 57
Hideyuki Sawada

5 **Electro-tactile Display: Principle and Hardware** 79
Hiroyuki Kajimoto

6 **Solid Ultrasonics Tactile Displays** 97
Masaya Takasaki

7 **Lateral-Force-Based Tactile Display** 111
Satoshi Saga

8 **Airborne Ultrasound Tactile Display** 121
Takayuki Hoshi and Hiroyuki Shinoda

9 **Tactile Sensing Techniques That Use Intrinsic Force Sensors** 139
Toshiaki Tsuji

10 **Reflection-Image-Based Tactile Sensor** ... 157
Satoshi Saga

11 **Thermal Displays and Sensors** ... 167
Katsunari Sato

Part III Solid Application

12 **TECHTILE Workshop for Creating Haptic Content** ... 185
Masashi Nakatani, Yasuaki Kakehi, Kouta Minamizawa, Soichiro Mihara, and Susumu Tachi

13 **Computational Aesthetics: From Tactile Score to Sensory Language** ... 201
Yasuhiro Suzuki and Rieko Suzuki

14 **Haptic Aids for the Visually Impaired** ... 221
Han SungMin

15 **Simple Tactile Technologies Utilizing Human Tactile and Haptic Characteristics** ... 231
Yoshihiro Tanaka and Akihito Sano

Part IV Search for New Fields

16 **Toward the Haptic Interaction in Daily Life** ... 249
Takuya Nojima

17 **Haptic Interfaces That Induce Motion and Emotion** ... 265
Hiroyuki Kajimoto

18 **Bidirectionality of Haptics** ... 275
Satoshi Saga

19 **Remote Transmission of Multiple Tactile Properties** ... 285
Masashi Konyo

Index ... 305

Part I
Fundamentals of Hapitcs

Chapter 1
Psychophysical Dimensions of Material Perception and Methods to Specify Textural Space

Shogo Okamoto, Hikaru Nagano, and Hsin-Ni Ho

Abstract This chapter explains the five types of perceptual dimensions in tactile perception of materials: namely, coarse and fine roughness, softness (hardness), warmness (coldness), and friction percepts. How these dimensions are specified is discussed, and the perceptual mechanisms of each dimension are outlined. Furthermore, experimental and analytical methods to specify these perceptual dimensions are introduced. Primarily, two types of analyses, factor analysis and multidimensional scaling, are described with appropriate experiments for data collection.

Keywords Texture • Roughness • Softness • Hardness • Friction • Warmness • Coldness • Perceptual dimension

1.1 Introduction

Human perceptual and affective experiences in touching products or materials are expressed by a semantically multilayered and multidimensional information space, as shown in Fig. 1.1. As discussed in the literature of perceptual and emotional responses of touch [15, 26, 53], the most primal layer is the psychophysical layer that is composed of the percepts of physical properties, comprising surface roughness and pliability of materials. Because this layer is regarded as an interface between human responses and physical stimuli, it is called the psychophysical layer or commonly the layer of texture. The higher layers are more cognitive and affective; they are mediated not only by physical stimuli but also personal background and the knowledge of objects to be touched. The preferential layer, which is composed of hedonic expressions, is more individual than the lower layers.

S. Okamoto (✉) • H. Nagano
Department of Mechanical Science and Engineering, Nagoya University, Nagoya, Japan
e-mail: okamoto-shogo@mech.nagoya-u.ac.jp

H.-N. Ho
NTT Communication Science Laboratories, Nippon Telegraph and Telephone Corporation, Tokyo, Japan

H. Kajimoto et al. (eds.), *Pervasive Haptics*, DOI 10.1007/978-4-431-55772-2_1

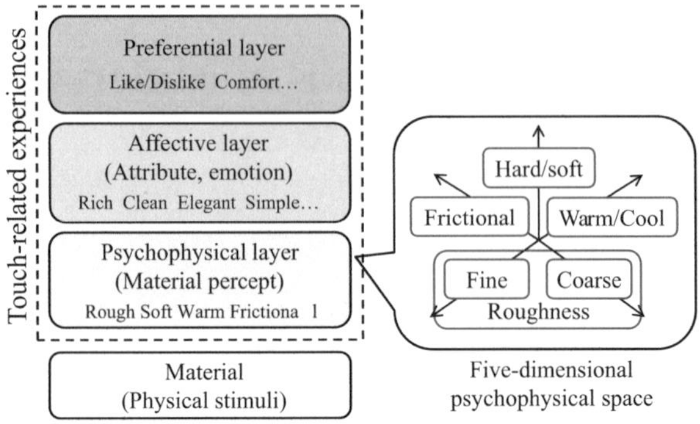

Fig. 1.1 Multilayered model of touch-related experiences. The *bottom layer* is composed of the percepts of physical properties of materials. The *middle layer* includes perceived attributes and emotional experiences. The *highest layer* pertains to personal preferences. The *lower layers* are more common among all people, whereas the *higher layers* are more individual

In this chapter, the structure of the psychophysical layer (texture) is discussed and explained in terms of their perceptual independence and mechanisms. The investigation of earlier studies on the textural dimensions leads us to a reasonable deduction that the space mainly comprises five types of perceptual dimensions [55]. The five perceptual dimensions are the percepts of fine roughness, coarse roughness, softness, warmness, and friction. We summarize the perceptual mechanisms of each dimension based on the psychophysics, contact mechanics, and neurophysiology in the field. Given that the texture is expressed in the multidimensional information space, we also introduce experimental and analytical methods to specify the dimensions that construct the textural space. The weakness of these methods are discussed, as well as their strength.

1.2 Five Types of Psychophysical Dimensions of Texture

1.2.1 Related Studies on Texture Dimensions

Thus far, the dimensional structure of tactile textures attracted the interest of many researchers. However, their conclusions are not necessarily consistent with each other. All studies differ in terms of materials, criteria (adjectives), and the modality involved in the research, which results in inconsistency between the studies.

The majority of the inconsistency may result from the differences in material samples and adjectives (criteria) used for judgment. Some studies used very limited number of samples that are insufficient to cover the wide aspects of human perception. Even if the number of material samples is adequate, an imbalanced sample set leads to an imbalanced result. Difference in the polarity of adjectives must be also considered. The majority of the studies used adjectival dyads such as "rough-smooth" and "hard-soft" for rating tasks. However, in some studies, unipolar descriptors were used and semantically opposite adjectives were not paired. The difference in modalities involved in psychological investigations is also a priority. In some studies, experimental participants were allowed to see their samples, whereas they were not in other studies. Furthermore, each study adopted different definitions of texture. The strictest definition of texture is the perception of topographic characteristics of the surface that do not include the senses of warmth or softness. Additionally, in some studies, the bending stiffness of fabrics or papers might have implicitly influenced the texture when they were actively held and explored by subjects. In contrast, in other studies, the material samples were attached to flat boards or the subjects' hand motions were controlled.

Table 1.1 summarizes the reports of related studies on textural dimensions. Because all studies used different but similar terms or descriptors, they are slightly changed from the original literature for consistency. Unipolar descriptors were also interpreted such that each dimension would become bipolar. Most studies specified the textural space as two- or three-dimensional. Clearly, they do not completely agree with each other. An important thing that we should try to accomplish is to reach a reasonable conclusion by conducting an overview of earlier studies. In the following part, these related studies are comprehensively discussed in order to rearrange the textural dimensions.

1.2.2 Sort-Out of Five Dimensions

Table 1.2 shows the number of dimensions reported in the abovementioned 20 studies. Based on these reports and the following aspects, the dimensionality of texture can be inferred.

- Rough-smooth, hard-soft, and warm-cold dimensions were robustly found under many different research conditions.
- Two types of roughness perceptions, namely coarse and fine roughness, were reported in some studies. Coarse roughness is described as voluminous, uneven, lumpy, coarse, and relief. This is distinguished from fine roughness that is typically described as harsh or rough. Furthermore, the perceptual mechanisms of these two types of roughness are clearly different, as described in Sect. 1.3.1.
- Percepts of roughness and friction were separately identified in many studies. These two types of percepts tend to be considered strongly linked with each other because friction is physically connected with its surface roughness. However, their perceptual mechanisms are different as described later.
- Moist-dry and sticky-slippery are likely to belong to the same (or very similar) dimension pertaining to friction. They have never been extracted as separate dimensions in single studies.

From these viewpoints, the textural space is conjectured to comprise five types of dimensions: fine roughness, coarse roughness, warmness (coldness), hardness (softness), and friction percepts. This classification also agrees with psychological and neurophysiological aspects [5, 7, 38].

Table 1.1 Early studies on perceptual dimensions of tactile textures in order of publication year

Literature	Dimension 1	Dimension 2	Dimension 3	Dimension 4
Howorth [35]	Rough/smooth, Hard/soft	Warm/cold		
Yoshida [81]	Hard/soft Warm/cold, Rough/smooth	Moist/dry, Rough/smooth	Hard/soft	
Lyne [47]	Hard/soft	Embossed (Roughness)		
Hollins [32]	Rough/smooth, Warm/cold, Sticky/slippery,	Hard/soft	Not specified (Stiff)	
Hollins [30]	Rough/smooth	Hard/soft	Sticky/slippery	
Tamura [73]	Rough/smooth Hard/soft	Warm/cold	Moist/dry	
Picard [58]	Hard/soft, Rough (Fine roughness)	Relief (Coarse roughness)	Hard/soft	
Picard [59]	Hard/soft	Rough/smooth		
Soufflet [67]	Rough/smooth Hard/soft	Warm/cold		
Shirado [63]	Rough/smooth	Warm/cold	Moist/dry	Hard/soft
Ballesteros [2, 3]	Rough/smooth	Hard/soft	Sticky/slippery	
Gescheider [24]	Coarse roughness	Rough/smooth	Fine roughness	
Bergmann Tiest [8]	Hard/soft	Rough/smooth	Not named	Not named
Tanaka [75]	Moist/dry	Hard/soft Rough/smooth	Warm/cold	
Yoshioka [82]	Hard/soft	Rough/smooth	Sticky/slippery	
Summers [72]	Rough/smooth			
Guest [27]	Sticky/slippery	Rough/smooth	Oily	
Guest [26]	Rough/smooth	Moist/dry	Hard/soft	
Baumgartner [4]	Rough/smooth Sticky/slippery	Hard/soft Warm/cold		
Ackerley [1]	Rough/smooth Hard/soft	Sticky/slippery (Moisture)	Warm/cold	

Table 1.2 Number of textural dimensions reported in 20 earlier studies

Rough/smooth		Hard/soft	Warm/cold	Frictional	
20		17	9	12	
Coarse	Fine			Moist/dry	Sticky/slipp.
2	2			6	6

1.3 Perceptual Mechanisms of Individual Dimensions

The perceptual mechanisms of each textural dimension have been investigated by many researchers. Here, they are outlined from psychophysical, contact-mechanical, and neurophysiological aspects as shown in Table 1.3; other comprehensive studies [5, 7, 38] reviewing such mechanisms are also recommended for reference.

1.3.1 Roughness

1.3.1.1 Coexistence of Two Types of Roughness

The percepts of surface roughness are mediated by two mechanisms that depend on the degree of surface roughness. In the case of grating scales and dotted scales, which have been frequently used for investigating the roughness percept, the width of two neighboring dots or ridges becomes a criterion to separate the two mechanisms. When it is larger than the range of a few hundreds of micrometers to 1 mm, the surface roughness is called coarse or macroscopic roughness. In contrary, a surface roughness with a value smaller than this range is called fine roughness. This threshold range is compared with the width of finger print ridges [21].

Although the perceptual mechanisms of the coarse and fine roughness clearly differ, not many studies on textural dimensions have reported them independently as described in the above section. This is partly because these two types of

Table 1.3 Summary of perceptual mechanisms of five textural dimensions

Dimension	Exploratory motion	Principle	Receptors	Physical dominance
Coarse roughness	Push	Pressure distribution	SA units	Surface roughness
Fine roughness	Rub	Skin vibration	FA units	Surface roughness
Softness	Push	Contact area/ Pressure distribution	SA units (FA units)	Elasticity, stiffness
Hardness	Tap	Vibration	FA units	Mechanical impedance
Friction	Rub	Skin stretch, Frictional vibration	–	Friction
Warmness/ coldness	Push, Soft touch	Heat transfer	Free nerves (TRP channels)	Temperature, Thermal properties

*SA and FA: slow and fast adaptive. TRP: transient receptor potential. Appropriate exploratory motions for each dimension are also discussed in [52]

roughness largely overlap, indicating that humans make use of the two types of perceptual mechanisms to deal with the wide range of surface roughness, and also to accurately estimate the common materials for which surface roughness lies within the overlapping range.

1.3.1.2 Coarse (Macroscopic) Roughness

Another definition of coarse roughness is that it is a type of roughness that humans can discern merely by pushing the surface with a finger without any relative motions between the finger and material. Hence, the percept of coarse roughness is presented by the pressure distribution within a contact area between the finger pad and asperities of material surface. The type of receptors that realize such perception are Merkel corpuscles, which are also known as SA I units; they exist densely in the hypodermis and respond to sustained skin deformation and pressure. Furthermore, some types of free nerve endings are also responsive to mechanical stimuli [61]. These receptors constitute dense pressure sensors spread across the skin surface and contribute to the percepts of coarse roughness [12, 18, 83]. When FA units that are other types of mechanical receptors in the skin are fully adapted, humans are virtually incapable of discriminating fine roughness, but the capacity to perceive coarse roughness remains intact [31, 33]. This also corroborates that SA and FA units govern the different types of surface roughness; coarse and fine roughness, respectively.

1.3.1.3 Fine Roughness

For the surface roughness of which asperities are finer than the density of Merkel corpuscles or the width of epidermal ridges, humans cannot discern the degree of roughness solely by the pressure information provided by pushing the surface. In this case, the temporal or vibratory information elicited by the relative motion between a finger pad and the surface contributes to the perception of roughness [6, 14, 43]. In short, fine roughness is coded by scanning the surface. In contrast, the percept of coarse roughness is scarcely influenced by such temporal information [34, 50]. The neural coding of skin vibrations is preferentially mediated by Meissner (FA I units) and Pacinian corpuscles (FA II units).

The vibration frequency of a finger pad relies on the fineness of surface asperities and relative speed of the finger. Finer roughness and faster exploration result in a higher frequency. However, mechanoreceptors in skin can respond to vibrations at a maximum of 1 kHz. For perception of finer roughness that leads to a much faster skin vibration, the degree of surface roughness (e.g., size of particles or height of asperities) rather than vibration frequency determines the roughness percepts. Miyaoka et al. [51] suggested this possibility from the evidence that humans can discriminate abrasive papers with particle sizes of several micrometers and rectangular single ridges with different heights of several micrometers with equal or comparable accuracies. This indicates that very fine roughness can be coded by the intensities of cutaneous deformations or resultant activities of vibration-sensitive mechanoreceptors.

1.3.2 Softness and Hardness

Softness percepts tend to be considered related to proprioception because the mechanical stiffness of an object is defined by the ratio of relative reaction force to relative surface displacement. However, in addition to other textural information, tactile cues play an important role [9, 69], and rather predominant over proprioceptive cues for soft materials whose Young's moduli are approximately 1 MPa [10]. A question that comes to mind is whether the object stiffness (e.g., spring constant) or material elasticity (e.g., Young's modulus) is more proximate to hardness that humans perceive. Bergmann Tiest et al. conducted an experiment where participants matched perceived softness of objects with different thickness (spring constant) and material (Young's modulus) [10]. As a result, the participants judged the perceived softness by integrating the two types of physical quantities or cutaneous and kinesthetic cues. Nonetheless, the cutaneous cue or material elasticity is perceptually dominant for softness percepts.

Skin deformation caused by the contact with materials provides their softness cues. It is not exactly specified what information gathered via skin deformation is used for softness percept; however, it is at least known that the contact area between the finger pad and material surface and pressure distribution on the contact area are deeply connected with perceived softness. Their relationships have been established decisively by psychophysical experiments using specific softness display devices [11, 23, 37, 62]. Specifically, when a finger pad is in contact with a soft material, the contact area is large and the contact load is widely spread over the area. In contrast, in the case of contact with a rigid material, the contact area is small with a large peak pressure. It is speculated that such differences in the pressure distribution contribute to the discrimination of material softness.

Regarding the receptive units related to softness percepts, slow adaptive units are considered to be potential mediators. However, during active contact between the finger pad and material surface, the skin dynamically deforms. From this viewpoint, fast adaptive units are also potentially concerned with softness perception.

As to the hardness percept, humans adopt another approach. Stiffness estimation based on finger pad deformation is not effective in the case of objects that are significantly stiffer than the finger pad. In such cases, the hardness of objects can be estimated by the damped natural vibration caused by tapping the object surface [42, 56]. The damped natural vibration with a single frequency component is characterized by its amplitude, vibration frequency, and decay coefficient. Although the mechanical impedance of an object influences all of these parameters, the most perceptually effective one is the vibration frequency. A higher frequency leads to greater perceived hardness of objects.

1.3.3 Friction

Sticky-slippery or moist-dry factors were found in some of the earlier studies as described above. These two factors are considered to be derived from the same factor because they have not been extracted together in a single study. Provided that both these factors are pertinent to friction, we comprehensively regard them as a single factor related to friction.

Roughness and friction percepts are sometimes considered to be dependent on each other [15, 49, 74] because the surface roughness physically influences its friction coefficients. In some studies, friction and roughness percepts were found to be correlated [20, 21, 65, 66]. However, roughness and friction percepts are actually independent. In a study using roughness samples with and without lubrication, the roughness percepts were not affected by the difference in friction [76]. Furthermore, in an experiment involving coated papers, the perceived roughness was negatively correlated with their friction coefficients [64]. Coated paper with larger surface roughness leads to a smaller real contact area with a finger pad, which causes weaker adhesion. As a result, its roughness perception negatively correlates with its friction perception. In many earlier studies on textural dimensions described in the previous section, the roughness and friction factors were separately extracted, and they were conjectured to be perceptually independent.

As a further classification of friction percepts, Guest et al. [27] suggested the separation of the percepts of watery and oily fluids from the results of experiments using liquid on a flat plate. The difference between these two types of percepts was attributed to frictional vibration caused by the repetition of stuck and slippery states between the finger pad and explored surfaces. Oily fluid is slippery with little frictional vibration whereas the purer water more frequently elicits such vibrations [54].

Because friction is a kind of force, a friction percept tends to be classified as proprioceptive; however, cutaneous cues are also a significant factor. In one report, when the force of 0.3 N was applied to a finger pad, cutaneous and proprioceptive cues equally contributed to the force perception [48]. For the perception of even smaller forces, cutaneous cues become more dominant. Although the perceptual mechanisms of friction remain to be studied, quasi-static shear deformation of finger pad is certainly used for the estimation of friction [60, 84]. The finger pad is deformed along its shear direction owing to the friction or interaction force during the exploration of material surfaces by a finger. Furthermore, the dynamic shear deformation of finger pad caused by the difference between static and kinetic friction coefficients is likely to provide human with frictional information of materials. Frictional vibrations caused by stick-slip phenomena between the finger and material present some aspects of friction [39, 54]. Additionally, when the finger pad transfers from a stuck state to slipping state, the contact area of finger pad rapidly decreases [77]. The friction-related dynamism of the finger pad has an intimate relationship with human percepts of friction. However, it is unknown whether humans can really estimate the static and kinetic properties of friction such as static and kinetic coefficients, based on such dynamism.

1.3.4 Warmness and Coldness

1.3.4.1 Physical Process

When the hand touches an object whose temperature is different from the skin temperature, heat transfer occurs between the two. For an object whose temperature is lower than the skin temperature, heat flows out of the skin during contact and coldness is perceived. On the other hand, for an object whose temperature is higher than the skin temperature, heat flows into the skin and warmness is perceived. In both cases, the degree of coldness or warmness is proportional to the amount of heat exchanged, which is in turn a function of the temperature difference between the skin and the object and the material properties of the object such as thermal conductivity and heat capacity.

In daily experience, we often find that a metal object feels colder than a wooden object of the same temperature. This difference comes from their difference in the material properties. The metal object feels colder because a large amount of heat flows out of the skin during contact due to its high thermal conductivity and heat capacity. The wooden object does not feel as cold because only a small amount of heat flows out of the skin during contact due to its relatively low thermal conductivity and heat capacity. In the case that the temperatures of a metal object and a wooden object are the same but higher than the skin temperature, the metal object would feel warmer than the wooden object because a larger amount of heat is exchanged with the metal object.

Several models have been proposed to predict the thermal interaction between the skin and an object during contact (for review, see [36]). Among them, Ho and Jones [28] proposed that the contact coefficient, which is the square root of thermal conductivity and heat capacity, can be used as an index to predict how much heat would be exchanged during contact. They also showed that people are able to discriminate between materials based on thermal cues, provided that the ratio of the contact coefficients of the materials is larger than 3. In the case of metal and wood, the contact coefficient of metal is 10 times larger than that of the wood, so people can discriminate between them reliably based on the difference in the perceived coldness.

1.3.4.2 Mechanisms of Temperature Sensing

Changes in skin temperature are encoded by warm and cold receptors. They are free nerve endings, with the warm receptors innervated by unmyelinated C fibers and the cold receptors innervated by small myelinated Aδ fibers. Because of the difference in fiber type, the conduction velocity of cold receptors is much faster (5–30 m/s) than that of warm receptors (0.5–2 m/s) [13]. The warm receptors respond to a temperature range between 30–50 °C with peak intensities around 45 °C. Cold receptors respond to a temperature range between 5–43 °C with peak intensities between 23–28 °C [19, 68]. In the neutral thermal zone between 30 and 36 °C, both warm and cold receptors discharge spontaneously at low rates and no thermal sensation is noted. When the skin temperature exceeds 45 °C or falls below 15 °C, responses from the nocioceptors result in the perception of pain.

The processes involved in converting thermal stimuli into electrical and chemical signals are mediated by the ion channels expressed in the thermoreceptors or the skin cells, called thermo-transient receptor potentials (thermoTRPs). To date, six ThermoTRP channels have been identified, which include four heat-activated channels and two cold-activated channels. Among heat-activated channels, TRPV1 and 2 are activated by painful levels of heat, while TRPV3 and 4 respond to non-painful warmth. Among the two cold-activated channels, TRPM8 is activated by non-painful cool temperatures, while TRPA1 responds to painful cold (for review, see [57]).

1.3.4.3 Psychophysical Characteristics of Temperature Perception

Humans can only perceive a thermal stimulus when it exceeds some specific threshold and this threshold depends on the stimulated site and whether the stimulus is warm or cold. In general the face is the most sensitive region and the extremities are the least sensitive. The sensitivity of the finger is in between and the warm and cold thresholds are about 0.5 and 0.3 °C, respectively [71]. Human thermal perception has been shown to have good spatial summation and poor localization for thermal stimuli at low intensities [70]. This is hardly noticed in the daily experience because concurrent tactile inputs can facilitate thermal localization. For example, when the hand makes contact with an object, the change in skin temperature and the deformation of the skin activate thermoreceptors and mechanoreceptors located in the skin. The cross-modal processing of the thermal and tactile inputs influences the localization and the perceived intensity of resulting thermal sensations [25, 29]. The time required to process a thermal stimulus depends on the response required. The simple reaction times to warming and cooling stimuli presented to the hand have been shown to be about 940 and 530 ms, respectively [22]. When the task is to respond to the thermal properties of a material based on the perceived coldness, it takes longer than the simple reaction time. It has been shown that it takes about 900 ms to discriminate a copper item from a wooded item, and the time increases with the number of the wooden items presented [44]. The time spent is considerably longer than the times required to respond to tactile properties. For example, it only took in average 400–500 ms to discriminating a hard item from a soft item.

1.4 Methods to Specify Dimensions of Material Perception

There are many variations in the methods available to specify the perceptual dimensions of tactile textures. Most of them are based on two types of multivariate

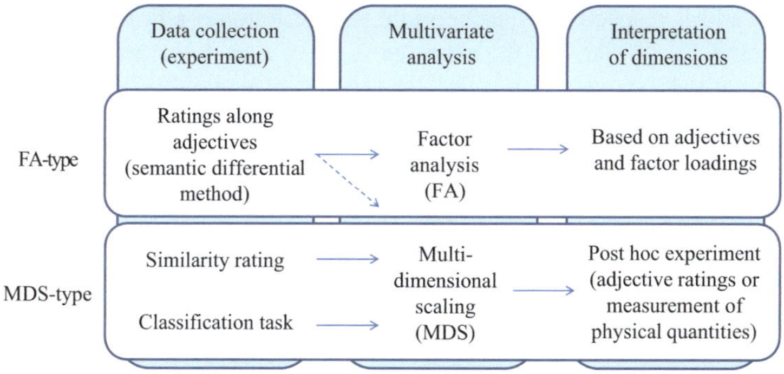

Fig. 1.2 Major approaches to specify textural dimensions. Methods are categorized into two approaches regarding multivariate analysis: factor analysis and multidimensional scaling. There are also three major experimental methods: semantic differential method, similarity rating, and classification task

analyses: factor analysis and multidimensional scaling, as shown in Fig. 1.2. In psychological experiments, subjective data must be collected such that the data conforms to either or both of these analyses. Here, these analyses and psychological experiments suitable to them are outlined.

1.4.1 Methods Based on Factor Analysis

Factor analysis, which is based on the eigenvalue decomposition of symmetric matrices in principle, constructs m types of components that are statistically orthogonal to each other from p types of variables that could co-vary ($m < p$). Each of the orthogonal components is considered as a dimensional axis of the space. Factor analysis is typically combined with a semantic differential method, in which a number of texture samples are rated one by one with adjective dyads as the criteria. For example, each material is judged in terms of multiple dyads including "rough/smooth" and "cold/warm."

For material sample i ($i = 1, \ldots, n$), the rating of a certain adjective dyad k ($k = 1, \ldots, p$) is assumed to be a linear combination of m types of latent factors and is expressed as

$$\begin{aligned} x_{ki} &= \mu_k + a_{k1} f_{i1} + \cdots + a_{km} f_{im} + e_{ki} \\ &= \mu_k + \boldsymbol{a}_k^T \boldsymbol{f}_i + e_{ki} \end{aligned}$$

where e_{ki} is a unique factor that is specific to x_{ki}, whereas a (factor loadings) is shared with x. $\boldsymbol{f}_i$ is the vector of factor score that corresponds to the coordinates of material i on the m-dimensional space. The average and variance of each element in $\boldsymbol{f}_i$ are 0 and 1, respectively. Furthermore, as restriction conditions, $e_i \perp e_j$ $(i, j = 1, \ldots, n)$ and $e \perp f$ hold, and the expected value of e_i is 0. For p types of adjective dyads, using p-variate vector $(\boldsymbol{x}_i \in \mathbb{R}^p)$, the scores of material i is noted as

$$\boldsymbol{x}_i = \boldsymbol{\mu} + \boldsymbol{A}\boldsymbol{f}_i + \boldsymbol{e}_i$$

where $\boldsymbol{\mu} = (\mu_1, \ldots, \mu_p)^T$, $\boldsymbol{A} = (\boldsymbol{a}_1, \ldots, \boldsymbol{a}_p)^T$, and $\boldsymbol{e}_i = (e_1, \ldots, e_p)^T$ are the average ratings of adjective scales, matrix of factor loadings, and uniqueness of material i, respectively. $\boldsymbol{A}$ comprises the structures of independent factors and is given by solving

$$\boldsymbol{\Sigma} = \boldsymbol{A}\boldsymbol{A}' + \boldsymbol{D}$$

where $\boldsymbol{\Sigma} \in \mathbb{R}^{p \times p}$ and $\boldsymbol{D} = \mathrm{diag}(\mathrm{Var}(e_1), \ldots, \mathrm{Var}(e_p))$ are the covariance matrices of x_{ki} and e_{ki}, respectively. In order to facilitate the interpretation of $\boldsymbol{A}$, rotational operation is usually applied to $\boldsymbol{A}$ and $\boldsymbol{f}_i$.

A benefit of the method based on a factor analysis is that it is easy to interpret the meanings of perceptual dimensions acquired through the analysis because $\boldsymbol{A}$ is mediated by adjective words. This also means that it does not require some post hoc experiments to clarify the dimensionality. In contrast, a caveat is imposed on the selection of adjective dyads; they should cover all of the aspects of the set of material samples.

1.4.2 Methods Based on Multidimensional Scaling

The multidimensional scaling method specifies the coordinates of events on the m-dimensional space using the distance matrix of n events. Depending on the level of distance scale and completeness of distance matrix, several methods are available. The most famous methods are Torgerson's metric method (Young-Householder's theorem) [78] and Kruskal's MDSCAL [40, 41] for ratio and ordinal scales, respectively. The latter method can deal with matrices from which some values are missing.

For the multidimensional scaling approaches, there are two major methods to collect distance data: the dissimilarity rating and the classification method. In dissimilarity rating, participants of experiments rate the dissimilarity between two material samples using grades or visual analog scales. By averaging the dissimilarity rates among the participants, the final distance values are obtained. In classification method, each participant categorizes all samples into several groups based on their subjective similarities. Samples with similar textures are labeled as belonging to the same group. For each pair of materials, the number of times the two materials in the pair are categorized into different groups is counted. From these counts, the distance between two materials is defined. When the two materials are assessed to belong to different groups by the majority of the participants, the distance between these two materials is considered to be large.

Strictly speaking, data collected through the abovementioned processes belong to ordinal scales (non-metric); however, multidimensional scaling methods for metric scales have been sometimes used for these data partly because the higher levels of scales are generally preferred. Among methods or mathematical models to establish ratio or interval scales, a similarity choice model [45, 46], an overlap model [79, 80], the constant ratio rule [16, 17], and a complete method of triads [78] are famous. However, they are hardly used in the study of textural dimensions because these methods require a larger number of trials or comparisons. In addition, there are few studies to which scales and theories should be applied in constructing the textural dimensions.

Here, a metric multidimensional scaling method is introduced, given that the distance information conforms to the ratio scale. Let $\boldsymbol{x}_i$ be the coordinate of material i on an m-dimensional space with arbitrary origin. The distance between materials i and j is noted by d_{ij}, and $d_{ij} = d_{ji}$. Through material k, the inner product of $\boldsymbol{x}_i$ and $\boldsymbol{x}_j$ is calculated by

$$\begin{aligned} z_{ij} &= (\boldsymbol{x}_i - \boldsymbol{x}_k) \cdot (\boldsymbol{x}_j - \boldsymbol{x}_k) \\ &= \frac{1}{2}(d_{ik}^2 + d_{jk}^2 - d_{ij}^2). \end{aligned}$$

Using this equation, a matrix of inner products $\boldsymbol{Z} \in \mathbb{R}^{n-1 \times n-1}$ is computed by using the distance matrix $\boldsymbol{D} \in \mathbb{R}^{n \times n}$ which contains the element d_{ij}. On the basis of the principle of eigenvalue decomposition, $\boldsymbol{Z}$ is expanded as

$$\begin{aligned} \boldsymbol{Z} &= \boldsymbol{A}\boldsymbol{\Lambda}\boldsymbol{A}^T \\ &= \boldsymbol{A}\boldsymbol{\Lambda}^{\frac{1}{2}}(\boldsymbol{A}\boldsymbol{\Lambda}^{\frac{1}{2}})^T \\ &= \boldsymbol{X}\boldsymbol{X}^T \end{aligned}$$

where $\boldsymbol{X}, \boldsymbol{A}$, and $\boldsymbol{\Lambda}$ are the matrix of x_{ij}, eigenvectors, and eigenvalues, respectively. These computations, where $\boldsymbol{x}_k$ is regarded as the origin of the space, are called Young-Householder's theorem. In Torgerson's method, the origin is set to the center of all the materials. As a result of expansion, n or $n-1$ dimensional coordinates are assigned to the materials. Considering the magnitudes of eigen values, these materials are located on the m-dimensional space ($m < n-1$).

In order to interpret the meanings of the dimensions of the information space acquired by multidimensional scaling, additional experiments are needed because the multidimensional scaling merely provides the coordinate sets for multiple materials. For this purpose, either of the two representative methods can be used. The first method is based on adjectives, where individual materials are rated according to adjectival labels. This process is same as the semantic differential method. The adjective scores and the coordinates of materials are then compared to interpret the meanings of the dimensions. For example, if the coordinates along a certain dimension show a larger correlation with rough-smooth scores, then this dimension is considered to be characterized by roughness percepts. Another method is based on the physical quantities of materials. In this method, quantities such as the average surface roughness or Young's moduli of the materials are measured and analyzed in terms of correlations with the coordinates of the materials to judge the meanings of dimensions.

1.4.3 Selection of Material Samples and Adjectives

An element shared by the two abovementioned analyses is the approximation of matrix by eigenvectors with large eigenvalues. An approximated matrix expresses the large part of sample variations. If the samples are primarily distributed along a certain dimension, then this dimension is successfully extracted. However, other dimensions along which few samples are observed are ignored. For example, when the majority of samples differ in their surface roughness and have similar compliance, the dimension of roughness percept becomes prominent whereas that of softness diminishes. The same holds true for the selection of adjectives in a semantic differential method. If adjectives relating to roughness, such as rough, uneven, coarse, voluminous, and harsh, are intensively used for the rating task, then the roughness dimension is easily found. Considering these characteristics of multivariate analyses, material samples and adjectives should be selected in a balanced manner.

1.5 Summary

In this chapter, the perceptual dimensions of tactile textures were focused on. From early studies on textural dimensions, the five types of dimensions were extracted. There are two types of roughness percepts (coarse and fine roughness), as well as softness (hardness), warmness (coldness), and friction percepts. The perceptual mechanisms of each dimension were then introduced. Although the perceptual mechanisms of each dimension have been well-studied thus far, the integration of information from multiple textural dimensions is still a work in progress. Further studies are expected to elucidate material recognition and tactile textures.

Acknowledgements The content of this chapter was in part supported by MEXT Kakenhi Shitsukan #23135514 and #25135717 and MIC SCOPE #142106003.

References

1. Ackerley, R., Saar, K., McGlone, F., Wasling, H.B.: Quantifying the sensory and emotional perception of touch: Differences between glabrous and hairy skin. Front. Behav. Neurosci. **8**(34) (2014). doi:10.3389/fnbeh.2014.00034
2. Ballesteros, S., Heller, M.A.: Haptic object identification. In: Grunwald, M. (ed.) Human Haptic Perception: Basics and Applications, pp. 212–213. Birkhauser, Basel (2008)
3. Ballesteros, S., Reales, J.M., Ponce de León, L., García, B.: The perception of ecological textures by touch: does the perceptual space change under bimodal visual and haptic exploration. In: Proceedings of IEEE World Haptics Conference, Pisa, pp. 635–638 (2005)
4. Baumgartner, E., Wiebel, C.B., Gegenfurtner, K.R.: Visual and haptic representations of material properties. Multisens. Res. **26**, 429–455 (2013)
5. Bensmaïa, S.J.: Texture from touch. Scholarpedia **4**(8), 7956 (2009)
6. Bensmaïa, S.J., Hollins, M.: The vibrations of texture. Somatosens. Motor Res. **20**(1), 33–43 (2003)
7. Bergmann Tiest, W.M.: Tactual perception of material properties. Vis. Res. **50**(24), 2775–2782 (2010)
8. Bergmann Tiest, W.M., Kappers, A.M.L.: Analysis of haptic perception of materials by multidimensional scaling and physical measurements of roughness and compressibility. Acta Psychol. **121**(1), 1–20 (2006)
9. Bergmann Tiest, W.M., Kappers, A.M.L.: Kinaesthetic and cutaneous contributions to the perception of compressibility. In: Ferre, M. (ed.) Haptics: Perception, Devices and Scenarios. Lecture Notes in Computer Science, vol. 5024, pp. 255–264. Springer, Berlin/New York (2008)
10. Bergmann Tiest, W.M., Kappers, A.M.L.: Cues for haptic perception of compliance. IEEE Trans. Haptics **2**(4), 189–199 (2009)
11. Bicchi, A., Schilingo, E.P., De Rossi, D.: Haptic discrimination of softness in teleoperation: the role of the contact area spread rate. IEEE Trans. Robot. Autom. **16**(5), 496–504 (2000)
12. Blake, D.T., Hsiao, S.S., Johnson, K.O.: Neural coding mechanisms in tactile pattern recognition: the relative contributions of slowly and rapidly adapting mechanoreceptors to perceived roughness. J. Neurosci. **17**(19), 7480–7489 (1997)
13. Brown, A.: Somatic sensation: peripheral aspects. In: Patton, H.D., Fuchs, A.F., Hille, B., Scher, A.M., and Steiner, R. (eds.) Textbook of Physiology, pp. 298–313. Saunders, Philadelphia (1989)
14. Cascio, C.J., Sathian, K.: Temporal cues contribute to tactile perception of roughness. J. Neurosci. **21**(14), 5289–5296 (2001)
15. Chen, X., Barnes, C.J., Childs, T.H.C., Henson, B., Shao, F.: Materials' tactile testing and characterization for consumer products' affective packaging design. Mater. Des. **30**, 4299–4310 (2009)
16. Clarke, F.R.: Constant ratio rule for confusion matrices in speech communication. J. Acoust. Soc. Am. **29**(6), 751–720 (1957)
17. Clarke, F.R., Anderson, C.D.: Further test of the constant ratio rule in speech communication. J. Acoust. Soc. Am. **29**(12), 1318–1230 (1957)
18. Connor, C.E., Hsiao, S.S., Phillips, J.R., Johnson, K.O.: Tactile roughness: neural codes that account for psychophysical magnitude estimates. J. Neurosci. **10**(12), 3823–3836 (1990)
19. Darian-Smith, I.: Thermal sensibility. In: Brookhart, J.M., Mountcastle, V.B., Darian-Smith, I., Geiger, S.R. (eds.) Handbook of Physiology: The Nervous System, pp. 879–913. American Physiological Society, Bethesda (1984)
20. Ekman, G., Hosman, J., Lindström, B.: Roughness, smoothness, and preference: a study of quantitative relations in individual subjects. J. Exp. Psychol. **70**(1), 18–26 (1965)
21. Fagiani, R., Massi, F., Chatelet, E., Berthier, Y., Akay, A.: Tactile perception by friction induced vibrations. Tribol. Int. **44**, 1100–1110 (2011)
22. Fowler, C., Sitzoglou, K., Ali, Z., Halonen, P.: The conduction velocities of peripheral nerve fibres conveying sensations of warming and cooling. J. Neurol. Neurosurg. Psychiatr. **51**, 1164–1170 (1988)

23. Fujita, K., Ohmori, H.: A new softness display interface by dynamic fingertip contact area control. In: Proceedings of 5th World Multiconference on Systemics, Cybernetics, and Informatics, Orlando, pp. 78–82 (2001)
24. Gescheider, G.A., Bolanowski, S.J., Greenfield, T.G., Brunette, K.E.: Perception of the tactile texture of raised-dot patterns: a multidimensional analysis. Somatosens. Mot. Res. **22**(3), 127–140 (2005)
25. Green, B.: Localization of thermal sensation: an illusion and synthetic heat. Percept. Psychophys. **22**, 331–337 (1977)
26. Guest, S., Dessirier, J.M., Mehrabyan, A., McGlone, F., Essick, G., Gescheider, G., Fontana, A., Xiong, R., Ackerley, R., Blot, K.: The development and validation of sensory and emotional scales of touch perception. Atten. Percept. Psychophys. **73**, 531–550 (2011)
27. Guest, S., Mehrabyan, A., Essick, G., Phillips, N., Hopkinson, A., Mcglone, F.: Physics and tactile perception of fluid-covered surfaces. J. Texture Stud. **43**(1), 77–93 (2012)
28. Ho, H.N., Jones, L.: Contribution of thermal cues to material discrimination and localization. Percept. Psychophys. **68**, 118–128 (2006)
29. Ho, H.N., Watanabe, J., Ando, H., Kashino, M.: Mechanisms underlying referral of thermal sensations to sites of tactile stimulation. J. Neurosci. **31**(1), 208–213 (2011)
30. Hollins, M., Bensmaïa, S.J., Karlof, K., Young, F.: Individual differences in perceptual space for tactile textures: evidence from multidimensional scaling. Percept. Psychophys. **62**(8), 1534–1544 (2000)
31. Hollins, M., Bensmaiä, S.J., Washburn, S.: Vibrotactile adaptation impairs discrimination of fine, but not coarse, textures. Somatosens. Mot. Res. **18**(4), 253–262 (2001)
32. Hollins, M., Faldowski, R., Rao, S., Young, F.: Perceptual dimensions of tactile surface texture: a multidimensional scaling analysis. Percept. Psychophys. **54**(6), 697–705 (1993)
33. Hollins, M., Fox, A., Bishop, C.: Imposed vibration influences perceived tactile smoothness. Perception **29**(12), 1455–1465 (2000)
34. Hollins, M., Rinser, S.R.: Evidence for the duplex theory of tactile texture perception. Atten. Percept. Psychophys. **62**(4), 695–705 (2000)
35. Howorth, W.S., Oliver, P.H.: The application of multiple factor analysis to the assessment of fabric hand. J. Text. Inst. Trans. **49**(11), T540–T553 (1958)
36. Jones, L., Ho, H.N.: Warm or cool, large or small? The challenge of thermal displays. IEEE Trans. Haptics **1**(1), 53–70 (2008)
37. Kimura, F., Yamamoto, A., Higuchi, T.: Development of a 2-DOF softness feeling display for tactile tele-presentation of deformable surfaces. In: Proceedings of IEEE International Conference on Robotics and Automation, Kobe, pp. 1822–1827 (2010)
38. Klatzky, R.L., Pawluk, D., Peer, A.: Haptic perception of material properties and implications for applications. Proc. IEEE **101**(9), 2081–2092 (2013)
39. Konyo, M., Yamada, H., Okamoto, S., Tadokoro, S.: Alternative display of friction represented by tactile stimulation without tangential force. In: Ferre, M. (ed.) Haptics: Perception, Devices and Scenarios. Lecture Notes in Computer Science, vol. 5024, pp. 619–629. Springer, Berlin/New York (2008)
40. Kruskal, J.B.: Multidimensional scaling by optimizing goodness of fit to a nonmetric hypothesis. Psychometrika **29**, 1–27 (1964)
41. Kruskal, J.B.: Nonmetric multidimensional scaling: a numerical method. Psychometrika **29**, 115–129 (1964)
42. Kuchenbecker, K., Fiene, J., Niemeyer, G.: Improving contact realism through event-based haptic feedback. IEEE Trans. Vis. Comput. Graphics **12**(2), 219–230 (2006)
43. Lederman, S.J.: Tactile roughness of grooved surfaces: the touching process and effects of macro-and microsurface structure. Percept. Psychophys. **16**(2), 385–395 (1974)
44. Lederman, S.J., Thorne, G., Jones, B.: Perception of texture by vision and touch: multidimensionality and intersensory integration. J. Exp. Psychol.: Human Percept. Perform. **12**, 169–180 (1986)
45. Luce, R.D.: Detection and recognition. In: Luce, R.D., Bush, R.R., Galanter, E. (eds.) Handbook of Mathematical Psychology, pp. 103–189. Wiley, New York (1963)

46. Luce, R.D.: Individual Choice Behavior: A Theoretical Analysis. Dover Publications, Mineola (2005)
47. Lyne, M.B., Whiteman, A., Donderi, D.C.: Multidimensional scaling of tissue quality. Pulp Pap. Can. **85**(10), 43–50 (1984)
48. Matsui, K., Okamoto, S., Yamada, Y.: Relative contribution ratios of skin and proprioceptive sensations in perception of force applied to fingertip. IEEE Trans. Haptics **7**(1), 78–85 (2014)
49. Matsuoka, T., Kanai, H., Tsuji, H., Shinya, T., Nishimatsu, T.: Predicting texture image of covering fabric for car seat by physical properties. J. Text. Eng. **54**(3), 63–74 (2008)
50. Meftah, E.M., Belingard, L., Chapman, C.E.: Relative effects of the spatial and temporal characteristics of scanned surfaces on human perception of tactile roughness using passive touch. Exp. Brain Res. **132**(3), 351–361 (2000)
51. Miyaoka, T., Mano, T., Ohka, M.: Mechanisms of fine-surface-texture discrimination in human tactile sensation. J. Acoust. Soc. Am. **105**, 2485–2492 (1999)
52. Nagano, H., Okamoto, S., Yamada, Y.: Haptic invitation of textures: perceptually prominent properties of materials determine human touch motions. IEEE Trans. Haptics **7**(3), 345–355 (2014)
53. Nagano, H., Okamoto, S., Yamada, Y.: Semantically layered structure of tactile textures. In: Auvray, M., Duriez, C. (eds.) Haptics: Neuroscience, Devices, Modeling, and Applications, Part I. Lecture Notes in Computer Science, vol. 8618, pp. 3–9. Springer, Berlin/Heidelberg (2014)
54. Nonomura, Y., Fujii, T., Arashi, Y., Miura, T., Maeno, T., Tashiro, K., Kamikawa, Y., Monchi, R.: Tactile impression and friction of water on human skin. Colloids Surf. B: Biointerfaces **69**, 264–267 (2009)
55. Okamoto, S., Nagano, H., Yamada, Y.: Psychophysical dimensions of tactile perception of textures. IEEE Trans. Haptics **6**(1), 81–93 (2013)
56. Okamura, A.M., Cutkosky, M.R., Dennerlein, J.T.: Reality-based models for vibration feedback in virtual environments. IEEE/ASME Trans. Mechatron. **6**(3), 245–252 (2001)
57. Patapoutian, A., Peier, A., Story, G., Viswanath, V.: ThermoTRPs and beyond: mechanisms of temperature sensation. Nat. Rev. Neurosci. **4**, 529–539 (2003)
58. Picard, D., Dacremont, C., Valentin, D., Giboreau, A.: Perceptual dimensions of tactile textures. Acta Psychol. **114**(2), 165–184 (2003)
59. Picard, D., Dacremont, G., Valentin, D., Giboreau, A.: About the salient perceptual dimensions of tactile textures space. In: S. Ballesteros, M.A. Heller (eds.) Touch, Blindness, and Neuroscience, pp. 165–174. UNED, Madrid (2004)
60. Provancher, W.R., Sylvester, N.D.: Fingerpad skin stretch increases the perception of virtual friction. IEEE Trans. Haptics **2**(4), 212–223 (2009)
61. Schepers, R.J., Ringkamp, M.: Thermoreceptors and thermosensitive afferents. Neurosci. Biobehav. Rev. **34**, 177–184 (2010)
62. Scilingo, E.P., Bianchi, M., Grioli, G., Bicchi, A.: Rendering softness: integration of kinesthetic and cutaneous information in a haptic device. IEEE Trans. Haptics **3**(2), 109–118 (2010)
63. Shirado, H., Maeno, T.: Modeling of texture perception mechanism for tactile display and sensor. J. Virtual Real. Soc. Jpn. **9**(3), 235–240 (2004)
64. Skedung, L., Danerlöv, K., Olofsson, U., Johannesson, C.M., Aikala, M., Kettle, J., Arvidsson, M., Berglund, B., Rutland, M.W.: Tactile perception: finger friction, surface roughness and perceived coarseness. Tribol. Int. **44**, 505–512 (2011)
65. Smith, A.M., Basile, G.: Roughness of simulated surfaces examined with a haptic tool: effects of spatial period, friction, and resistance amplitude. Exp. Brain Res. **202**(1), 33–43 (2010)
66. Smith, A.M., Chapman, C.E., Deslandes, M., Langlais, J.S., Thibodeau, M.P.: Role of friction and tangential force variation in the subjective scaling of tactile roughness. Exp. Brain Res. **144**(2), 211–223 (2002)
67. Soufflet, I., Calonnier, M., Dacremont, C.: A comparison between industrial experts' and novices' haptic perceptual organization: a tool to identify descriptors of the handle of fabrics. Food Qual. Prefer. **15**, 689–699 (2004)
68. Spray, D.: Cutaneous temperature receptors. Ann. Rev. Physiol. **48**, 625–638 (1986)

69. Srinivasan, M.A., LaMotte, R.H.: Tactual discrimination of softness. J. Neurophysiol. **73**(1), 88–101 (1995)
70. Stevens, J.: Thermal sensibility. In: Heller, M.A., Schiff, W. (eds.) The Psychology of Touch, pp. 61–90. Lawrence Erlbaum, Mahwah (1991)
71. Stevens, J., Choo, K.: Temperature sensitivity of the body surface over the life span. Somatosen. Mot. Res. **15**, 13–28 (1998)
72. Summers, I.R., Irwin, R.J., Brady, A.C.: Haptic discrimination of paper. In: Grunwald, M. (ed.) Human Haptic Perception: Basics and Applications, pp. 525–535. Birkhäuser, Berlin/Heidelberg (2008)
73. Tamura, K., Oyama, O., Yamada, H.: Study on feeling evaluation applied to material recognition. In: Proceedings of JSME Dynamics and Design Conference, Tokyo, no. 709 (2000)
74. Tanaka, Y., Sukigara, S.: Evaluation of "shittori" characteristic for fabrics. J. Text. Eng. **54**(3), 75–81 (2008)
75. Tanaka, Y., Tanaka, M., Chonan, S.: Development of a sensor system for measuring tactile sensation. In: Proceedings of the 2006 IEEE Sensors, Houston, pp. 554–557 (2006)
76. Taylor, M.M., Lederman, S.J.: Tactile roughness of grooved surfaces: a model and the effect of friction. Percept. Psychophys. **17**(1), 23–36 (1975)
77. Terekhov, A.V., Hayward, V.: Minimal adhesion surface area in tangentially loaded digital contacts. J. Biomech. **44**(13), 2508–2510 (2011)
78. Torgerson, W.S.: Multidimensional scaling: I. Theory and method. Psychometrika **17**, 401–419 (1952)
79. Townsend, J.T.: Alphabetic confusion: a test of models for individuals. Percept. Psychophys. **9**, 449–454 (1971)
80. Townsend, J.T., Landon, D.E.: An experimental and theoretical investigation of the constant-ratio rule and other models of visual letter confusion. J. Math. Psychol. **25**, 119–162 (1982)
81. Yoshida, M.: Dimensions of tactual impressions (1). Jpn. Psycholog. Res. **10**(3), 123–137 (1968)
82. Yoshioka, T., Bensmaïa, S.J., Craig, J.C., Hsiao, S.S.: Texture perception through direct and indirect touch: an analysis of perceptual space for tactile textures in two modes of exploration. Somatosens. Mot. Res. **24**(1–2), 53–70 (2007)
83. Yoshioka, T., Gibb, B., Dorsch, A., Hsiao, S.S., Johnson, K.O.: Neural coding mechanisms underlying perceived roughness of finely textured surfaces. J. Neurosci. **21**(17), 6905–6916 (2001)
84. Zigler, M.J.: An experimental study of the perception of stickiness. Am. J. Psychol. **34**(1), 73–84 (1923)

Chapter 2
The Brain Network for Haptic Object Recognition

Ryo Kitada

Abstract Humans can haptically identify common three-dimensional objects surprisingly well. What are the neural mechanisms underlying this ability? Previous neuroimaging studies have shown that haptic object recognition involves a distributed network of brain regions beyond the conventional somatosensory cortices. However, the relative contributions of these regions to haptic object recognition are not well understood. In this chapter, I discuss three key hypotheses concerning the brain network underlying haptic object processing and its interaction with visual object processing. The first is that the occipito-temporal cortex, which has been considered to be part of the conventional visual cortex, plays a critical role in the haptic identification of common objects. The second is that distinct brain regions are involved in the haptic processing of two types of feature used for object identification: macro-geometric (e.g., shape) and material (e.g., roughness) properties. The third is that different brain regions are also involved in the visuo-haptic interaction of macro-geometric and material properties. Finally, I discuss some issues that remain to be addressed in future studies.

Keywords fMRI • Neuroimaging • Object recognition • Tactile • Touch

2.1 Introduction

In daily life, we frequently touch and manipulate objects such as keyboards, cups, and coins. When we put our hands inside a bag to find a wallet, it is relatively simple to identify, even in the absence of vision. Klatzky et al. (1985) demonstrated that humans can identify around 100 common inanimate objects in the absence of vision at well over 90 % accuracy [1]. Moreover, most objects were identified within 5 s. This indicates that touch is an effective sensory channel for recognizing common

---

R. Kitada (✉)
Division of Cerebral Integration, National Institute for Physiological Sciences, Okazaki 444-8585, Japan

Department of Physiological Sciences, SOKENDAI (The Graduate University for Advanced Studies), Hayama 240-0193, Japan
e-mail: kitada@nips.ac.jp

H. Kajimoto et al. (eds.), *Pervasive Haptics*, DOI 10.1007/978-4-431-55772-2_2

three-dimensional (3D) objects. More recent studies have shown that humans are also capable of haptically recognizing animate objects, such as facial identity [2, 3], facial expressions of basic emotions [3–6], body part identity [7, 8] and body expressions [9, 10]. These findings indicate that haptic recognition allows humans to identify both animate and inanimate objects to such an extent that it could be used for interpersonal communication. This raises fundamental questions concerning the nervous systems that underlie the haptic recognition of common objects. Recent advances in neuroimaging techniques have led to an accumulation of evidence regarding these brain networks. In this chapter, I discuss the findings to date in relation to the three key hypotheses introduced above.

2.2 A Distributed Brain Network Underlies Haptic Object Recognition

When we touch a familiar object (e.g., an orange), we initially discern properties such as its shape, roughness, softness, and temperature (in this case, the object might be spherical, smooth, slightly soft, and cold). We then use this information to identify the object. Thus, the brain network underlying haptic recognition realizes both the extraction of object properties and the identification of the object based on them (Fig. 2.1b).

Neuroscience textbooks often highlight the function of the somatosensory cortices in tactile processing [11]. These are the primary regions that receive haptic inputs from peripheral receptors. They consist of the postcentral gyrus (PostCG), which contains the primary somatosensory cortex (SI), and the parietal operculum (PO), which contains the secondary somatosensory cortex (SII) (Fig. 2.1a left). Electrophysiological studies on non-human primates indicate that neurons in the SI can encode several properties of stimuli such as roughness [12, 13], orientation [14, 15], and curvature [16]. These findings indicate that the SI is involved in the extraction of object properties. However, it is not only the somatosensory cortices that are involved in such processing. It is now widely accepted that a distributed network of brain regions beyond the conventional somatosensory cortices is involved in haptic object recognition [17–20] (Fig. 2.1a). My review focuses mainly on the posterior parts of this brain network, which contain several important nodes for haptic object recognition (Fig. 2.1a left).

2.2.1 Involvement of the "Visual Cortex" in Haptic Object Recognition

The occipital cortex was once thought to be involved exclusively in the processing of visual information. In the late 1990s, the primary visual cortex was found to be active during the tactile recognition of Braille by early blind individuals [21, 22].

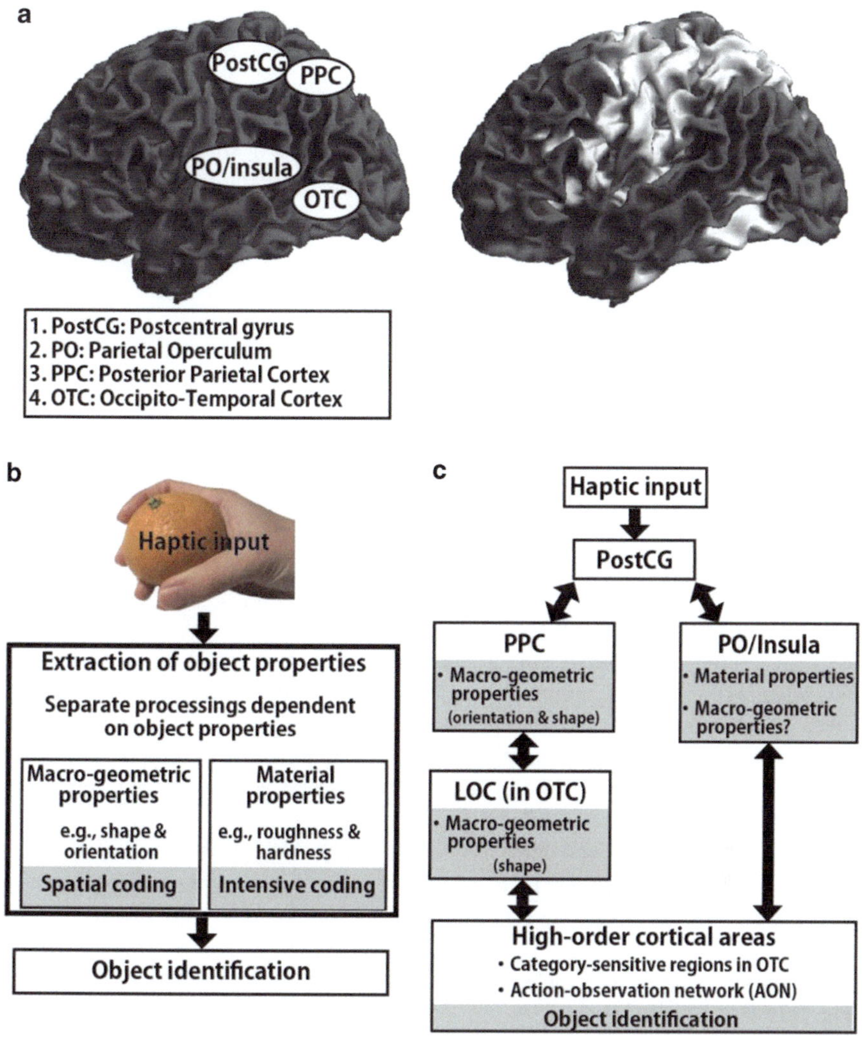

Fig. 2.1 Model of the brain network for haptic recognition of common objects. (**a**) Brain regions involved in haptic object recognition. *Left*, The regions used in the proposed model. PostCG, postcentral gyrus containing the primary somatosensory cortex (SI); PO/Insula, parietal operculum including the secondary somatosensory cortex (SII) and adjacent insula; PPC, posterior parietal cortex including the intraparietal sulcus (IPS); OTC, occipito-temporal cortex. *Right*, Brain regions that are active during haptic object recognition with the right hand (relative to rest condition) (Kitada et al. unpublished data; $Z > 2.58$ corrected for multiple comparisons at cluster level). Only the left side of the brain is shown. (**b**) Model of haptic recognition of common objects. (**c**) Model of the brain network for haptic recognition of common objects

This activation of the primary visual cortex in the early blind is now regarded as a consequence of plastic change of its functional organization due to early visual deprivation. However, subsequent studies showed that the occipital cortex in the sighted individuals was more active when paying attention to the orientation of gratings than to the spacing of gratings [23], and when haptically identifying objects relative to the judgment of perceived roughness [24]. These findings indicate that the occipital cortex, which was once considered as the "visual cortex", is also involved in haptic object processing.

2.2.1.1 Vision and Touch Share Category-Sensitivity in the Ventral Visual Pathway

The visual cortex is characterized by its unique functional organization. It consists of anatomically different structures and functionally distinct regions. For instance, neuroimaging studies have consistently shown that the occipito-temporal cortex (OTC), known as the high-order visual cortex, contains regions that are distinctively responsive to different categories of objects: face-sensitive regions, such as the occipital face area (OFA) and the fusiform face area (FFA) [25–27]; body-sensitive regions, such as the extrastriate body area (EBA) and the fusiform body area (FBA) [28, 29]; a scene-sensitive region known as the parahippocampal place area (PPA) [30]; and a word-sensitive region known as the visual word-form area (VWFA) [31]. These category-sensitive regions appear to play critical roles in the identification of objects within the corresponding groupings. Previous neuroimaging studies on sighted individuals have found that haptics and vision share the same category-sensitivity in some of these regions, such as the FFA [5, 7], EBA [7, 10, 32], PPA [33], and VWFA [34]. For instance, Kitada et al. (2009) showed that the FFA in the sighted individuals had a greater response to faces identified haptically than to other categories of objects (hands, feet, and bottles) [7]. These findings indicate that the category-sensitive regions in the OTC are critical nodes of the brain network underlying the haptic, as well as the visual, recognition of common objects (Fig. 2.1c).

2.2.1.2 Development of Functional Organization in the OTC of Early Blind and Sighted Individuals

A common criticism of these findings is that the involvement of this region is not essential for haptic object recognition. For instance, this region might be activated due to visual imagery, which could help, but not be indispensable for, haptic object recognition. To what extent is this area critical for haptic object recognition? One approach to addressing this question is to investigate the effect of visual deprivation on the development of the functional organization of the OTC. If this region is essential for the haptic recognition of common objects, its unique functional organization should develop regardless of visual experience. Alternatively, if it is not essential for haptic object recognition, its functional organization might be subject to plastic change like the primary visual cortex [21].

Previous neuroimaging studies have investigated the brain activation during haptic object recognition in early-blind individuals, who cannot recall seeing familiar objects [6, 10, 33–35]. These studies have revealed that some category-sensitive regions, such as the VWFA, PPA, and part of the EBA, develop object sensitivity in both early-blind and sighted individuals. Although it is necessary to examine the effect of visual deprivation on other category-sensitive regions (e.g., the FFA), these findings support the view that the functional organization of the OTC develops even without visual experience. Thus, the OTC, which has been considered as the high-order visual cortex, might play an essential role in haptic object recognition.

2.2.2 Are Object Properties Processed Separately in the Brain?

In order to identify the category of an object, we need to determine properties such as its shape, roughness, and softness (Fig. 2.1c). Recognizing additional properties can improve the accuracy of haptic object identification [2]. Thus, one possible mechanism underlying haptic object identification involves the OTC receiving information on the extracted object properties from other brain regions (Fig. 2.1c). Submodalities of visual information (e.g., motion, disparity, contrast, color, and orientation) are thought to be processed in a parallel-distributed manner in the brain [36]. Is the same principle applicable to touch?

Object properties can be broadly categorized as macro-geometric or material. Macro-geometric properties, such as shape and orientation, can be characterized by spatial coding that involves reference frames beyond somatotopic representation (e.g., allocentric reference frame) (spatial coding) [37]. By contrast, material properties indicate physical object information that is characterized by intensity data (intensity coding) [37]. So, are macro-geometric and material properties processed in a parallel-distributed manner in the brain?

2.2.2.1 Involvement of the Posterior Parietal Cortex in Processing Macro-geometric Properties

Previous neuroimaging studies have indicated that the posterior parietal cortex (PPC) is critical for the haptic recognition of macro-geometric properties [38–42]. Roland et al. (1998) found that the PPC was more active during discrimination of the shape of ellipsoids than the roughness of cylinders [38]. This finding was supported by subsequent studies [39, 42].

However, the difference in activation between macro-geometric and material properties in these studies can be explained by confounding factors such as variations in the stimuli (e.g., ellipsoids vs. cylinders) and in the hand movements

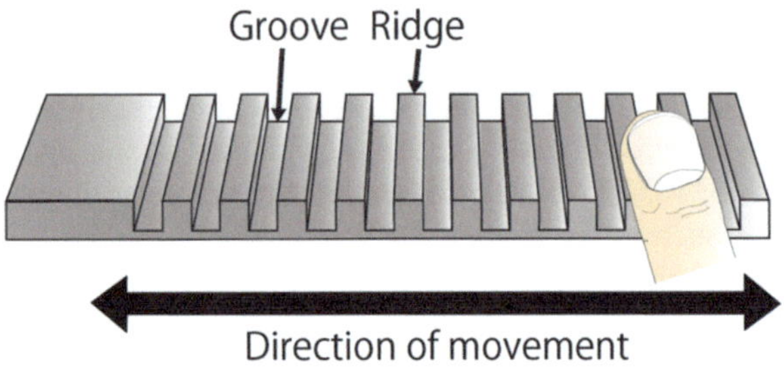

Fig. 2.2 Linear gratings. These comprise a series of ridges and grooves. Changing the groove width changes the perceived magnitude of roughness, whereas changing the orientation of ridges and grooves changes the orientation perception. Different brain activation patterns between macro-geometric and material properties can be tested by controlling factors such as hand movement and stimuli

used to extract each object property (the exploratory procedures) [43]. In order to eliminate these potentially confounding factors, Kitada et al. (2006) conducted a functional magnetic resonance imaging (fMRI) study in which the subject's finger was passively stimulated by a surface of linear gratings [40]. Roughly speaking, these comprised a series of bars that were aligned in parallel on a plate (Fig. 2.2). The advantage of using linear gratings is that we can control the amount of stimulation between macro-geometric and material properties. More specifically, linear gratings can be used for orientation perception by changing the orientation of the bars, and can also be used for roughness perception by changing the distance between bars (i.e., the groove width). The authors found that the intraparietal sulcus, which is a part of the PPC, and the caudal part of the PostCG were more strongly activated by the condition of orientation than by the roughness classification. This result indicates that the intraparietal sulcus is involved in haptic orientation classification. The PPC might have a role in employing spatial reference frames to represent macro-geometric properties [44].

In addition to the PPC, previous neuroimaging studies have also shown that the haptic perception of an object's shape activates a part of the OTC called the lateral occipital complex (LOC) [45–48]. More specifically, this region is functionally defined by showing a greater response to visually recognized shapes (relative to textures). In previous studies, the LOC was not activated during the haptic perception of object orientation [40, 41, 49] or the perception of object location [50]. Thus, it is possible that the contributions of the PPC and LOC to haptic object processing differ, with the former being involved in spatial coding and the latter in the processing of shape among the macro-geometric properties.

2.2.2.2 Involvement of the PO and Insula in Processing Material Properties

Is any brain region more involved in processing material properties than macro-geometric properties? Previous studies showed that, compared to the discrimination of the shape of objects, roughness discrimination of different types of object produced stronger activation in the PO [38, 42]. Kitada et al. (2005) revealed that the PO and insula showed activation that was inversely related to the magnitude estimate of roughness: the rougher the linear gratings felt, the smaller the activation in this region became [51]. Other than roughness perception, Craig et al. (2000) showed that the insula and possibly the PO showed activation dependent on thermal perception [52]. Servos et al. (2001) showed evidence that the haptic classification of hardness activated the PO as compared to motor control (i.e., gripping movement without an object) [53]. These studies indicate that the PO and insula play critical roles in the processing of material properties.

However, these findings do not provide conclusive evidence that the PO and insula are more important for processing material properties than macro-geometric properties. Again, the activation of the PO and insula in previous studies [38, 42] can be explained by confounding factors such as differences in stimuli and patterns of hand movements. In fMRI studies that controlled for these factors, the PO and insula showed relatively little difference between roughness and orientation classification [40, 41]. In electrophysiological studies in non-human primates, the PO contained neurons that are sensitive to macro-geometric properties such as object orientation [54]. Accordingly, future studies are needed to determine the relative contributions of the PO and the insula to the processing of material properties and macro-geometric properties.

Collectively, the previous findings of neuroimaging experiments provide partial support for the perspective that haptically-perceived object properties are processed in a distributed manner in the brain.

2.3 Brain Networks Involved in Visuo-Haptic Interactions of Object Properties

In daily life, we frequently recognize common objects using both vision and touch. According to the two stages of object identification (i.e., the extraction of object properties and object identification), information that originates from touch and vision can interact both at the level of the processing of object properties and at the level of object identification (Fig. 2.3a). In this section, I review the brain network in which visuo-haptic object interaction occurs in relation to these two levels.

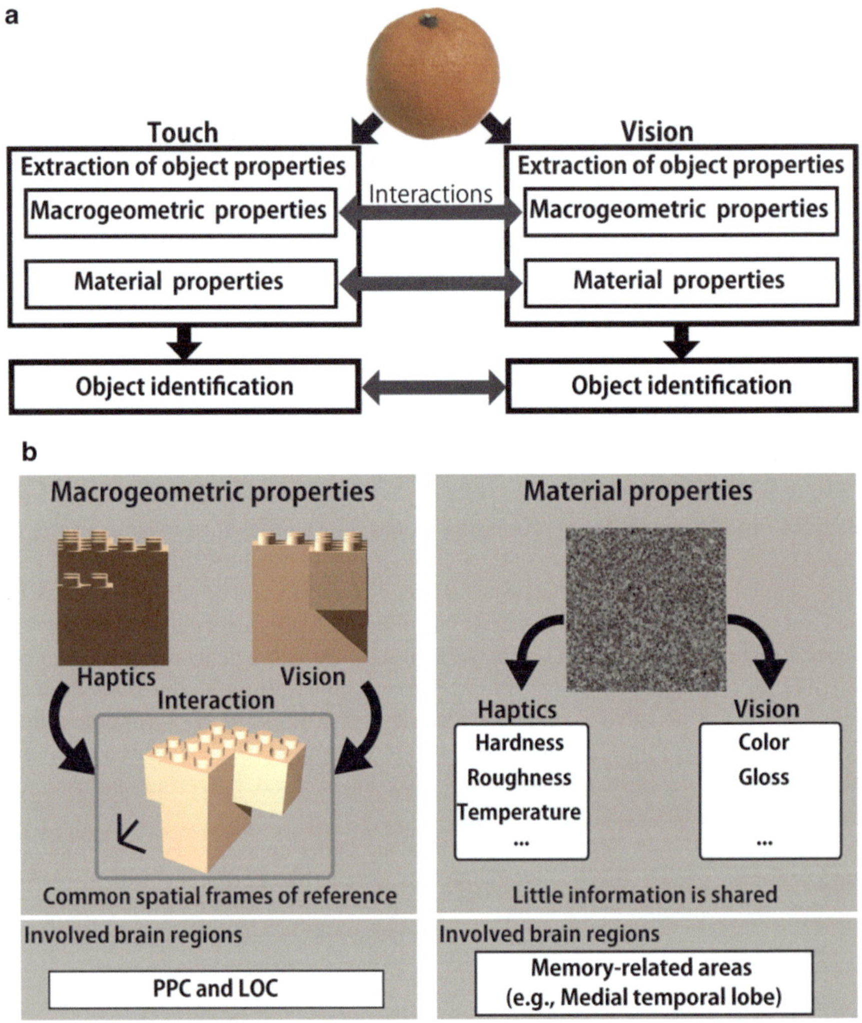

Fig. 2.3 Model for visuo-haptic interactions. (**a**) Stages of visuo-haptic object interactions. We frequently recognize objects using both vision and touch. Information originating from the two sensory modalities can interact at the level of extraction of object properties and at the level of object identification. (**b**) Visuo-haptic object interaction of object properties. Difference in visuo-tactile interactions between macro-geometric and material properties. By employing common frames of reference, spatial information between the two sensory modalities can be directly compared. Previous studies suggest that the posterior parietal cortex (PPC) and lateral occipital complex (LOC) are involved in the visuo-haptic interaction of macro-geometric properties. By contrast, little information on material properties can be directly integrated between touch and vision. One possible heuristic is to compare physical object information extracted by touch and vision in the same "format", after translating it from one sensory modality to the corresponding other. As previously learned physical associations between vision and touch are necessary for this translation, we can hypothesize that memory-related regions (such as the medial temporal lobe) are involved in visuo-haptic interactions of material properties

2.3.1 *Supramodal Representation of Common Objects in the Brain*

In the previous section, I explained that the OTC is involved in visual and haptic object recognition. In other words, this region plays a critical role in supramodally representing common objects in the brain. However, it is not the sole region contributing to supramodal object representation. For instance, we can recognize others' facial and bodily expressions by touch [3, 4, 9, 10]. It is well known that the brain contains a distributed network called the "action observation network" (AON). This network involves not only the OTC, including the EBA [28], but also the inferior parietal lobule and the inferior frontal gyrus. Some of these regions are considered to constitute the human homologue of the mirror-neuron system, in which neurons discharge both when individuals perform a particular action and when they observe the same action of others [55]. The AON is activated during the visual recognition of others' body actions and facial expressions [56, 57]. However, a part of the AON is also activated by the haptic recognition of other's facial expressions [5] and hand gestures [10]. Moreover, the presence of such activity is independent of visual experience [6, 10]. These findings indicate that the AON is related to the recognition of others' actions supramodally.

A limitation of the previous findings on supramodal object representation should be noted. Previous studies have demonstrated a convergence of activation between touch and vision in the OTC and AON. If there is supramodal representation, there should be mechanisms that integrate haptic and visual information in these structures. Neuroimaging studies are needed to examine this integration process by investigating the interaction effects between visual and tactile information: supra-additive effects ([58, 59] but see [18]) and cross-modal repetitive suppression [60] or enhancement [61]. Moreover, it would be useful to employ a cross-modal multi-voxel pattern (MVPA) approach, in which the differences of activation patterns between recognized objects in one modality can explain those that are recognized by the other modality [35, 62]. Such an approach could provide support for the hypothesis that the visuo-haptic interaction of common objects occurs at the level of object identification in these regions.

2.3.2 *Visuo-Tactile Interaction of Macro-geometric Properties*

Touch and vision share spatial information regarding macro-geometric properties. By employing common frames of reference, spatial information can be directly compared between the two sensory modalities (Fig. 2.3b left). I have discussed the involvement of the PPC in the haptic processing of macro-geometric properties and of the LOC in the haptic processing of shape in Sect. 2.2.1. Previous neuroimaging studies have shown that these regions are activated by the visual recognition of macro-geometric properties, and show evidence of interactions between vision and touch [40, 63–66]. For instance, Kitada et al. (2006) showed that the intraparietal sulcus is activated by both tactile and visual classification of object orientation [40]. Saito et al. (2003) compared brain activity during the visuo-tactile comparison of two-dimensional (2D) spatial patterns (Mahjong tiles) with that during the uni-sensory (i.e., either visual or tactile) comparison of the same stimuli. They found that a posterior part of the IPS showed greater activation in a cross-sensory condition than in uni-sensory conditions [65].

In the LOC, the haptic and visual perception of object shapes relative to textures showed overlapping activation [45–47]. Moreover, other studies showed evidence of the visuo-haptic interaction of shape in this region [46, 48, 66]. For instance, viewing visually and haptically primed objects produced more activation than viewing non-primed objects in this region [46]. These findings indicate that both the LOC and the IPS play a critical role in the visuo-haptic integration of object shape.

An additional region that has been implicated in the visuo-tactile interaction of object shape is the claustrum/insula [67, 68]. However, it is unclear whether this region is involved in the visuo-tactile interaction of macro-geometric properties rather than material properties or is involved with all object properties.

2.3.3 Visuo-Tactile Interaction of Material Properties

Unlike macro-geometric properties, relatively little is known about the brain network involved in the visuo-tactile interaction of material properties. There is limited information on material properties that can be directly integrated between touch and vision. More specifically, the physical properties of the object material itself (e.g., roughness, softness, and temperature) are perceived by touch, whereas the properties of surface reflectance (e.g., color and gloss) are accessible only by vision. As the physical object information that is extracted by vision and touch differs substantially, it has been argued that they might contribute to the perception of material properties in an independent, rather than an integrated, manner [69]. How then can different types of physical information (from touch and vision) be compared in the brain?

One possible heuristic is to compare the physical object information extracted by touch and vision in the same "format", after translating it from one sensory modality to the corresponding other. This comparison might be implemented by interactions between visual and tactile physical object information in the brain. For instance, we can retrieve tactile information that was previously associated with the visual appearance of an object (e.g., the high thermal conductivity of gold) and compare it with incoming tactile information (e.g., the low thermal conductivity of plastic). In order for this heuristic to be implemented, previously learned physical associations between vision and touch must be retrieved [70], otherwise there is no link for this translation between the two modalities. Accordingly, comparing visuo-tactile information about material properties involves neural mechanisms that can retrieve and then utilize previously learned vision-touch associations.

Several neuroimaging studies have reported that parts of the occipital cortex are active during the tactile, as well as visual, perception of material properties [42, 50, 71]. Thus, like the category-sensitive regions, these visual regions might be involved in the interaction of visual and tactile material information. However, if the retrieval of information that is shared between touch and vision is critical, additional cortical regions, such as memory-related areas, should be involved. Previous studies have identified the neural substrates underlying the retrieval of stimulus pairs during paired-association tasks such as the medial temporal lobe [72–76], lateral prefrontal cortex [72, 74, 77, 78], and precuneus [72, 74, 79, 80].

Given this background, Kitada et al. (2014) investigated whether these regions are involved in the visuo-tactile comparison of material properties rather than macro-geometric properties [41]. The stimuli consisted of surfaces on which an oriented plastic bar was placed on a background texture (Fig. 2.4). The subjects determined whether the orientations of visually- and tactually-presented bar stimuli were congruent in the orientation conditions, and whether the visually- and tactually-presented background textures were congruent in the texture conditions. The texture conditions revealed greater activation of not only the occipital cortex, but also the medial temporal lobe and lateral prefrontal cortex compared with the orientation conditions. In the texture conditions, the precuneus showed a greater response to incongruent stimuli than to congruent stimuli. This incongruency effect was greater for the texture conditions than for the orientation conditions. These results suggest that the precuneus is involved in detecting incongruency between tactile and visual texture information in concert with the medial temporal lobe, which is tightly linked with long-term memory.

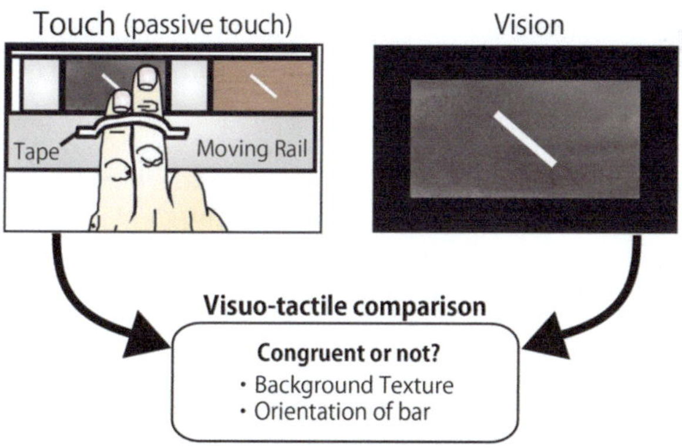

Fig. 2.4 Task design of the visuo-tactile comparisons in Kitada et al. [41]. The stimuli consisted of surfaces on which an oriented plastic bar was placed on a background texture. The subject determined whether the orientations of visually- and tactually-presented bar stimuli were congruent in the orientation conditions, and whether visually- and tactually-presented background textures were congruent in the texture conditions. Brain activity in the texture conditions was compared with that in the orientation conditions

A limitation of this finding is that it is unclear to what extent it can be generalized to the dissociation of the neural substrates involved in the visuo-tactile interaction of macro-geometric and material properties. For instance, the visuo-tactile comparison of familiar shape might involve not only common spatial frames of reference, but also cross-modal association [81]. Thus, it remains critical to examine the dissociation by employing different types of material and macro-geometric stimuli.

2.4 Future Questions

Three key points should be addressed in future studies. First, I have discussed the contributions of several nodes of the brain to haptic object identification. However, it is clear that other brain regions (e.g., frontal cortex) are involved in haptic object processing and its interaction with vision (Fig. 2.1 right). It is necessary to clarify the contribution of these brain regions such that the process can be modelled. Second, I have explained the relative contributions of the brain regions to haptic object recognition (functional specialization). A critical next step is to consider how these brain regions interact with each other to achieve haptic object recognition (functional integration).

Third and finally, when we touch objects, we not only identify them (discriminative touch), but also experience associated affective sensations such as pleasantness and unpleasantness (affective touch). A fundamental question in tactile research is how affective touch and discriminative touch are related. Their relationship has been psychophysically investigated for material properties such as roughness [82–85] and temperature [86–89]. These studies showed that the perceived magnitudes between discriminative and affective touch were different in thermal perception, but not in roughness perception. For instance, Kitada et al. (2012) conducted a psychophysical experiment wherein the participants estimated the magnitude of roughness and unpleasantness when surfaces consisting of 2D raised-dot patterns moved on the subjects' skin. The patterns of perceived estimates of roughness and unpleasantness as a function of inter-element spacing were highly similar when the speed of movement was held constant (Pearson's $r > 0.98$) [84]. Previous studies have shown that regions in and around the limbic system are related to affective touch, including the insula [90–92], the orbitofrontal cortex [93, 94] and the ventral striatum [94]. Thus, the insula, which is related to the processing of roughness perception [51], could also be related to the processing of unpleasantness and pleasantness. Further studies are necessary to investigate the nature of affective touch and how it is related to discriminative touch.

2.5 Conclusion

A wide cortical network underlies haptic object recognition. In this chapter, I have discussed its underlying functional organization. Under the current working hypothesis, haptic object processing and its interaction with vision might be conducted differently for macro-geometric and material properties. These separately processed properties are combined in supramodal regions such as the occipito-temporal cortex. A fundamental question that remains to be addressed is how haptic object processing is related to the affective processing of touch in the brain. Answering this question might also provide clues as to how contact from others (e.g., child–parent interactions) influences and maintains homeostasis in the mind.

Acknowledgment This work was supported by a Grant-in-aid for Young Scientists (B) (#23700326) from the Japan Society for the Promotion of Science and by a Grant-in-Aid for Scientific Research on Innovative Areas, "Brain and Information Science on SHITSUKAN" (#25135734) from the Ministry of Education, Culture, Sports, Science and Technology (MEXT) of Japan to R.K.

References

1. Klatzky, R.L., Lederman, S.J., Metzger, V.A.: Identifying objects by touch: an "expert system". Percept. Psychophys. **37**, 299–302 (1985)
2. Kilgour, A.R., Lederman, S.J.: Face recognition by hand. Percept. Psychophys. **64**, 339–352 (2002)
3. Lederman, S.J., Kilgour, A., Kitada, R., Klatzky, R.L., Hamilton, C.: Haptic face processing. Can. J. Exp. Psychol. **61**, 230–241 (2007)
4. Lederman, S.J., Klatzky, R.L., Abramowicz, A., Salsman, K., Kitada, R., Hamilton, C.: Haptic recognition of static and dynamic expressions of emotion in the live face. Psychol. Sci. **18**, 158–164 (2007)
5. Kitada, R., Johnsrude, I.S., Kochiyama, T., Lederman, S.J.: Brain networks involved in haptic and visual identification of facial expressions of emotion: an fMRI study. Neuroimage **49**, 1677–1689 (2010)
6. Kitada, R., Okamoto, Y., Sasaki, A.T., Kochiyama, T., Miyahara, M., Lederman, S.J., Sadato, N.: Early visual experience and the recognition of basic facial expressions: involvement of the middle temporal and inferior frontal gyri during haptic identification by the early blind. Front. Hum. Neurosci. **7**, 7 (2013)
7. Kitada, R., Johnsrude, I.S., Kochiyama, T., Lederman, S.J.: Functional specialization and convergence in the occipito-temporal cortex supporting haptic and visual identification of human faces and body parts: an fMRI study. J. Cogn. Neurosci. **21**, 2027–2045 (2009)
8. Kitada, R., Dijkerman, H.C., Soo, G., Lederman, S.J.: Representing human hands haptically or visually from first-person versus third-person perspectives. Perception **39**, 236–254 (2010)
9. Hertenstein, M.J., Keltner, D., App, B., Bulleit, B.A., Jaskolka, A.R.: Touch communicates distinct emotions. Emotion **6**, 528–533 (2006)
10. Kitada, R., Yoshihara, K., Sasaki, A.T., Hashiguchi, M., Kochiyama, T., Sadato, N.: The brain network underlying the recognition of hand gestures in the blind: the supramodal role of the extrastriate body area. J. Neurosci. **34**, 10096–10108 (2014)
11. Gardner, E.P., Johnson, K.O.: 23. Touch. In: Kandel, E.R., Schwartz, J.H., Jessell, T.M., Siegelbaum, S.A., Hudspeth, A.J. (eds.) Principles of Neural Science, 5th edn, pp. 498–529. McGraw-Hill, New York (2012)

12. Sinclair, R.J., Burton, H.: Neuronal activity in the primary somatosensory cortex in monkeys (*Macaca mulatta*) during active touch of textured surface gratings: responses to groove width, applied force, and velocity of motion. J. Neurophysiol. **66**, 153–169 (1991)
13. Chapman, C.E., Tremblay, F., Jiang, W., Belingard, L., el Meftah, M.: Central neural mechanisms contributing to the perception of tactile roughness. Behav. Brain Res. **135**, 225–233 (2002)
14. Warren, S., Hamalainen, H.A., Gardner, E.P.: Objective classification of motion- and direction-sensitive neurons in primary somatosensory cortex of awake monkeys. J. Neurophysiol. **56**, 598–622 (1986)
15. Pei, Y.C., Bensmaia, S.J.: The neural basis of tactile motion perception. J. Neurophysiol. **112**, 3023–3032 (2014)
16. Yau, J.M., Connor, C.E., Hsiao, S.S.: Representation of tactile curvature in macaque somatosensory area 2. J. Neurophysiol. **109**, 2999–3012 (2013)
17. Amedi, A., von Kriegstein, K., van Atteveldt, N.M., Beauchamp, M.S., Naumer, M.J.: Functional imaging of human crossmodal identification and object recognition. Exp. Brain Res. **166**, 559–571 (2005)
18. Beauchamp, M.S.: See me, hear me, touch me: multisensory integration in lateral occipital-temporal cortex. Curr. Opin. Neurobiol. **15**, 145–153 (2005)
19. James, T.W., Kim, S., Fisher, J.S.: The neural basis of haptic object processing. Can. J. Exp. Psychol. **61**, 219–229 (2007)
20. Lacey, S., Sathian, K.: Visuo-haptic multisensory object recognition, categorization, and representation. Front. Psychol. **5**, 730 (2014)
21. Sadato, N., Pascual-Leone, A., Grafman, J., Ibanez, V., Deiber, M.P., Dold, G., Hallett, M.: Activation of the primary visual cortex by Braille reading in blind subjects. Nature **380**, 526–528 (1996)
22. Cohen, L.G., Celnik, P., Pascual-Leone, A., Corwell, B., Falz, L., Dambrosia, J., Honda, M., Sadato, N., Gerloff, C., Catala, M.D., Hallett, M.: Functional relevance of cross-modal plasticity in blind humans. Nature **389**, 180–183 (1997)
23. Sathian, K., Zangaladze, A., Hoffman, J.M., Grafton, S.T.: Feeling with the mind's eye. Neuroreport **8**, 3877–3881 (1997)
24. Deibert, E., Kraut, M., Kremen, S., Hart Jr., J.: Neural pathways in tactile object recognition. Neurology **52**, 1413–1417 (1999)
25. Puce, A., Allison, T., Asgari, M., Gore, J.C., McCarthy, G.: Differential sensitivity of human visual cortex to faces, letterstrings, and textures: a functional magnetic resonance imaging study. J. Neurosci. **16**, 5205–5215 (1996)
26. Kanwisher, N., McDermott, J., Chun, M.M.: The fusiform face area: a module in human extrastriate cortex specialized for face perception. J. Neurosci. **17**, 4302–4311 (1997)
27. Gauthier, I., Tarr, M.J., Moylan, J., Skudlarski, P., Gore, J.C., Anderson, A.W.: The fusiform "face area" is part of a network that processes faces at the individual level. J. Cogn. Neurosci. **12**, 495–504 (2000)
28. Downing, P.E., Jiang, Y., Shuman, M., Kanwisher, N.: A cortical area selective for visual processing of the human body. Science **293**, 2470–2473 (2001)
29. Peelen, M.V., Downing, P.E.: Selectivity for the human body in the fusiform gyrus. J. Neurophysiol. **93**, 603–608 (2005)
30. Epstein, R., Kanwisher, N.: A cortical representation of the local visual environment. Nature **392**, 598–601 (1998)
31. Cohen, L., Dehaene, S., Naccache, L., Lehericy, S., Dehaene-Lambertz, G., Henaff, M.A., Michel, F.: The visual word form area: spatial and temporal characterization of an initial stage of reading in normal subjects and posterior split-brain patients. Brain **123**, 291–307 (2000)
32. Costantini, M., Urgesi, C., Galati, G., Romani, G.L., Aglioti, S.M.: Haptic perception and body representation in lateral and medial occipito-temporal cortices. Neuropsychologia **49**, 821–829 (2011)
33. Wolbers, T., Klatzky, R.L., Loomis, J.M., Wutte, M.G., Giudice, N.A.: Modality-independent coding of spatial layout in the human brain. Curr. Biol. **21**, 984–989 (2011)

34. Reich, L., Szwed, M., Cohen, L., Amedi, A.: A ventral visual stream reading center independent of visual experience. Curr. Biol. **21**, 363–368 (2011)
35. Pietrini, P., Furey, M.L., Ricciardi, E., Gobbini, M.I., Wu, W.H., Cohen, L., Guazzelli, M., Haxby, J.V.: Beyond sensory images: object-based representation in the human ventral pathway. Proc. Natl. Acad. Sci. U. S. A. **101**, 5658–5663 (2004)
36. Gilbert, C.D.: 25. The constructive nature of visual processing. In: Kandel, E.R., Schwartz, J.H., Jessell, T.M., Siegelbaum, S.A., Hudspeth, A.J. (eds.) Principles of Neural Science, 5th edn, pp. 556–576. McGraw-Hill, New York (2012)
37. Lederman, S.J., Klatzky, R.L.: Relative availability of surface and object properties during early haptic processing. J. Exp. Psychol. Hum. Percept. Perform. **23**, 1680–1707 (1997)
38. Roland, P.E., O'Sullivan, B., Kawashima, R.: Shape and roughness activate different somatosensory areas in the human brain. Proc. Natl. Acad. Sci. U. S. A. **95**, 3295–3300 (1998)
39. Bodegård, A., Geyer, S., Grefkes, C., Zilles, K., Roland, P.E.: Hierarchical processing of tactile shape in the human brain. Neuron **31**, 317–328 (2001)
40. Kitada, R., Kito, T., Saito, D.N., Kochiyama, T., Matsumura, M., Sadato, N., Lederman, S.J.: Multisensory activation of the intraparietal area when classifying grating orientation: a functional magnetic resonance imaging study. J. Neurosci. **26**, 7491–7501 (2006)
41. Kitada, R., Sasaki, A.T., Okamoto, Y., Kochiyama, T., Sadato, N.: Role of the precuneus in the detection of incongruency between tactile and visual texture information: a functional MRI study. Neuropsychologia **64C**, 252–262 (2014)
42. Stilla, R., Sathian, K.: Selective visuo-haptic processing of shape and texture. Hum. Brain Mapp. **29**, 1123–1138 (2008)
43. Lederman, S.J., Klatzky, R.L.: Hand movements: a window into haptic object recognition. Cogn. Psychol. **19**, 342–368 (1987)
44. Andersen, R.A., Snyder, L.H., Bradley, D.C., Xing, J.: Multimodal representation of space in the posterior parietal cortex and its use in planning movements. Annu. Rev. Neurosci. **20**, 303–330 (1997)
45. Amedi, A., Malach, R., Hendler, T., Peled, S., Zohary, E.: Visuo-haptic object-related activation in the ventral visual pathway. Nat. Neurosci. **4**, 324–330 (2001)
46. James, T.W., Humphrey, G.K., Gati, J.S., Servos, P., Menon, R.S., Goodale, M.A.: Haptic study of three-dimensional objects activates extrastriate visual areas. Neuropsychologia **40**, 1706–1714 (2002)
47. Peltier, S., Stilla, R., Mariola, E., LaConte, S., Hu, X., Sathian, K.: Activity and effective connectivity of parietal and occipital cortical regions during haptic shape perception. Neuropsychologia **45**, 476–483 (2007)
48. Kim, S., James, T.W.: Enhanced effectiveness in visuo-haptic object-selective brain regions with increasing stimulus salience. Hum. Brain Mapp. **31**, 678–693 (2010)
49. Zhang, M., Mariola, E., Stilla, R., Stoesz, M., Mao, H., Hu, X., Sathian, K.: Tactile discrimination of grating orientation: fMRI activation patterns. Hum. Brain Mapp. **25**, 370–377 (2005)
50. Sathian, K., Lacey, S., Stilla, R., Gibson, G.O., Deshpande, G., Hu, X., Laconte, S., Glielmi, C.: Dual pathways for haptic and visual perception of spatial and texture information. Neuroimage **57**, 462–475 (2011)
51. Kitada, R., Hashimoto, T., Kochiyama, T., Kito, T., Okada, T., Matsumura, M., Lederman, S.J., Sadato, N.: Tactile estimation of the roughness of gratings yields a graded response in the human brain: an fMRI study. Neuroimage **25**, 90–100 (2005)
52. Craig, A.D., Chen, K., Bandy, D., Reiman, E.M.: Thermosensory activation of insular cortex. Nat. Neurosci. **3**, 184–190 (2000)
53. Servos, P., Lederman, S., Wilson, D., Gati, J.: fMRI-derived cortical maps for haptic shape, texture, and hardness. Brain Res. Cogn. Brain Res. **12**, 307–313 (2001)
54. Fitzgerald, P.J., Lane, J.W., Thakur, P.H., Hsiao, S.S.: Receptive field properties of the macaque second somatosensory cortex: representation of orientation on different finger pads. J. Neurosci. **26**, 6473–6484 (2006)

55. Rizzolatti, G., Craighero, L.: The mirror-neuron system. Annu. Rev. Neurosci. **27**, 169–192 (2004)
56. Iacoboni, M., Woods, R.P., Brass, M., Bekkering, H., Mazziotta, J.C., Rizzolatti, G.: Cortical mechanisms of human imitation. Science **286**, 2526–2528 (1999)
57. Carr, L., Iacoboni, M., Dubeau, M.C., Mazziotta, J.C., Lenzi, G.L.: Neural mechanisms of empathy in humans: a relay from neural systems for imitation to limbic areas. Proc. Natl. Acad. Sci. U. S. A. **100**, 5497–5502 (2003)
58. Calvert, G.A., Campbell, R., Brammer, M.J.: Evidence from functional magnetic resonance imaging of crossmodal binding in the human heteromodal cortex. Curr. Biol. **10**, 649–657 (2000)
59. Raij, T., Uutela, K., Hari, R.: Audiovisual integration of letters in the human brain. Neuron **28**, 617–625 (2000)
60. Grill-Spector, K., Malach, R.: fMR-adaptation: a tool for studying the functional properties of human cortical neurons. Acta. Psychol. (Amst.) **107**, 293–321 (2001)
61. Segaert, K., Weber, K., de Lange, F.P., Petersson, K.M., Hagoort, P.: The suppression of repetition enhancement: a review of fMRI studies. Neuropsychologia **51**, 59–66 (2013)
62. Oosterhof, N.N., Tipper, S.P., Downing, P.E.: Crossmodal and action-specific: neuroimaging the human mirror neuron system. Trends Cogn. Sci. **17**, 311–318 (2013)
63. Grefkes, C., Weiss, P.H., Zilles, K., Fink, G.R.: Crossmodal processing of object features in human anterior intraparietal cortex: an fMRI study implies equivalencies between humans and monkeys. Neuron **35**, 173–184 (2002)
64. Nakashita, S., Saito, D.N., Kochiyama, T., Honda, M., Tanabe, H.C., Sadato, N.: Tactile-visual integration in the posterior parietal cortex: a functional magnetic resonance imaging study. Brain Res. Bull. **75**, 513–525 (2008)
65. Saito, D.N., Okada, T., Morita, Y., Yonekura, Y., Sadato, N.: Tactile-visual cross-modal shape matching: a functional MRI study. Brain Res. Cogn. Brain Res. **17**, 14–25 (2003)
66. Tal, N., Amedi, A.: Multisensory visual-tactile object related network in humans: insights gained using a novel crossmodal adaptation approach. Exp. Brain Res. **198**, 165–182 (2009)
67. Hadjikhani, N., Roland, P.E.: Cross-modal transfer of information between the tactile and the visual representations in the human brain: a positron emission tomographic study. J. Neurosci. **18**, 1072–1084 (1998)
68. Kassuba, T., Klinge, C., Holig, C., Roder, B., Siebner, H.R.: Vision holds a greater share in visuo-haptic object recognition than touch. Neuroimage **65**, 59–68 (2013)
69. Whitaker, T.A., Simoes-Franklin, C., Newell, F.N.: Vision and touch: independent or integrated systems for the perception of texture? Brain Res. **1242**, 59–72 (2008)
70. Fleming, R.W.: Visual perception of materials and their properties. Vision Res. **94**, 62–75 (2014)
71. Eck, J., Kaas, A.L., Goebel, R.: Crossmodal interactions of haptic and visual texture information in early sensory cortex. Neuroimage **75**, 123–135 (2013)
72. Gonzalo, D., Shallice, T., Dolan, R.: Time-dependent changes in learning audiovisual associations: a single-trial fMRI study. Neuroimage **11**, 243–255 (2000)
73. Naya, Y., Yoshida, M., Miyashita, Y.: Backward spreading of memory-retrieval signal in the primate temporal cortex. Science **291**, 661–664 (2001)
74. Ranganath, C., Cohen, M.X., Dam, C., D'Esposito, M.: Inferior temporal, prefrontal, and hippocampal contributions to visual working memory maintenance and associative memory retrieval. J. Neurosci. **24**, 3917–3925 (2004)
75. Tanabe, H.C., Honda, M., Sadato, N.: Functionally segregated neural substrates for arbitrary audiovisual paired-association learning. J. Neurosci. **25**, 6409–6418 (2005)
76. Weniger, G., Boucsein, K., Irle, E.: Impaired associative memory in temporal lobe epilepsy subjects after lesions of hippocampus, parahippocampal gyrus, and amygdala. Hippocampus **14**, 785–796 (2004)
77. Fuster, J.M., Bodner, M., Kroger, J.K.: Cross-modal and cross-temporal association in neurons of frontal cortex. Nature **405**, 347–351 (2000)

78. Hasegawa, I., Fukushima, T., Ihara, T., Miyashita, Y.: Callosal window between prefrontal cortices: cognitive interaction to retrieve long-term memory. Science **281**, 814–818 (1998)
79. Krause, B.J., Schmidt, D., Mottaghy, F.M., Taylor, J., Halsband, U., Herzog, H., Tellmann, L., Muller-Gartner, H.W.: Episodic retrieval activates the precuneus irrespective of the imagery content of word pair associates. A PET study. Brain **122**, 255–263 (1999)
80. Tanabe, H.C., Sadato, N.: Ventrolateral prefrontal cortex activity associated with individual differences in arbitrary delayed paired-association learning performance: a functional magnetic resonance imaging study. Neuroscience **160**, 688–697 (2009)
81. Holdstock, J.S., Hocking, J., Notley, P., Devlin, J.T., Price, C.J.: Integrating visual and tactile information in the perirhinal cortex. Cereb. Cortex **19**, 2993–3000 (2009)
82. Ekman, G., Hosman, J., Lindstroem, B.: Roughness, smoothness, and preference: a study of quantitative relations in individual subjects. J. Exp. Psychol. **70**, 18–26 (1965)
83. Verrillo, R.T., Bolanowski, S.J., McGlone, F.P.: Subjective magnitude of tactile roughness. Somatosens. Mot. Res. **16**, 352–360 (1999)
84. Kitada, R., Sadato, N., Lederman, S.J.: Tactile perception of nonpainful unpleasantness in relation to perceived roughness: effects of inter-element spacing and speed of relative motion of rigid 2-D raised-dot patterns at two body loci. Perception **41**, 204–220 (2012)
85. Klocker, A., Oddo, C.M., Camboni, D., Penta, M., Thonnard, J.L.: Physical factors influencing pleasant touch during passive fingertip stimulation. PLoS One **9**, e101361 (2014)
86. Chatonnet, J., Cabanac, M.: The perception of thermal comfort. Int. J. Biometeorol. **9**, 183–193 (1965)
87. Mower, G.D.: Perceived intensity of peripheral thermal stimuli is independent of internal body temperature. J. Comp. Physiol. Psychol. **90**, 1152–1155 (1976)
88. Attia, M., Engel, P.: Thermal pleasantness sensation: an indicator of thermal stress. Eur. J. Appl. Physiol. **50**, 55–70 (1982)
89. Nakamura, M., Yoda, T., Crawshaw, L.I., Yasuhara, S., Saito, Y., Kasuga, M., Nagashima, K., Kanosue, K.: Regional differences in temperature sensation and thermal comfort in humans. J. Appl. Physiol. **105**, 1897–1906 (2008)
90. Craig, A.D.: How do you feel? Interoception: the sense of the physiological condition of the body. Nat. Rev. Neurosci. **3**, 655–666 (2002)
91. Olausson, H., Lamarre, Y., Backlund, H., Morin, C., Wallin, B.G., Starck, G., Ekholm, S., Strigo, I., Worsley, K., Vallbo, A.B., Bushnell, M.C.: Unmyelinated tactile afferents signal touch and project to insular cortex. Nat. Neurosci. **5**, 900–904 (2002)
92. McGlone, F., Wessberg, J., Olausson, H.: Discriminative and affective touch: sensing and feeling. Neuron **82**, 737–755 (2014)
93. Rolls, E.T., O'Doherty, J., Kringelbach, M.L., Francis, S., Bowtell, R., McGlone, F.: Representations of pleasant and painful touch in the human orbitofrontal and cingulate cortices. Cereb. Cortex **13**, 308–317 (2003)
94. Rolls, E.T., Grabenhorst, F., Parris, B.A.: Warm pleasant feelings in the brain. Neuroimage **41**, 1504–1513 (2008)

Chapter 3
Understanding Force Perception Characteristics of a Human and Its Applications

Yuichi Kurita

Abstract Many assistive devices designed to reduce fatigue on muscles during work are in development. Building models of human sensory and motor functions is helpful when creating devices to effectively and safely assist us in daily life. There is a sense of force associated with voluntary muscular exertion, which contributes to the perception of weight. Researchers in the field of perceptual psychology have investigated human perception characteristics of force over the years. These models of perception characteristics of force are helpful to evaluate subjective efforts associated with intuitive, safe, and easy-to-use products. In this chapter, two research studies are addressed in detail. The first study aims to model the perception characteristics of force, and explore its application in developing a perception prediction method during steering. This study estimates the muscle activity when using a steering wheel based on a three-dimensional musculoskeletal model, develops a perception conversion model that follows the Weber-Fechner law, and describes the experimental results that confirm the feasibility of the proposed force perception prediction. The second study aims to explore the challenges in developing wearable assistive technology that can enhance humans' force perception capability by unloading voluntary muscle activation. This study measures the force perception capability of participants with different postures to investigate how differences in voluntary muscle activation affect force perception capability. Based on the experiments, a muscle-assistive wearable device that unloads the upper limb is developed, and the improvement in sensorimotor capability when voluntary muscle activation is reduced while wearing the device is evaluated.

Keywords Force perception • Sensorimotor function • Musculoskeletal model • Steering wheel operation • Wearable assistive device

Y. Kurita (✉)
Hiroshima University, 1-4-1 Kagamiyama, Higashi-Hiroshima, Hiroshima 739-8527, Japan
e-mail: kurita@bsys.hiroshima-u.ac.jp

H. Kajimoto et al. (eds.), *Pervasive Haptics*, DOI 10.1007/978-4-431-55772-2_3

3.1 Introduction

In recent years, many assistive devices for individuals with injury or disabilities have been developed as auxiliary rehabilitation and assistive exercise equipment with electric and pneumatic actuators [1–5]. To reduce the maintenance cost, assistive wearable devices made of stretchable fabric have also been developed [6, 7]. These devices and clothing were designed to reduce fatigue on muscles during work. To evaluate the performance of the assistive devices or human factors in product usability, the intensity of electromyography (EMG) signals is usually focused on. Kong et al. measured the maximum pulling force, the EMG signals, and the distribution of grasping pressure when pulling seven different meat hooks, and reported a positive relationship between subjective usability and EMGs [8–10]. Iwamoto et al. proposed a method to characterize the physiological response of a driver during the dynamic operation of a vehicle [11]. Measuring EMG signals is useful for evaluating human effort; however, evaluation methods based on EMG measurements require human experiments where an actual human participant uses or manipulates an actual physical object. The limitation of available muscles whose activities can be measured with a certain level of accuracy is also a big issue.

In this context, simulation-based evaluation methods of motion effort have been explored by taking human physical capacity into consideration. Some studies in biomechanics have proposed an accurate musculoskeletal model of the human body. An et al. established a three-dimensional normative hand model based on X-ray image analysis [12]. Holzbaur et al. developed a biomechanical model of the upper extremity [13]. Valero-Cuevas proposed a detailed model of the human finger, including neuro-musculoskeletal interactions [14]. Flanagan et al. discussed control strategies for the human fingertips[15]. Sueda et al. developed a method for generating the motions of tendons and muscles for a hand animation [16]. Also, the robotics field, contains studies related to skin and muscle modeling. For example, Nakamura et al. proposed a method of computing somatosensory information from motion-capture data [17]. Delp et al. developed a freely available, open-source software system (OpenSim) that lets users create models of musculoskeletal structures and dynamic simulations of a wide variety of movements [18]. Endo et al. developed a simulator that can evaluate the stability of grasp when a user holds an object, such as a digital camera, without real participants and physical objects [19]. Tada and Pai developed a simple finger shell model that can quickly and efficiently simulate finger surface deformation [20]. Sugiyama et al. proposed a grasp criterion that yields the best grasping position for a robot hand with two soft fingers [21]. Murai et al. developed a system that estimates and visualizes muscle tensions in real time using optical motion capture data and electromyography (EMG) signals [22]. Kurita and Ikeda developed a tendon skeletal model of the fingers, and proposed the evaluation method of product usability from the viewpoint of the grasp effort by calculating muscle activity when pinching an object [23]. They also developed robotic fingers to measure the torque during a pinch motion and evaluated the pinching effort through the estimation of muscle force [24].

These studies evaluate the body posture in terms of the muscle activity or joint torque estimated using a musculoskeletal model, whereas there are few studies that consider the characteristics of human perception. Human skeletal muscles contain sensory systems, such as the muscle spindle and Golgi tendon organ. It is known that the muscle spindles primarily detect changes in the length of the muscle, and that the Golgi tendon organs primarily detect changes in muscle tension. These receptors are believed to be responsible for the proprioceptive sense of body posture and muscle effort. There is a sense of force associated with voluntary muscular exertion, which contributes to the perception of weight. Over the years, researchers in the field of perceptual psychology have investigated human force perception characteristics. Some researchers have reported that the sense of force is overestimated after fatigue [25–27]. Brockett et al. reported that both the sense of joint position and the sense of force are disturbed over a number of days following eccentric exercise [28]. Vuillerme et al. reported that muscle fatigue degrades the sense of force at the ankle joint [29]. van Beek et al. reported that humans seem to possess a sense of effort rather than a sense of force, which might be more helpful when performing tasks in everyday life [30]. Takemura et al. investigated human force perception characteristics when manipulating the steering wheel of a vehicle [31], and reported that the perception characteristics of force follow the Weber-Fechner law [32] with a constant posture, where the magnitude of the force sense is a logarithmic function of the stimulus intensity.

These models of the perception characteristics of force are helpful in evaluating the subjective efforts associated with intuitive, safe, and easy-to-use products. In this chapter, two research studies are addressed in detail. The first study aims to model the perception characteristics of force, and explore its application in developing a perception prediction method during steering. This study estimates the muscle activity when using a steering wheel based on a three-dimensional musculoskeletal model, develops a perception conversion model that follows the Weber-Fechner law, and describes the experimental results that confirm the feasibility of the proposed force perception prediction [33]. The second study aims to explore the challenges in developing assistive clothing that can enhance human force perception capability by unloading voluntary muscle activation. This study measures the force perception capability of human participants in different postures to investigate how differences in voluntary muscle activation affect human force perception capability. Based on the experiments, a muscle-assistive wearable device that unloads the upper limb is developed, and the improvement in sensorimotor capability when voluntary muscle activation is reduced by wearing the developed suit is evaluated [34].

3.2 Prediction of Subjective Force Perception During a Steering Operation

3.2.1 Estimation Model of the Perceived Force

Takemura and colleagues measured human force perception characteristics during the operation of a vehicle's steering wheel [31]. Figure 3.1 shows the measured force perception characteristics of a representative participant. The horizontal axis is the given force, the vertical axis is the perceived force, and the solid curve is the approximated curve obtained from the following equation:

$$F_p = a \log(F_t) + b \tag{3.1}$$

where the parameters $a = 11.21$ and $b = 13.42$ are determined by the least-squares method. In the experiment, the experimenter informed the participants of the intensity of the standard load (= 20 N), but not the other loads. The results show that the perception characteristics of force in the steering operation follow the Weber-Fechner law. This indicates that the magnitude of the sensed force perception is a logarithmic function of the stimulus intensity.

Figure 3.2 shows the perceived load when a 20 N load is given to the participants via the steering wheel. The horizontal axis shows the steering wheel angle, and the vertical axis shows the perceived force. These results strongly suggest that the perception characteristics of force depend on both body posture and stimulus direction. Although the Weber-Fechner law models the perception characteristics against the stimulus intensity, it does not take the effect of body posture and stimulus direction into consideration. Based on Fig. 3.1, the change ratio of the perceived force, which is determined by the perceived force F_p and the given external force F_t, can be calculated using the following equation:

$$Z = \frac{F_p}{F_t} = \frac{a \log F_t + b}{F_t}. \tag{3.2}$$

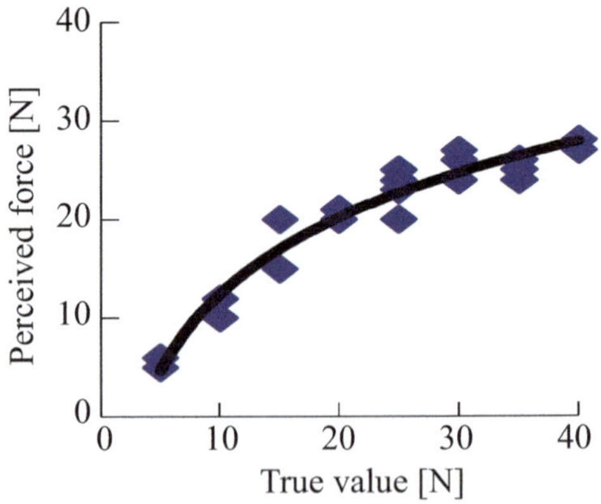

Fig. 3.1 Relationship between the given and perceived force

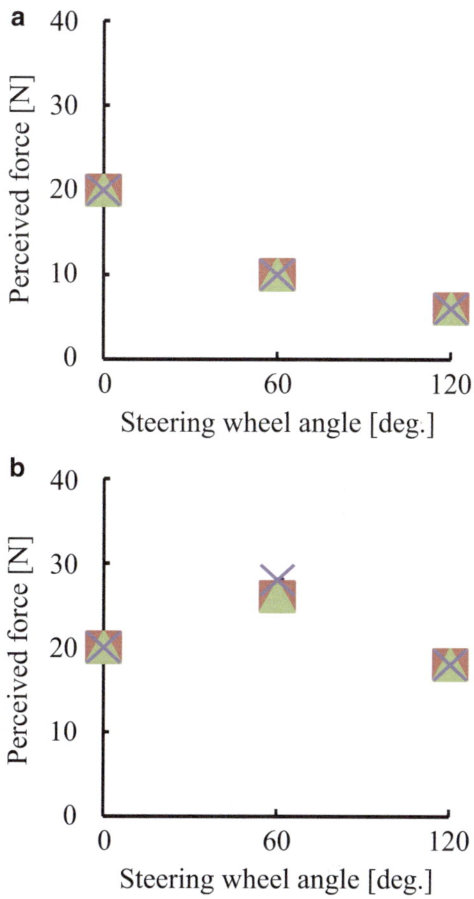

Fig. 3.2 Perceived force when operating a steering wheel of a vehicle. The results of the three subjects are shown as square, triangle, and cross marks, respectively. (**a**) Leftward load. (**b**) Rightward load

By determining the muscle activation that generates the joint torque to balance the external force, the obtained change ratio of the perceived force against the external force can be converted to that against the muscle activation.

Three-dimensional musculoskeletal models for left ant right arms were developed based on an upper extremity model [13]. The developed model of the right arm is shown in Fig. 3.3. The muscle force was calculated based on a muscle contraction model [35] composed of elastic and contractile components. The maximum isometric force, the optimal muscle fiber length, and the penetration angle of the muscles were determined according to the parameters reported by [36]. Same parameters were used for the left and right arms. Muscle activations were calculated using OpenSim, which is an open source software system for biomechanical modeling, simulation and analysis [18].

Figure 3.4 shows the relationship between the mean muscle activation of all the muscles and the exerted force at the wrist at the posture shown in Fig. 3.5. From this figure, there is a linear relationship between the muscle activation and the exerted

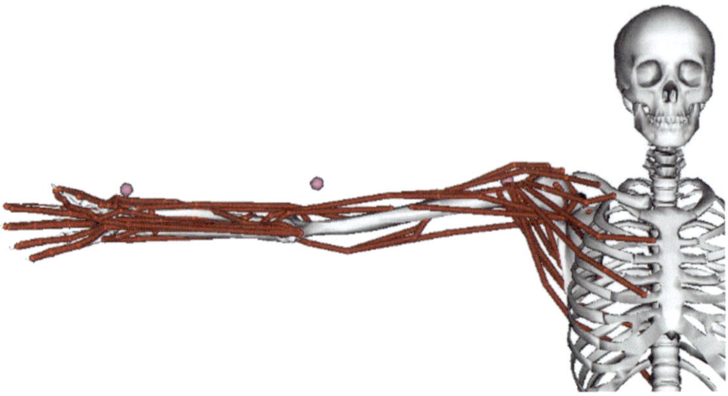

Fig. 3.3 Musculoskeletal model

Fig. 3.4 Relationship between the muscle activity and the external force

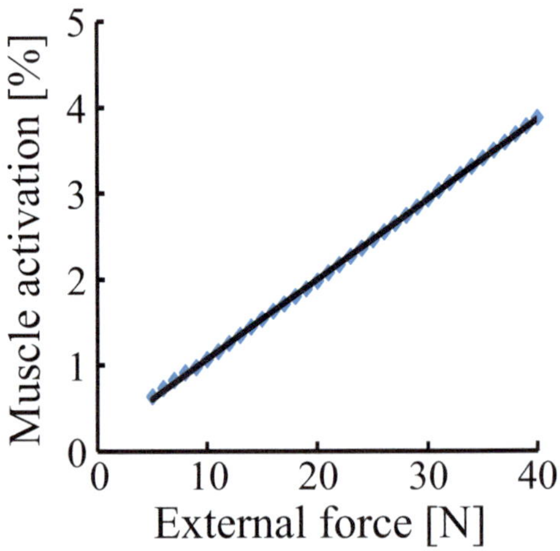

force. The relationship between the intensity of the force F_t and the mean muscle activation $\bar{\alpha}^{all}$ is modeled by the following linear approximation:

$$\bar{\alpha}^{all} = kF_t + m \tag{3.3}$$

where the coefficients $k = 0.00093$ and $m = 0.0014$ are obtained by the least-squares method. Substituting this into Eq. (3.2) gives:

$$Z'(\bar{\alpha}^{all}) = \frac{k\left(a \log\left(\frac{\bar{\alpha}^{all} - m}{k}\right) + b\right)}{\bar{\alpha}^{all} - m}. \tag{3.4}$$

This equation expresses the perception change ratio as a function of the muscle activation.

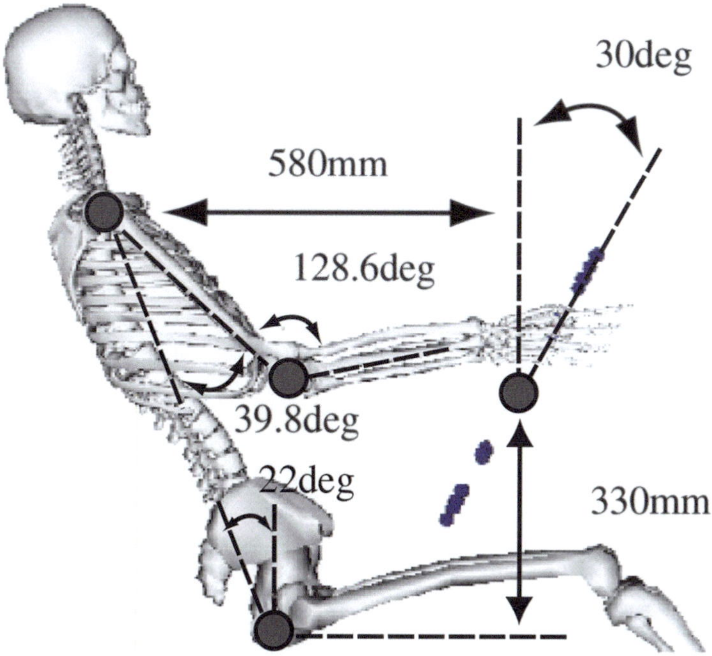

Fig. 3.5 Body posture for the simulation

Fig. 3.6 Rotation angle of the steering for the simulation

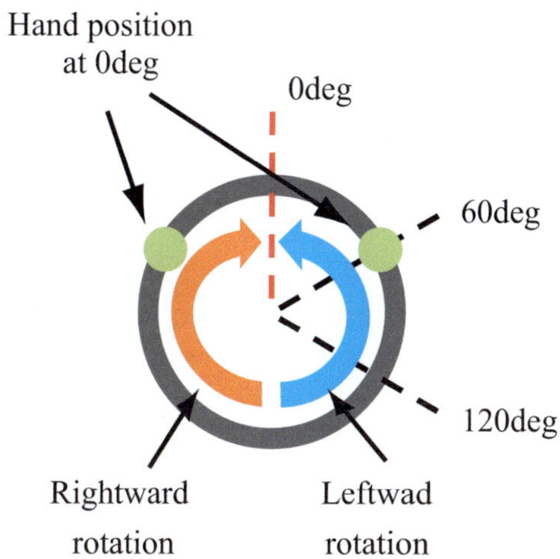

3.2.2 *Prediction of the Perceived Force*

The perception characteristics of force when the body posture changes during steering wheel operation were investigated using the proposed model. The experimental conditions are shown in Fig. 3.6. In this study, the rotation motions of the steering

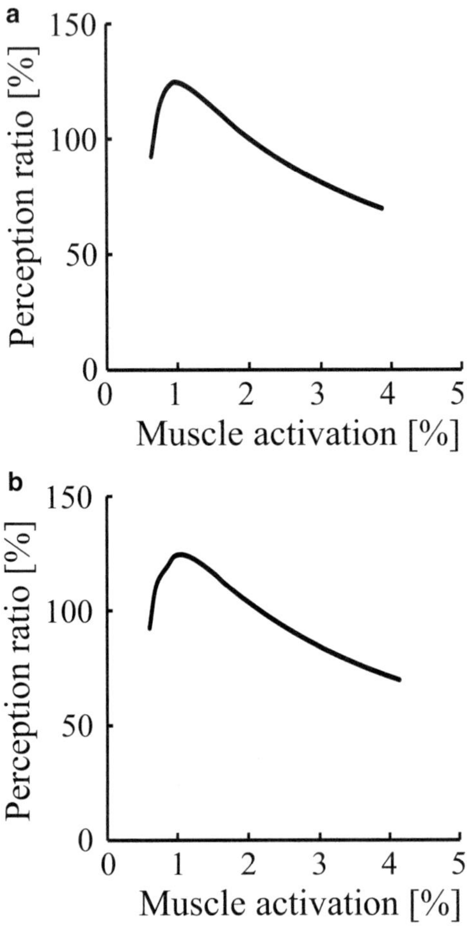

Fig. 3.7 Perception change ratio against the muscle activations. (**a**) Leftward rotation. (**b**) Rightward rotation

wheel at 0, 60, and 120° in the clockwise direction were discussed. The joint angles of the body at each steering angle were measured using a motion capture system. The length, mass, center of mass, and moment of inertia of the upper extremity were determined according to the method of Ae et al. [37]. The external stimuli of 10 N were given to both right and left hands (20 N in total).

Figure 3.7 shows the relationship between the perception change ratio and the mean muscle activation calculated using Eq. (3.4) at a posture of 0°. (a) and (b) are the results when the leftward and rightward stimuli are given, respectively. The horizontal axis shows the mean muscle activation, and the vertical axis shows the perception change ratio. These figures indicate that humans underestimate the stimuli when the mean muscle activation is smaller than 0.8 % or larger than 2 %, whereas they overestimate when the muscle activation is between 0.8 % and 2 %. Figure 3.8 shows the predicted perceived force calculated based on the perception change ratio. (a) and (b) are the results for the leftward and rightward stimuli,

Fig. 3.8 Predicted perceived force. (**a**) Leftward rotation. (**b**) Rightward rotation

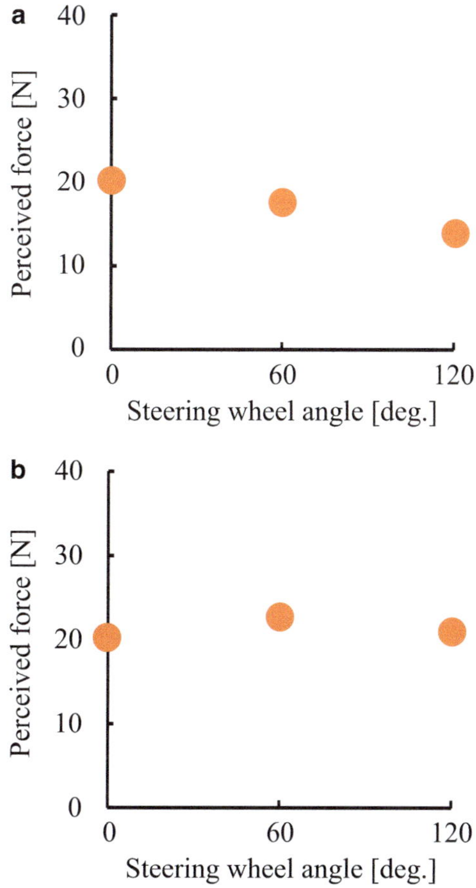

respectively. The horizontal axis shows the steering wheel angle, and the vertical axis shows the predicted perceived force. (a) indicates that when a leftward stimulus is given, the perceived force decreases as the steering wheel angle increases. Conversely, (b) suggests that when a rightward stimulus is given, the perceived force is not significantly affected by the change of the steering wheel angle. The obtained perception characteristics of force predicted by the perception change ratio have a similar tendency to those obtained by human participants as shown in Fig. 3.8. This indicates that the biomechanical model has the potential to simulate the phenomenon and that this comes not only from the mechanism of the motor system, but also from the nature of the sensory system.

The simulation done in this study contains some assumptions for simplicity. First, the distribution of the external stimuli is not based on the actual measurement. Because of the limitations of the experimental system, the grasping force on the steering wheel was not measured. In our simulation, we assume that the same grasping force was exerted at the left and right hands. However, the distribution of force is a very important factor for the accurate estimation of human muscle activity. Second, the internal force is completely neglected in the muscle force and muscle activity estimation process. In the current estimation, the muscle activity was determined by an optimization calculation where the total intensity of the muscle activity is minimized. However, humans sometimes increase their joint stiffness by the co-contraction of muscles for precise and dexterous movements. More accurate evaluation of the perception characteristics of force will be possible by taking these factors into consideration.

3.3 Muscle-Assistive Wear That Can Enhance Sensorimotor Performance

3.3.1 Muscle-Force Assistive Suit

A previously mentioned study [33] revealed that the perception characteristics of force changes depend on the physical capacity of the human body. In particular, the results suggested that humans can sense the intensity of external stimuli more accurately when voluntary muscle activation is lower. For better understanding of human sensory capabilities, we measured force perception capability in various postures to investigate how differences in voluntary muscle activations affect force perception capability [34]. The experimental results show that there are postures in which humans can perceive loads more accurately, and these postures require relatively low voluntary muscle activation.

Taking this into consideration, we developed muscle-assistive equipment that unloads the upper limb, and enhances the sensorimotor capability of humans. The developed equipment is shown in Fig. 3.9. This equipment is made of stretchy fabric

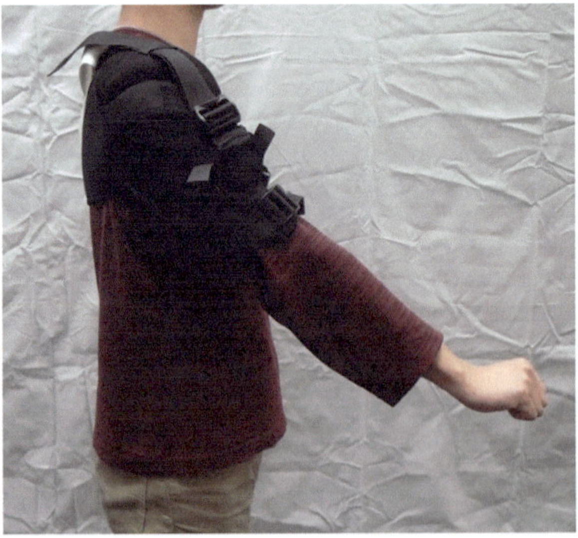

Fig. 3.9 Muscle assistive equipment

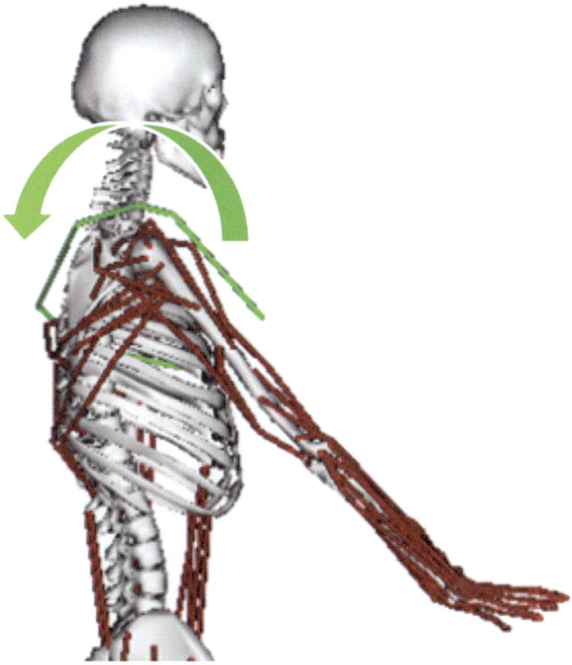

Fig. 3.10 Assisting torque by the developed assistive equipment

and employs passive assistance to reduce maintenance cost. One side of the stretchy fabric is fixed at the waist and the other side is fixed at the upper arm. The stretchy fabric has a spring constant of 392 N/m, and it generates a flexion moment at the shoulder joint as shown in Fig. 3.10. This shoulder flexion moment generated by the equipment unloads part of the shoulder extension moment generated by the weight of the arm.

3.3.2 Experimental Results

We conducted human experiments to confirm the improvement of the force perception capability using the proposed assistive equipment. Ten male volunteers aged 22–24 years participated in the experiment. The participants were instructed to stretch their arm forward with their elbow extended and close their eyes. First, the experimenter hung an empty container on the wrist of the participant, put a weight of 1500 gf into the container as shown in Fig. 3.11, and asked the participant to remember the perception of holding the weight. Second, the experimenter detached the container and put a force transducer at the participant's wrist as shown in Fig. 3.12. The participants were asked to exert a force upward (shoulder flexion torque) to match the force with the weight that was previously hung on their

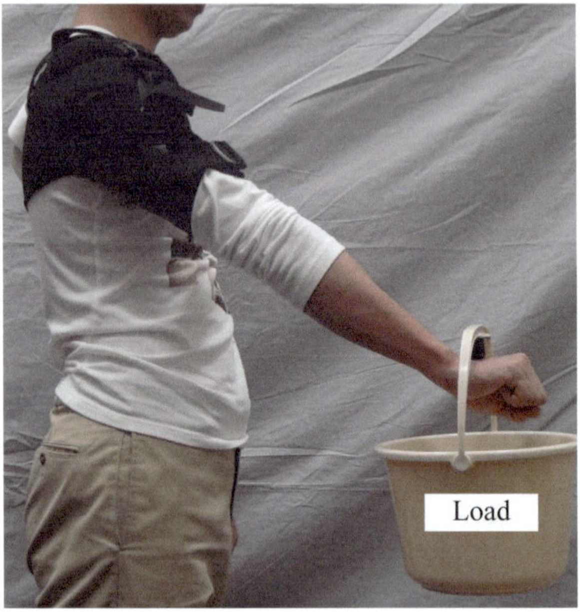

Fig. 3.11 Experimental condition: the load is given at the wrist

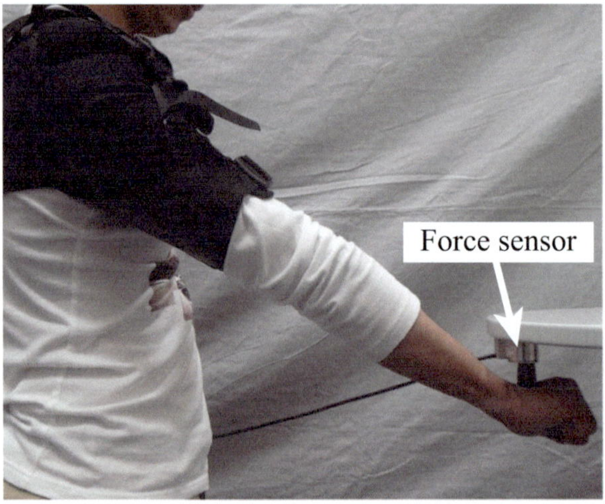

Fig. 3.12 Experimental condition: the participant exert force on the force sensor

arm. Each participant alternately underwent three trials for the conditions with and without the assistive equipment. Rest periods or 30-s long were provided between trials. Informed consent was obtained from all participants before the experiments based on the Declaration of Helsinki.

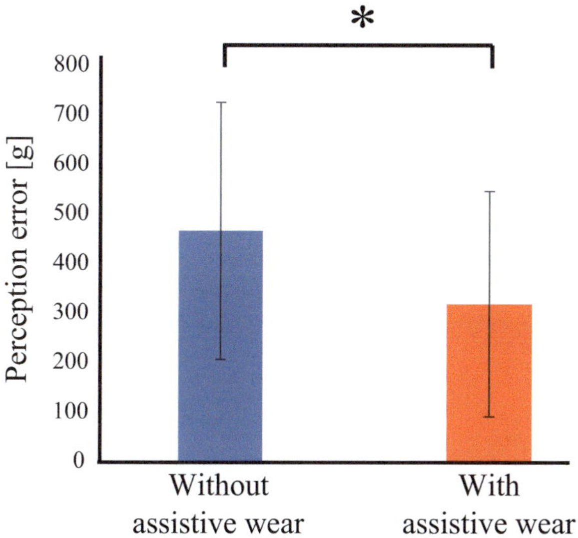

Fig. 3.13 Error between the remembered and exerted force. Statistically significant differences was found by the Student's T-test ($* : p < 0.05$)

Figure 3.13 shows the difference between the remembered load (1500 gf) and the force exerted by participants based on their memory. On average, 450 gf and 350 gf errors were observed without and with the assistance of the equipment, respectively. A statistically significant difference was found using Student's t-test. This result suggests that the assistive equipment can significantly improve the sensorimotor function of the participants. To sense a weight and to exert force at the wrist with a stretched arm posture, the shoulder muscles, e.g., the deltoid muscles, infraspinatus muscle, and biceps brachii muscles, play an important role. The muscle-assistive equipment mainly supports the flexion moment about the shoulder. Counteracting the shoulder extension moment unloads the voluntary muscle activation related to the flexion of the shoulder.

The results of the force perception capability experiments show that errors of perceived force can be significantly reduced by supporting the weight of the relevant upper limb, which suggests that force perception capability can be improved by mitigating loads from voluntary muscle activation. The outcomes also suggest that force perception capability improvement is greater with heavier loads. The central nervous system, sensory nerves and motor nerves in the brain and spinal cord play a role in controlling muscle activation, with motor units being made up of one alpha motor neuron and a muscle fiber group innervated by it. Muscle tension is controlled by the number of mobilized motor units. Henneman et al. introduced the concept of orderly motor unit recruitment by which units are recruited in order of increasing size (called the size principle; [38]). In this size principle, as the requirement for force grows, motor units are recruited depending on the magnitude of their force output, with small units being recruited first and large ones being recruited later. It is known that there is variability in motor output; as the standard deviation of noise is proportional to the magnitude of motor output, it is known as signal-dependent noise. This biological noise may be caused by the size principle of motor unit recruitment [39, 40]. Signal-dependent noise increases the standard deviation of noise as force increases, which makes it difficult to precisely control force output when a large force is exerted. As assisting muscle load helps to mitigate muscle activation, it can help to reduce noise due to the recruitment of large motor units. As the sensory system inside human skeletal muscles is believed to be associated with a sense of force, it is reasonable to assume that reducing unnecessary voluntary muscle activation may contribute to improved sensory performance.

3.4 Conclusions

Building models of human sensory and motor function is helpful for effective and safe assistance in our daily life. The first study explored the challenge of predicting perceived force during steering wheel operation based on the estimation of muscle activation using a three-dimensional musculoskeletal model. The experimental results revealed that the perception characteristics of force change depending on the physical capacity of the human body. The second study examined the sensorimotor capability of human participants when using muscle-assistive equipment. The results of the sensorimotor capability test show that unloading voluntary muscle activation is an effective way to improve sensorimotor performance. Continued research in this area may also lead to development of muscle-assistive equipment to allow the unloading of larger forces on a broader ranges of muscles.

The author hopes that this chapter will be useful to anybody interested in understanding human force perception characteristics and their applications.

References

1. Suzuki, K., Mito, G., Kawamoto, H., Hasegawa, Y., Sankai, Y.: Intention-based walking support for paraplegia patients with robot suit HAL. Adv. Robot. **21**(12), 1441–1469 (2007)
2. Tanaka, T., Satoh, Y., Kaneko, S., Suzuki, Y.: Smart suit: soft power suit with semi-active assist mechanism -prototype for supporting waist and knee joint-. In: International Conference on Control, Automation and Systems, Seoul (2008). doi:10.1109/ICCAS.2008.4694428
3. Yeh, T.I., Wu, M.J., Lu, T.J., Wu, F.K., Huang, C.R.: Control of McKibben pneumatic muscles for a power-assist, lower-limb orthosis. Mechatronics **20**(6), 686–697 (2010)
4. Muramatsu, Y., Kobayashi, H., Sato, Y., Jiaou, H., Hashimoto, T., Kobayashi, H.: Quantitative performance analysis of exoskeleton augmenting devices -muscle suit- for manual worker. Int. J. Autom. Technol. **5**(4), 559–567 (2011)
5. Gordon, K.E., Kinnaird, C.R., Ferris, D.P.: Locomotor adaptation to a soleus EMG-controlled antagonistic exoskeleton. J. Neurophysiol. **109**(7), 1804–1814 (2013)

6. Imamura, Y., Tanaka, T., Suzuki, Y., Takizawa, K., Yamanaka, M.: Motion-based-design of elastic material for passive assistive device using musculoskeletal model. J. Robot. Mechatron. **23**(6), 978–990 (2011)
7. Takahashi, K., Tanaka, T., Nara, H., Kaneko, S., Yoshida, E.: A model of burden sense from psychophysical factors in lifting action with and without power assist device. In: International Conference on Advanced Cognitive Technologies and Applications, Valencia, pp. 27–33 (2013)
8. Kong, Y.K., Freivalds, A.: Evaluation of meat-hook handle shapes. Int. J. Ind. Ergon. **32**(2), 13–23 (2003)
9. Kong, Y.K., Freivalds, A.: Evaluation of hook handles in a pulling task. Int. J. Occup. Saf. Ergon. **11**(3), 303–313 (2005)
10. Kong, Y.K., Lowe, B.D., Lee, S.J., Krieg, E.F.: Evaluation of handle design characteristics in a maximum screwdriving torque task. Ergonomics **50**(9), 1404–1418 (2007)
11. Iwamoto, Y., Umetsu, D., Ozaki, S.: Evaluation of driver-vehicle matching using neck muscle activity and vehicle dynamics. Kansei Eng. Int. J. **11**(3), 147–154 (2012)
12. An, K.N., Chao, E.Y., Cooney, W.P., Linscheid, R.L.: Normative model of human hand for biomechanical analysis. J. Biomech. **12**(10), 775–788 (1979)
13. Holzbaur, K.R.S., Murray, W.M., Delp, S.L.: A model of the upper extremity for simulating musculoskeletal surgery and analyzing neuromuscular control. Ann. Biomed. Eng. **33**(6), 829–840 (2005)
14. Valero-Cuevas, F.J.: An integrative approach to the biomechanical function and neuromuscular control of the fingers. J. Biomech. **38**(4), 673–684 (2005)
15. Flanagan, J.R., Bowman, M.C., Johansson, R.S.: Control strategies in object manipulation tasks. Curr. Opin. Neurobiol. **16**(6), 650–659 (2006)
16. Sueda, S., Kaufman, A., Pai, D.K.: Musculotendon simulation for hand animation. ACM Trans. Graphics (Proceedings SIGGRAPH) **27**(3), 775–788 (2008)
17. Nakamura, Y., Yamane, K., Fujita, Y., Suzuki, I.: Somatosensory computation for man-machine interface from motion capture data and musculoskeletal human model. IEEE Trans. Robot. **21**(1), 58–66 (2005)
18. Delp, S.L., Anderson, F.C., Arnold, A.S., Loan, P., Habib, A., John, C.T., Guendelman, E., Thelen, D.G.: Opensim: open-source software to create and analyze dynamic simulations of movement. IEEE Trans. Biomed. Eng. **54**(11), 1940–1950 (2007)
19. Endo, Y., Kanai, S., Kishinami, T., Miyata, N., Kouchi, M., Mochimaru, M.: Virtual grasping assessment using 3D digital hand model. In: Proceedings of 10th Annual Applied Ergonomics Conference, Texas (2007)
20. Tada, M., Pai, D.K.: Finger shell: predicting finger pad deformation under line loading. In: Proceedings of the Symposium on Haptic Interfaces for Virtual Environment and Teleoperator Systems, Reno, pp. 107–112 (2008)
21. Sugiyama, S., Koeda, M., Fujimoto, H., Yoshikawa, T.: Measurement of grasp position by human hands and grasp criterion for two soft-fingered robot hands. In: IEEE International Conference on Robotics and Automation, Kobe, pp. 1127–1132 (2009)
22. Murai, A., Kurosaki, K., Yamane, K., Nakamura, Y.: Musculoskeletal-see-through mirror: computational modeling and algorithm for whole-body muscle activity visualization in real time. Prog. Biophys. Mol. Biol. **103**, 310–317 (2010)
23. Ikeda, A., Kurita, Y., Ogasawara, T.: Product usability estimation using musculoskeletal model. In: IEEE RAS/EMBS International Conference on Biomedical Robotics and Biomechatronics, Tokyo, pp. 307–312 (2010)
24. Kurita, Y., Ikeda, A., Matsumoto, T., Ogasawara, T.: Evaluation of grasp efficiency based on muscle activity estimation by anthropomorphic robot fingers. In: International Symposium on Micro-NanoMechatronics and Human Science, Nagoya, pp. 466–468 (2011)
25. Gandevia, S.C., McClosky, D.I.: Sensations of heaviness. Brain **100**(2), 345–354 (1977)
26. Jones, L.A.: Role of central and peripheral signals in force sensation during fatigue. Exp. Neurol. **81**(2), 497–503 (1983)

27. Jones, L.A.: Perception of force and weight: theory and research. Psycholog. Bull. **100**(1), 29–42 (1986)
28. Brockett, C., Warren, N., Gregory, J.E., Morgan, D.L., Proske, U.: A comparison of the effects of concentric versus eccentric exercise on force and position sense at the human elbow joint. Brain Res. **771**(2), 251–258 (1997)
29. Vuillerme, N., Boisgontier, M.: Muscle fatigue degrades force sense at the ankle joint. Gait Posture **28**(3), 521–524 (2008)
30. van Beek, F.E., Tiest, W.M.B., Kappers, A.M.L.: Anisotropy in the haptic perception of force direction and magnitude. IEEE Trans. Haptics **6**(4), 399–407 (2013)
31. Takemura, K., Yamada, N., Kishi, A., Nishikawa, K., Nouzawa, T., Kurita, Y., Tsuji, T.: A subjective force perception model of humans and its application to a steering operation system of a vehicle. In: IEEE International Conference on Systems, Man, and Cybernetics, Manchester, pp. 3675–3680 (2013)
32. Fechner, G.T.: Elements of Psychophysics. Handbook of Physiology Section 12 Exercise: Regulation and Integration of Multiple Systems. Holt, Rinehart and Winston, New York (1966)
33. Sato, J., Takemura, K., Yamada, N., Kishi, A., Nishikawa, K., Nouzawa, T., Tsuji, T., Kurita, Y.: Investigation of subjective force perception based on estimation of muscle activities during steering operation. In: IEEE/SICE International Symposium on System Integration, Kobe, pp. 76–81 (2013)
34. Kurita, Y., Sato, J., Tanaka, T., Shinohara, M., Tsuji, T.: Unloading muscle activation enhances force perception. Augmented Human (2014). doi:10.1145/2582051.2582055
35. Thelen, D.G.: Adjustment of muscle mechanics model parameters to simulate dynamic contractions in older adults. J. Biomech. Eng. **25**, 70–77 (2003)
36. Menegolo, A.: Simtk.org: upper and lower body model: overview (2011). https://simtk.org/home/ulb_project
37. Ae, M., Tang, H.P., Yokoi, T.: Estimation of inertia properties of the body segments in Japanese athletes. Biomechanism **11**, 23–33 (1992)
38. Henneman, E.: Relation between size of neurons and their susceptibility to discharge. Science **126**(3287), 1345–1347 (1957)
39. Harris, C.M., Wolpert, D.M.: Signal-dependent noise determines motor planning. Nature **394**(6695), 780–784 (1998)
40. Jones, K.E., Hamilton, A.F., Wolpert, D.M.: Sources of signal-dependent noise during isometric force production. J. Neurophysiol. **88**, 1533–1544 (2002)

Part II
Component Design

Chapter 4
Tactile Display Using the Micro-vibration of Shape-Memory Alloy Wires and Its Application to Tactile Interaction Systems

Hideyuki Sawada

Abstract A compact device for the presentation of tactile sensation is introduced in this chapter. By employing shape-memory alloy (SMA) wires, a novel tactile information transmission device is developed for displaying various tactile sensations. A tactile actuator consists of a thin SMA wire with the diameter of 50 μm, and generates micro-vibrations with various frequencies and amplitudes in accordance with a driving pulse signal. The actuator features compactness and low energy consumption, and is applied to tactile displays and tactile interaction systems.

Keywords Tactile sensation • Shape memory alloy • Micro vibration • Tactile higher-level psychological perception • Tactile interaction • Texture • Image processing • Braille display • Diabetes mellitus

4.1 Introduction

A novel micro-vibration actuator using a shape-memory alloy (SMA) for the presentation of various tactile sensations to human skin will be introduced in this chapter. SMAs are metals, which have a shape-memory effect. The alloy has a peculiar temperature T_p, and the shape memory effect is observed when the temperature is cooled to below T_p. The effect has been widely applied in different fields such as robotic muscles, medical surgeries, actuators and elastic wires. SMAs also have a unique characteristic to shrink at a certain temperature. The authors found that by making the SMA into a thin wire, it accepts a pulse-signal to generate a vibration in accordance with the pulse frequency of the signal, and can be developed into a micro-vibration actuator for presenting tactile information. By coupling the actuators as a pair, an information transmission system is created for presenting higher-level tactile perceptions such as the phantom sensation and the apparent movement to transmit novel tactile sensations to human skin. The apparent

H. Sawada (✉)
Department of Intelligent Mechanical Systems Engineering, Faculty of Engineering, Kagawa University, Takamatsu, Japan
e-mail: sawada@eng.kagawa-u.ac.jp

H. Kajimoto et al. (eds.), *Pervasive Haptics*, DOI 10.1007/978-4-431-55772-2_4

movement was especially well perceived by users as a sensation of something running across their fingers, or as being tapped by an object, according to well-determined signals given to the devices. The tactile actuators employing SMA wires are applied to various tactile displays and tactile interaction systems [1–9].

4.2 Micro-vibration Actuator for the Presentation of Tactile Sensation

4.2.1 Human Tactile Sensory System and the Higher-Level Psychological Perception

In order to facilitate the sense of touch, the human skin includes four main tactile receptors, which are Merkel discs, Meissner corpuscles, Ruffini endings and Pacinian corpuscles. Merkel discs are located in the epidermis approximately 10 μm in diameter, and they can sense pressure and texture. Meissner corpuscles are primarily located just beneath the epidermis within the dermis and between 30–140 μm in length and 40–60 μm in diameter, and sense vibration and fluttering stimuli. Ruffini endings are also located in the dermis around 0.5–2 mm in the length, and they can sense skin stretch. Pacinian corpuscles are located in the subcutis, and they are about 0.5–2 mm in the length and 0.7 mm in the diameter. According to the response time and size of the receptors, the four receptors can be classified into four categories, which are the fast adapting I and II (FAI and FAII) and the slow adapting I and II (SAI and SAII). The receptors are distributed with different densities in different regions of a human body, and it is known that the receptors are densely distributed in human fingers, especially at the tip of the fingers.

In this study, we pay attention to the higher-level psychological perception of tactile sensation for the presentation of tactile information, which are the phantom sensation (PS), the apparent movement (AM) and the cutaneous rabbit illusion (CRI) [10–12].

4.2.2 Higher-Level Psychological Perception of Tactile Sensation

Apparent movement is known as one of the higher-level psychological perceptions of human tactile sensation. When two locations on our skin surface are excited by two mechanical vibratory stimuli with transient time delay, we perceive an illusory sensation which continuously moves from the first location to the other, as shown in Fig. 4.1a.

Phantom sensation is also the psychological perception of tactile sensation. A variable sensation appears between two locations when they are stimulated simultaneously with arbitrary intensity. If two stimuli have the same intensity, the phantom sensation is perceived in the middle of them. If one stimulus is

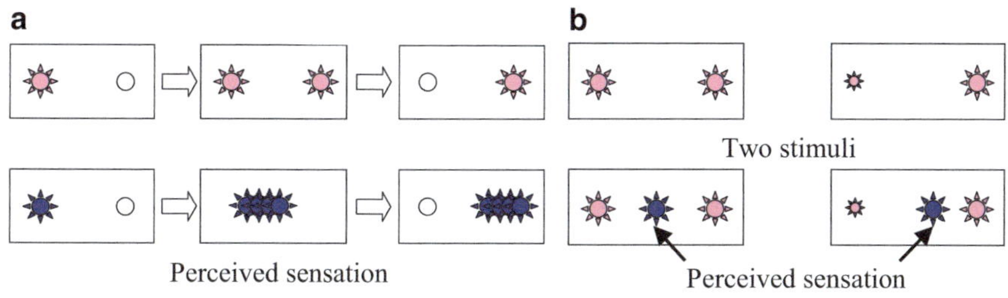

Fig. 4.1 Tactile higher-level psychological perception. (**a**) Apparent movement (AM). (**b**) Phantom sensation (PS)

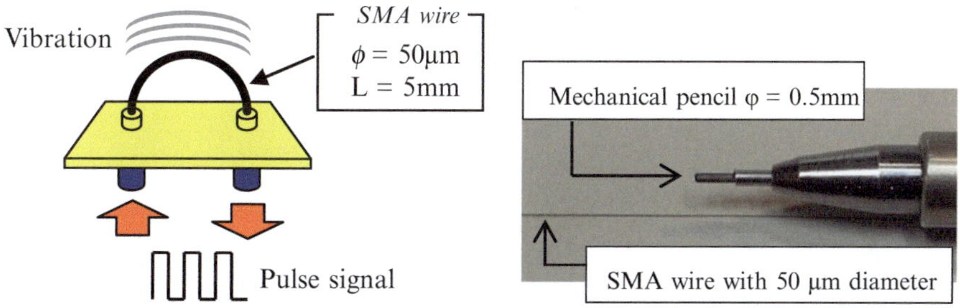

Fig. 4.2 SMA wire and vibration actuator

stronger than the other, the illusory sensation appears closer to the stronger location, according to the strength ratio. Figure 4.1b shows the schematic figure of the phantom sensation, which appears between two mechanical stimuli.

The cutaneous rabbit illusion was discovered by Geldard and Sherrick in 1972 [12]. When rapid sequential taps are delivered from one location on the skin to another, it creates a tactile illusion like sequential taps hopping along the skin from the first location to the second. Some reports have shown that the cutaneous rabbit effect appears not only on the body, but also "hopped out of the body" and onto a stick held by the subject's index fingers.

4.2.3 Design of Micro-vibration Actuators

The authors have developed a micro-vibration actuator electrically driven by periodic signals generated by a current-control circuit for tactile information transmission. Figure 4.2 shows a vibration actuator composed with a 5 mm-long SMA wire with a diameter of 0.05 mm. Applying a weak current to the alloy, the temperature rises to T_2 due to the generated heat inside it and the alloy shrinks up to 5 % lengthwise of the original length. When the current stops and the temperature drops to T_1, the alloy returns to its original length. Figure 4.3 shows the temperature characteristics of the SMA wire employed in this study having the temperatures $T_1 = 68$ and $T_2 = 72°$.

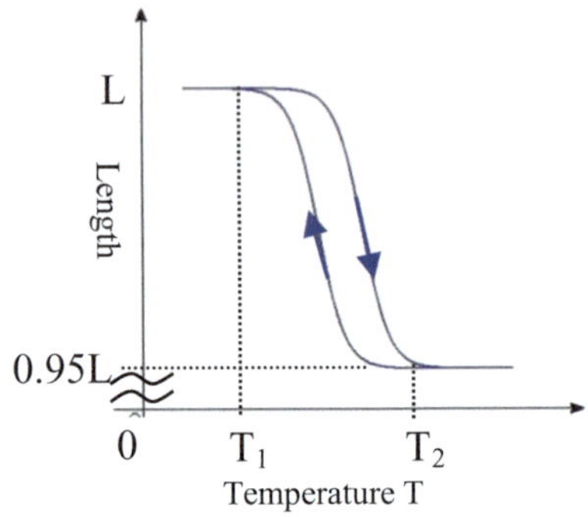

Fig. 4.3 Temperature characteristics of SMA

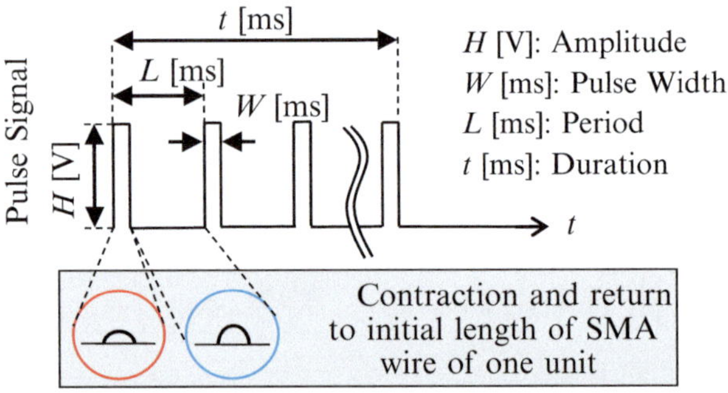

Fig. 4.4 Pulse current

The SMA wire is so thin that it rapidly cools after the current stops, and returns to its original shape when the temperature shifts from T_2 to T_1. This means that the shrinkage and the return to initial length of the SMA wire can be controlled by the pulse current. By driving the SMA wire with a pulse current as shown in Fig. 4.4, micro-vibration with the amplitude of several micrometers is generated, which is perceived by the human body as tactile sensation. We employ pulse-width modulated (PWM) current to control the vibration mode of the SMA wire generated from a specially designed amplifier. The pulse has the amplitude of H [V] and the width of W msec. The duty ratio W/L determines the heating and cooling time of the SMA. $W \times H$, which is equivalent to the calories exchanged, determines the amplitude of a vibration, and the vibration frequency is completely controlled by regulating L. When the tactile actuator generated micro vibrations of 50 Hz and 100 Hz, a high-speed camera verified that the SMA wire perfectly synchronized with the ON/OFF pulse current, and shrunk in the contraction state to about 2 μm toward the length.

The actuator has the advantage of compactness to give a vibration stimulus to a small spot on the skin, and low energy consumption of approximately 10 mW to quickly respond to generating the vibration.

The vibration generated from the SMA wire is able to transmit various tactile sensations to human hands and feet, especially to hairless parts of the skin. However, the greatest deformation of skin given by the wire is approximately 10 μm. A pin-type tactile actuator was developed to amplify the vibration given to any part of

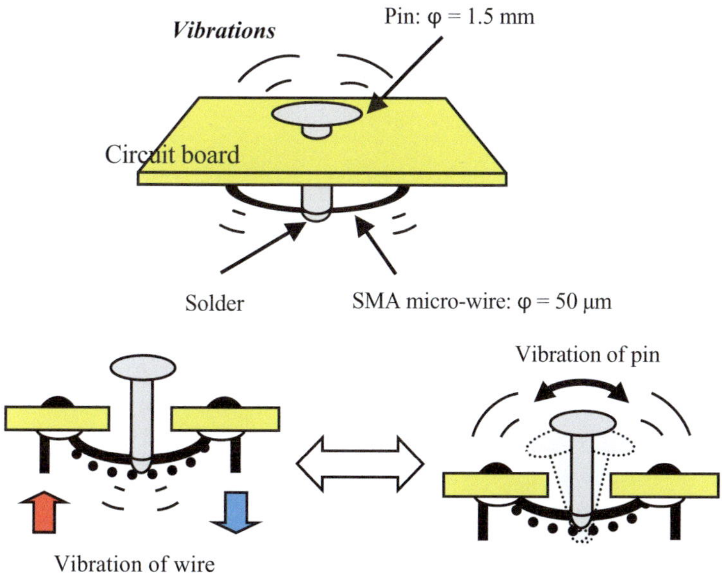

Fig. 4.5 Pin-type vibration actuator

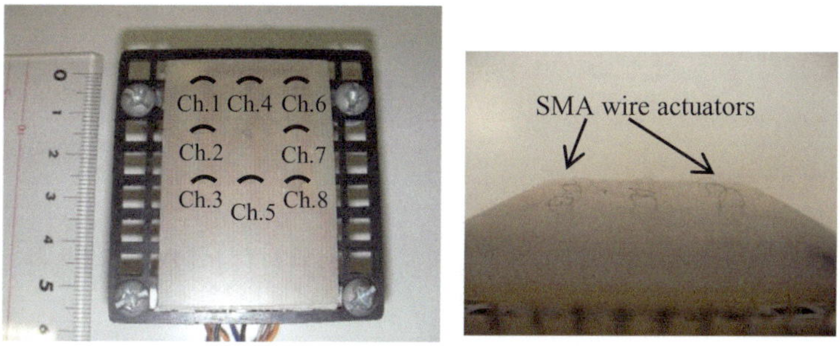

Fig. 4.6 Tactile display with eight actuators

the human body as shown in Fig. 4.5. Micro-vibration generated by the SMA wire generates greater vibratory stimuli to any part of the human skin by a pin soldered to the wire. We confirm that up to 300 Hz is generated by synchronizing properly prepared pulse-signals, and the mechanical vibrations are preferably perceived by tactile mechano-receptors under the skin, such as Meissner corpuscles and Pacinian corpuscles, which respond to frequencies lower than 100 Hz, and also from 50 to 300 Hz, respectively.

By arranging plural vibration actuators, a tactile display is constructed. Figure 4.6 shows one example of a tactile display, which is equipped with eight actuators arranged in a 3×3 matrix. The eight actuators are independently driven by the current-control amplifier to present human higher-level perceptions. A user puts his palm on the display to feel the tactile sensations. With AM, one could perceive a stroking sensation moving from one location to the other on the palm, under

Fig. 4.7 Different types of current-control amplifiers. (**a**) 32-channel amplifier. (**b**) 16-channel amplifier with Bluetooth. (**c**) 16-channel amplifier with LCD controller

the control of the time delay between two vibratory stimuli. In this manner, by using eight actuators, the tactile display presents tactile phantom images in arbitrary locations among multiple stimuli [1].

4.2.4 Current-Control Amplifier

Considering the calories exchanged in the SMA wire body to generate necessary vibration for the tactile presentation, the pulse current given to actuators should be properly controlled. The authors have developed several types of current-control amplifiers with different channel numbers and communication functions as shown in Fig. 4.7. All the amplifiers are driven by the DC supply of 5 V, together with all the SMA actuators.

4.3 Presentation of Texture Sensations by the Control of Pulse-Signal Density

4.3.1 Presentation of Stroking Sensation Using the Apparent Movement

Figure 4.8 shows an example of a tactile stroking sensation moving from left to right on a palm by driving the actuators in the order of (Ch. 1, Ch. 3) – (Ch. 4,

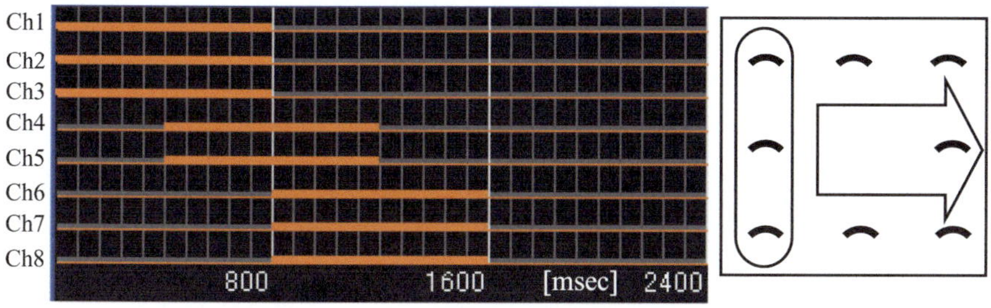

Fig. 4.8 Stroking sensation moving from left to right

Ch. 5) – (Ch. 6, Ch. 8) using the display shown in Fig. 4.6. In the left figure, the driving signals of each channel are shown in bold lines, and the perceived sensation is schematically shown on the right. Multiple AM and PS simultaneously arises to generate continuous motion of a phantom image moving across the palm. We found that the tactile presentation gives clear phantom images to be used for displaying various tactile sensations.

We conducted an experiment to evaluate stroking sensations. The sensation of a stroke moving randomly in four directions was presented to the right palm of five able-bodied subjects. The frequency for all the vibration actuators was set to 50 Hz, and the stimuli were presented to each subject 30 times randomly. The subjects correctly recognized 90 % of the randomly presented stimuli, and most of the subjects gave positive preferences to the tactile presentation by the display.

4.3.2 Presentation by Driving Actuators Randomly with a Probability Density Function

We further examined the presentation of an object's texture and its rubbing sensation using randomly generated pulse signals for driving actuators. The actuator is basically driven by pulse signals, but we paid attention to the pulse density in order to present the texture of a virtual object's surface. The greater the density of the pulse, the greater the roughness of a surface was expected to be generated. We also noted that by changing the pulse density, a rubbing stroking motion could be presented. With these assumptions, we constructed a tactile presentation system to control the signals using a pulse-signal probability density function (PPDF).

The probability density of a pulse occurrence is determined by PPDF using the Gaussian distribution as

$$p(t) = \alpha + \beta \exp\left\{\frac{-(t-m)^2}{2\sigma^2}\right\} \tag{4.1}$$

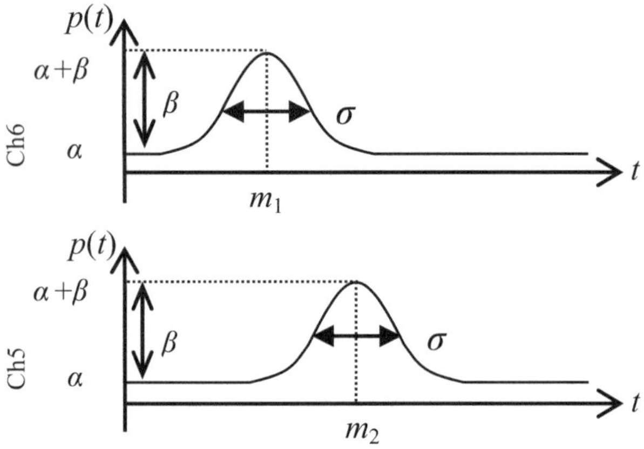

Fig. 4.9 Pulse-signal probability density functions

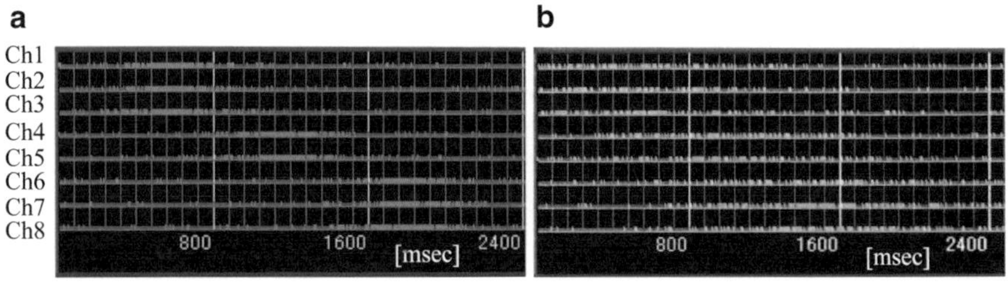

Fig. 4.10 Examples of signals generated by PPDF. (**a**) $\alpha=0.1$, $\beta=0.6$, $m=600$, $\sigma=200$. (**b**) $\alpha=0.1$, $\beta=0.4$, $m=500$, $\sigma=500$

where m: average, σ: variance, α: offset, β: gain, $\alpha+\beta<1.0$, and the control signal for each channel is generated as shown in Fig. 4.9. Two examples of control signals generated by PPDF are shown in Fig. 4.10, where high-density pulses are presented by bold lines.

4.3.3 Experiment of Tactile Presentation with PPDF

As a first step to evaluating tactile sensations presented by the display, an experiment was conducted by changing the values of the four parameters m, σ, α, and β. We found that subjects could perceive different rubbing sensations and also could roughly discriminate a moving direction. By changing the parameter β with the fixed $\alpha=0.1$, most stimuli were successfully perceived as a stroking sensation. With a frequency of 20 Hz, the subject perceived a stroking sensation with disconnected motions. With frequencies higher than 50 Hz, better sensations with different tactile textures were recognized.

Based on the examination, we selected the parameter values for presenting rubbing sensations. Three subjects evaluated the left-to-right rubbing strokes generated by PPDF where $\alpha = 0.1$, $\beta = 0.6$, $m = 400$ and $\sigma = 150$, 300 and 450, by comparing the results with the fixed pulse signals. The frequency was set to 50 Hz, and the driving duration of each actuator was 800 ms. Using a scale of 1 (Negative) to 5 (Positive), the subjects evaluated the sensations according to (A) Smooth motion, (B) Texture of smoothness, (C) Texture of roughness, and (D) Disconnection of motion.

The subjects felt coarse or rough textures when the σ value was small, and smooth textures when σ was greater. At $\sigma = 150$, the subjects evaluated the presented sensations better and could perceive a sensation of touching or rubbing a real object with a coarse surface. In addition, greatly different and unique sensations were perceived with $\sigma = 300$, even though the parameter values were close with each other.

4.3.4 Presentation of Different Tactile Sensations

For the presentation of different tactile sensations, the parameters to drive the tactile display were examined. Driving parameters m_i-s, σ, α, and β described in Fig. 4.9 were changed in a pre-determined range, and the generated tactile stimuli were presented to seven subjects to evaluate the sensations. We selected four couples of evaluation points as listed below, which are often used for the assessment of textures. The subjects scored them in seven levels.

1. Rough (1) – Smooth (7)
2. Bumpy (1) – Flat (7)
3. Hard (1) – Soft (7)
4. Elastic (1) – Non-elastic (7)

The duty ratio of pulsed current was fixed at 1:20, and the pulse frequencies were set to 50 or 100 Hz. The time difference $|m_i - m_j|$ and the variance σ were changed to 300, 500 and 800, respectively. Eighteen different conditions in total were randomly changed to present different stimuli to seven subjects to be evaluated in seven levels. The greater the scores were given, the smoother, flatter, softer and less elastic the object was evaluated.

Throughout the experiments, a frequency of 100 Hz gave greater scores compared to 50 Hz, which means that the frequency presents fine and minute tactile sensations. For the evaluation of smoothness and flatness, the conditions of 50 Hz and $\sigma = 300$ were scored with less values. As an example, a signal with the above conditions is shown in Fig. 4.11. The overlap time among actuators is small, and the vibration stimuli tend to be localized spatially and temporally to present rough sensations. For the sensation of the softness, $\sigma = 500$ gave the greatest scores for both frequencies. In order to judge the elasticity of an object, we pushed the object

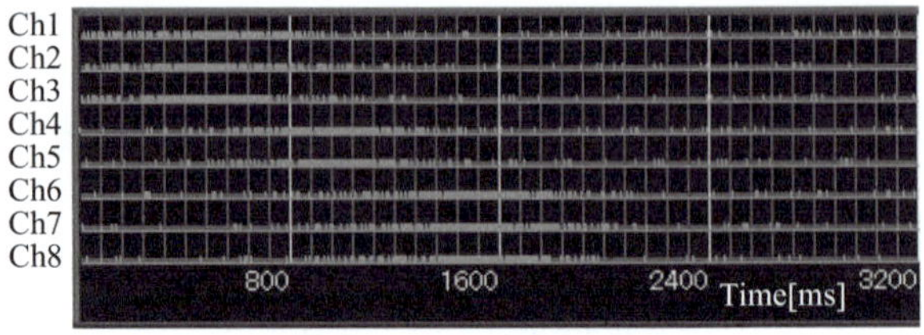

Fig. 4.11 Stimulus condition with $|m_i - m_j| = 500$ and $\sigma = 300$

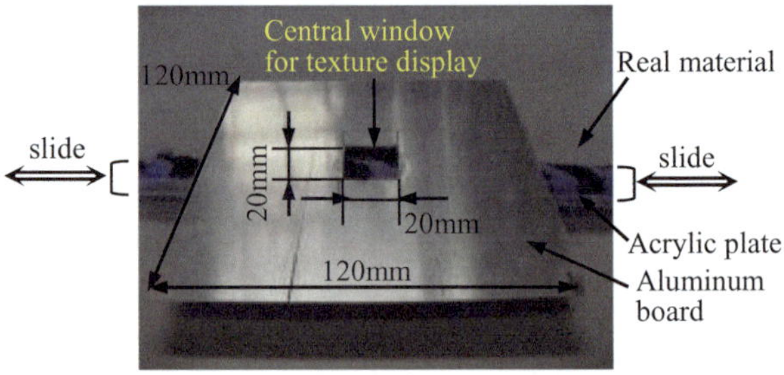

Fig. 4.12 Experimental equipment for the evaluation of human tactile sensation

to sense the force of repulsion. As it seemed difficult to push the tactile display, there was some difficulty evaluating the objects for elasticity.

In summary, with the tactile display, we found that sensations such as smoothness, flatness and softness are possible to present by changing the parameters to generate pulse-signal probability density functions (PPDF) for giving pulse signals.

4.3.5 Evaluation of Human Tactile Sensation Against Real Object's Textures

To compare the sensations presented by the tactile display with the human tactile sensible ability, we examined human skin sensations and how the textures of real objects were perceived without seeing them. Five objects, which were a polishing sponge, a felt cloth, a cotton towel, a plywood board, and a rubber sheet, were selected to test the different sensations in the experiment. Six male subjects aged from 21 to 30 years old touched five different objects with a bandage over their eyes, and selected the object they felt.

Before this experiment, the subjects touched all five objects freely to remember the tactile sensations of each object presented. Then in the experiment, one object out of five was presented blindly to the subject using the display equipment shown in Fig. 4.12, and the subject selected what it was.

In the experiment, a subject put his hand on the equipment to feel the texture presented from the central window. The objects were presented by sliding an acrylic plate into the central window with the speed of approximately 3 cm/s by a person

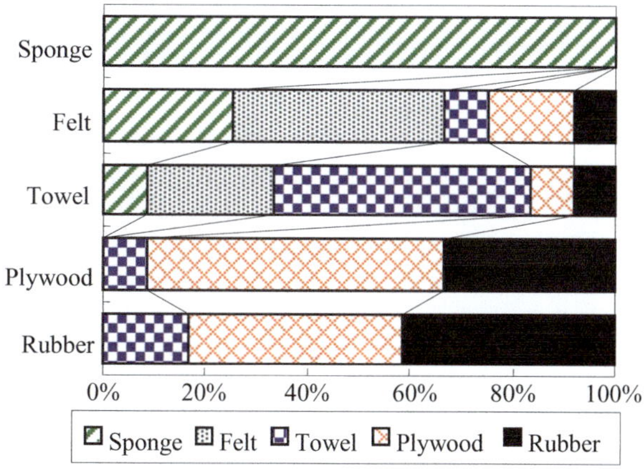

Fig. 4.13 Relationship between presented objects and answers

different than the subjects. The experiment was conducted in a random order, and the trial was repeated ten times, in which each object was presented twice.

The results are shown in Fig. 4.13. Each bar represents the five presented objects, and the ratios in the bars show the answers given by subjects. For example, all the subjects recognized the polishing sponge correctly. However, the rubber was correctly recognized by only 40 % of the subjects, and was frequently understood as the plywood. What was interesting in the experiment is that four objects except the polish sponge were correctly recognized less than 60 % in the blind condition. Furthermore, the cotton towel was often misrecognized as a felt cloth, and the plywood was frequently misunderstood as a rubber sheet. This experiment implies that in the recognition process of an object's texture, visual information helps to understand the object. Misrecognition would occur among the textures having close roughness or smoothness. The experiment also revealed the ambiguity of human tactile sensations.

From this experiment, we found a novel aspect of human tactile perception to object sensation in a blind condition. For a more detailed examination of tactile perception, further study and experiments should be expected.

4.3.6 *Comparison of the Presented Tactile Sensation with Real Objects*

Based on the experimental results described in the above chapters, we found that the display could present various different tactile sensations by changing the driving parameters. We conducted an experiment to compare presented sensations with real objects, by employing nine subjects with normal tactile sensitivity.

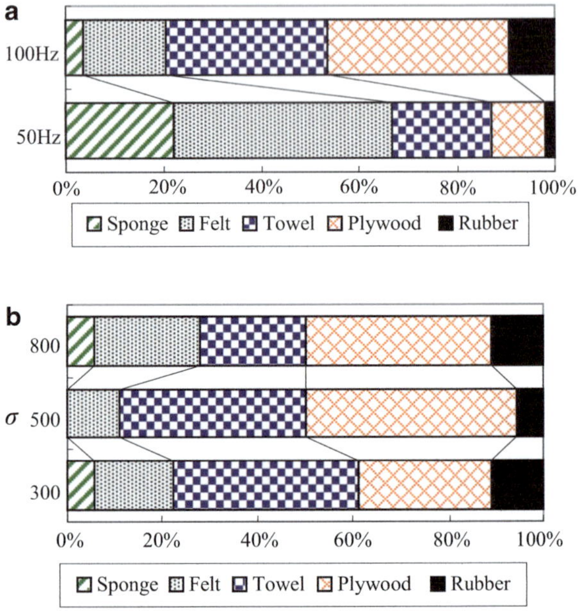

Fig. 4.14 Results of comparison experiment. (**a**) Result according to frequencies. (**b**) Difference by σ with 100 Hz

Five real objects, which were a polishing sponge, a felt cloth, a cotton towel, a plywood board, and a rubber sheet, as employed in the preliminary experiments described above, were also prepared in this experiment, and subjects answered which object had the same tactile texture as a displayed sensation. The duty ratio of pulsed current was fixed at 1:20, and the pulse frequencies were set to 50 or 100 Hz. The parameters α and β were fixed to 0.1 and 0.6, respectively, and the time difference $|m_i - m_j|$ and the variance σ were both changed to 300, 500 and 800 ms.

Figure 4.14 shows the summary of the results. The stimuli with the frequency of 50 Hz were mostly answered as materials with rough textures such as a polish sponge and a felt cloth, on the other hand, the stimuli of 100 Hz were recognized as comparatively smooth textures such as a cotton towel and plywood. Figure 4.14b shows the difference given by the parameter σ with the same frequency of 100 Hz. At $\sigma = 500$ ms, the sensations of a cotton towel and a plywood were mainly perceived. At $\sigma = 800$, the perception of a cotton towel was decreased and the answers of a felt cloth and a rubber increased. From the results, we could imply that the parameter σ would contribute to the sensations of softness and smoothness.

Figure 4.15 shows two selected parameter conditions, which gave the closest sensations to actual objects such as a felt cloth and plywood. As presented in the figures, by carefully selecting the driving parameters, the tactile display would present the tactile sensations given by actual objects. We also found that if one stimulus was presented to different subjects by the display, the perceived tactile sensations were occasionally different for each individual. Adversely, two different

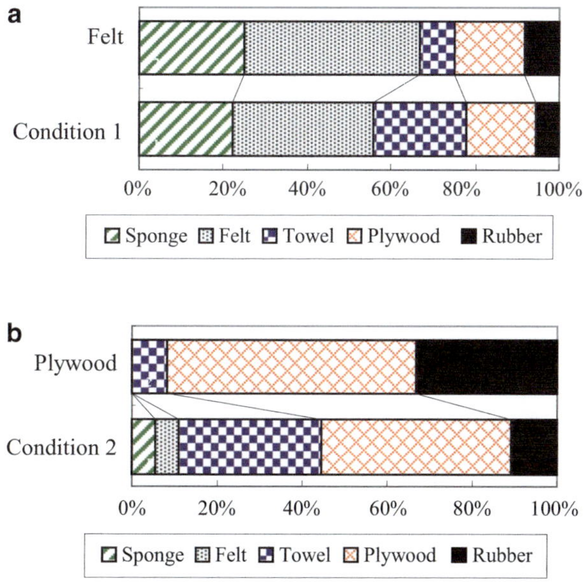

Fig. 4.15 Comparison with the results of Sect. 4.3.4. (**a**) Condition 1: 50 Hz, $|m_i - m_j| = 300$. (**b**) Condition 2: 100 Hz, $|m_i - m_j| = 300$

stimuli from the display would be recognized as the same sensations. We are still conducting experiments, and trying to justify the hypotheses of human tactile perception.

4.4 Applications to Tactile Displays and Tactile Interaction Systems

Tactile displays and tactile interaction systems are being developed by employing the SMA actuators. Since SMA wires with the diameter of 50 μm are thin enough to be attached to any surface of a conventional interface devices and also stitched in a cloth, various tactile devices will be realized. In this section, several devices and systems to present tactile sensation synchronizing with visual and auditory information are introduced.

4.4.1 Presentation of Texture Sensation from Pictures

Different tactile display devices have been developed so far as shown in Fig. 4.16, which are a pen device, a mouse and a specially-designed pad [9, 13]. For example, the tactile pen system consists of a tactile pen connected with a tablet device, a tactile controller, a PC and a visual display. Two pin-type tactile actuators are

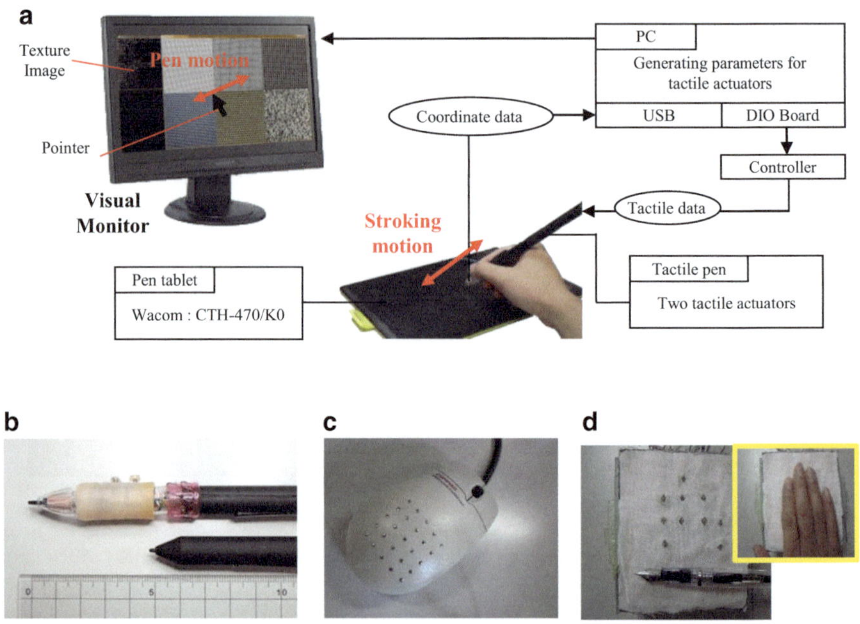

Fig. 4.16 Tactile display devices and tactile presentation systems. (**a**) Tactile pen system to present texture sensation synchronizing with pictures in a visual display. (**b**) Tactile pen. (**c**) Tactile mouse. (**d**) Tactile pad

mounted in the grip of the tactile pen, so that the tactile stimuli are presented to the user's thumb or the index finger. Tactile actuators present different tactile sensations based on the driving signals generated from the tactile controller. A tablet-device connected to a PC via USB is used for tracking the user's pen motions. A pointer in a visual display presents the trajectory in the tablet device. A texture picture is, at the same time, displayed in the visual display, and a user strokes the tactile pen on the tablet device, as if he rubs the texture picture in the display. Synchronizing with the stroking motions, he feels the tactile sensation presented by the tactile actuators in the pen grip.

4.4.2 Tactile Glove for Displaying Tactile Sensations in Tacto-gestural Interaction

The augmented reality (AR) system recognizes human gestural actions, and realizes visual, auditory and tactile interactions with a three-dimensional (3D) virtual object superimposed in a real environment as shown in Fig. 4.17. The system consists of a computer, a head mounted display (HMD), a pair of webcam, stereo speakers and a pair of Tactile gloves. User's gestures are captured by the stereo cameras mounted

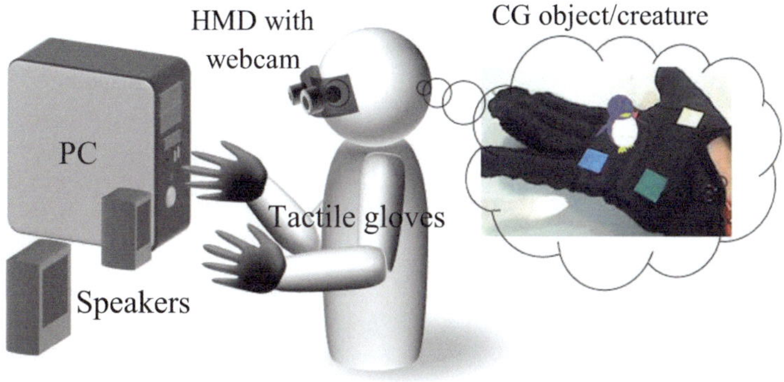

Fig. 4.17 AR interaction system with tactile gloves

Fig. 4.18 On-palm CG characters and the gestural interaction with tactile feedback

on the HMD, and the gestural actions are recognized in real-time. The tactile glove presents tactile sensations of a virtual object displayed in the HMD.

For presenting tactile sensations from a glove, 19 SMA wires in total are stitched inside the glove, where 9 wires form a 3×3 array to contact to the palm and 2 wires are placed at the tip of each finger.

We constructed an augmented reality system "PhantomTouch", in which a virtual penguin-like character or a robot character is walking on a user's palm as shown in Fig. 4.18. A user wears the HMD and the Tactile gloves, and interacts with the CG character by the hand gestures and gestural actions, while watching and listening the reactions of the character. In the initial state, the CG character is walking on the palm randomly in the area surrounded by the four color markers, and the user is able to feel the walking and stepping sensations and hears the sounds synchronizing with the character's location and walking behaviors.

The user is able to interact with the CG character by his gestural actions such as tilting the right hand, shaking the right hand and pinching the character with his left hand fingers. For example, if he tilts his right hand, the penguin starts to slip off from the palm as shown in Fig. 4.18, and the slipping sensations are presented from the glove. When he moves his hand up and down largely, the penguin bounces up and down on the palm, and at the same time the jumping and hitting sensations are presented. A user is also able to pinch the character up in the air using his left hand fingers. While he picks it up, he feels a pressing sensation on the inner surface of the thumb and index fingers. Then, in a couple of seconds, the character starts to slip off from the fingers, and he feels slipping sensations on the fingers.

4.4.3 Presentation of Pseudo-tactile Sensations in VR Space

We consider the method of presentation of a tactile feeling for users who can explore virtual objects by user's hands in a VR application, and develop a system that can improve interaction with VR objects enabling the user's hands to move freely and touch virtual objects in a VR environment. A tactile glove with SMA wires is employed to present pseudo-tactile sensations, and a VR application has been developed.

The system consists of a PC, an immersive HMD (Oculus Rift), a hand-motion sensor (Leap Motion), tactile gloves and an amplifier for tactile actuators. A user wears the tactile gloves and Oculus Rift that has Leap Motion mounted in the front. Oculus Rift has a wide viewing angle, and is equipped with a head angle measuring instrument so that the user can see a 360° view in virtual space. The motion sensor recognizes human hand motion in front of a face by two infrared sensors. These sensors are used to judge when a user's hand touches the VR object in this system. When the user touches the VR object, the tactile gloves present tactile sensation to the user's hand.

In this study, a tactile glove with a total of nine actuators in one's hand is employed. Four actuators are set in a pattern of 2×2 arrays facing to the palm inside the glove, and one actuator is set in each user's fingertip. First, the inside, white glove with the four actuators is worn. Then the user puts on the outside, black glove with finger actuators over the white one, so that the camera sensors can easily recognize the human hand, as shown in Fig. 4.19.

The system presents pseudo-tactile sensations by sending pulse current for controlling the SMA wires as described in the previous sections, when the user touches a virtual object in the virtual space. We firstly examined the difference of tactile feeling by different frequencies and amplitudes to associate the tactile

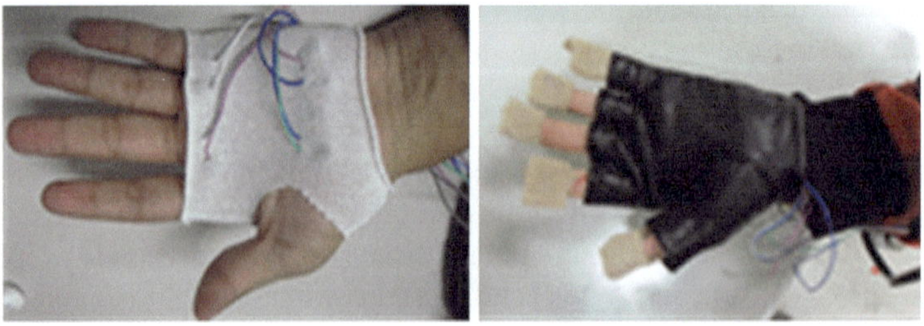

Fig. 4.19 Tactile glove and its nine tactile actuators inside

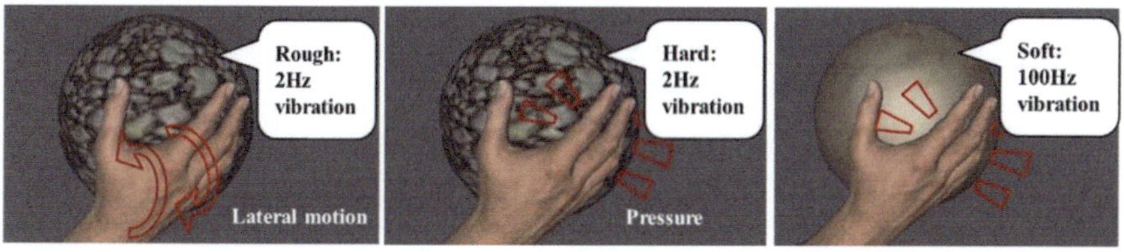

Fig. 4.20 Hand motions against virtual objects and their vibrating stimuli

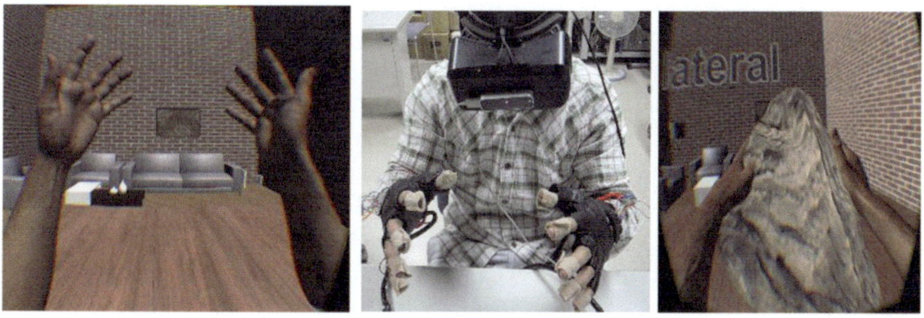

Fig. 4.21 System for presenting pseudo-tactile sensations in VR Space

parameters of controlling the SMA wires with virtual CG objects by presenting pseudo-tactile sensations. Examples of tactile presentation to the touch of different CG objects are shown in Fig. 4.20.

We have developed a virtual environment that can synchronize the movement of real hands with a hand CG as shown in Fig. 4.21. A user feels the pseudo-tactile sensations while looking of the visual appearance of his or her touching the virtual objects by the hand CG. Moreover, this system can present the realistic texture feeling by tactile presentation of the tactile actuators with each motion of users' hands.

4.4.4 A Novel Braille Display Using the Vibration of SMA Actuators

The visually impaired usually get the information through Braille and speech synthesizers. Since Braille is currently the most widely used communicative tool by visually impaired people, electronic Braille displays with actively sliding surface, consisting of 20–40 characters per line have been widely used. Such displays have been developed by researchers and developers to replace traditional Braille printed on the paper, and information in Braille can be easily accessed using these devices. For a normal Braille display, a user moves their fingers on top of each dot in a Braille cell in order to read the Braille character, which is called active touching.

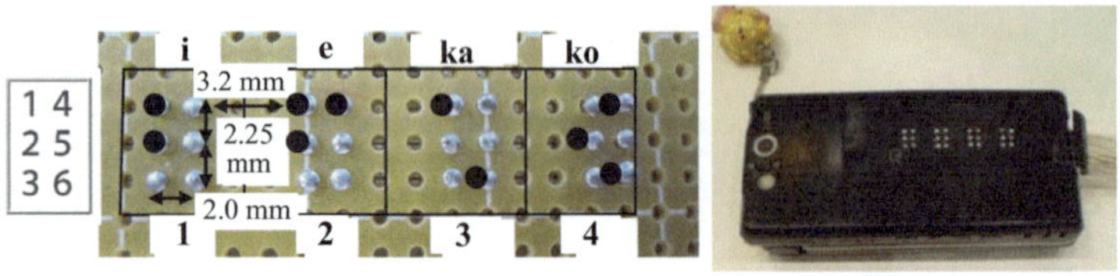

Fig. 4.22 Braille display with four cells and its installation to a mobile phone

Such Braille displays are constructed by using mechanical actuators such as small solenoid electromagnetic vibrators, small motors, piezoelectric actuators, pneumatic actuators, phase-change micro-actuators, ultrasonic actuators and electrostatic actuators. Small motors, for example, can easily be adapted as vibration actuators, and have been applied to mobile phones, PDAs, and Braille displays for visually impaired people in previous studies. However, the disadvantage is that the electromagnetic device is large, and the power consumption is greater than the energy provided by a mobile battery. Other actuators also have disadvantages such as the requirement of high-voltage and complex structure for mobile use.

Considering the above problems, we are developing a Braille display which has low power consumption, small size and passively touched features and can easily be combined with portable mobile devices. By using the vibration actuators, a compact Braille display with four Braille cells has been constructed as shown in Fig. 4.22. A user just puts the fingertip on top of the Braille cell in order to read the presented character.

4.4.5 A Screening Device of Diabetic Peripheral Neuropathy Based on the Perception of Micro-vibration Patterns

Diabetes mellitus is a group of metabolic diseases, which causes high blood sugar to a person, due to the functional problems of the pancreas or the metabolism. Patients of untreated diabetes would be damaged by the high blood sugar in vessels, and this starts to destroy capillary vessels to lower the sensitivity of tactile sensations, then effects to various organs and nerve systems. Diabetic peripheral neuropathy is one of the complications of diabetes mellitus, and we pay attention to the decline of the sensitivity of tactile sensations, and develop a non-invasive screening method of the level of diabetes using a novel micro-vibration actuator that employs shape-memory alloy wires.

In this study, a new quantitative micro-vibration sensation measurement device is developed to present variety of tactile stimuli by employing SMA actuators, and a pilot study using the device in diabetes patients has been performed to confirm a significant reduction in the vibratory sensation. The micro-vibration actuators using

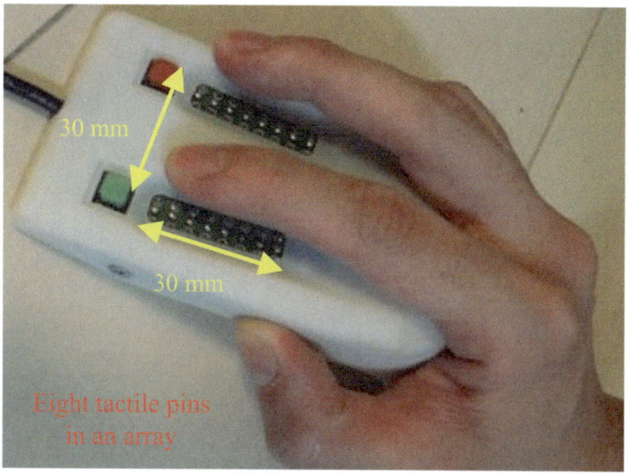

Fig. 4.23 Tactile display for screening diabetes

the SMA wires are arranged in an array, and various tactile stimuli are generated by just controlling the driving current signals.

As shown in Fig. 4.23, two tactile arrays are settled in a tactile display part, on which a patient places his index and middle fingers along with the indicators, so that the tips of the two fingers touch to the arrays. The presentation of vibratory stimuli generated by the tactile pins is made using tactile higher-level perceptual processes [6]. When the pins in an array are driven by pulse current signals with time delays, a subject perceives that a vibratory object continuously moves from Ch. 1 (fingertip) to Ch. 8 (second joint of a finger), due to the effects of tactile apparent movement perception. In this manner, moving tactile stimuli are presented from different directions, according to the time delay among tactile pins.

Tactile presentation using apparent movement perception was measured by using micro-vibration sensation measurement device at different frequencies and amplitudes. The amplitude of the vibration was divided into 30 steps from 1, with the lowest amplitude being difficult for healthy people with normal tactile sensitivity to perceive the movement, to 30, which has the greatest amplitude and is perceived even by a subject with severely damaged tactile sensation.

To examine the lowest threshold of tactile sensitivity of the index and middle fingers, we propose the Peripheral Neuropathy Vibration (PNV) score. A subject places his or her index and middle fingers on the pin arrays and is presented with tactile stimuli at different vibration frequencies and amplitudes in random directions as shown in Fig. 4.24. Through "yes" or "no" responses to questions about tactile perception, the system allows subjects for the measurement of threshold of tactile presentation, which is related to the severity of DPN.

Using this new device, we investigated the differences in the sensory threshold of index and middle fingers to compare DPN and non-DPN patients divided according to the American Diabetes Association diagnostic criteria. In the DPN group, PNV score measured by the finger method using the new device was significantly higher than that in the non-DPN group.

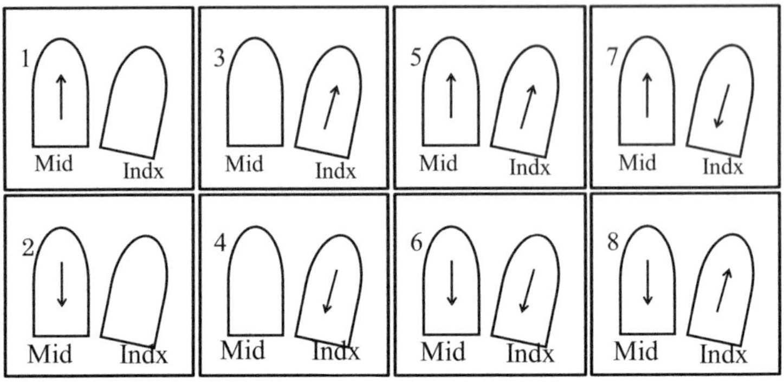

Fig. 4.24 Eight patterns of moving directions

4.5 Conclusions

In this chapter, a novel micro-vibration actuator using a shape-memory alloy for the presentation of various tactile sensations to human skin was introduced. The tactile actuator consists of a thin SMA wire with the diameter of 50 μm, and generates micro-vibrations with various frequencies and amplitudes in accordance with a driving pulse signal. The actuator features compactness and low energy consumption, and is applied to tactile displays and tactile interaction systems by attaching or stitching SMA wires in to any surface of a conventional interface devices. As examples of tactile presentations, several devices and systems for displaying tactile sensation synchronizing with visual and auditory information were introduced.

References

1. Mizukami, Y., Sawada, H.: Tactile information transmission by apparent movement phenomenon using shape-memory alloy device. Int. J. Disabil. Hum. Dev. **5**(3), 277–284 (2006)
2. Fukuyama, K., Takahashi, N., Zhao, F., Sawada, H.: Tactile display using the vibration of SMA wires and the evaluation of perceived sensations. Proc. IEEE Int. Conf. Hum. Syst. Interact. 685–690 (2009)
3. Hafiz, M., Sawada, H.: Presentation of button repulsive sensations on touch screen using SMA wires. Proc. IEEE Int. Conf. Mechatron. Autom. 1–6 (2011)
4. Zhao, F., Jiang, C., Sawada, H.: A novel Braille display using the vibration of SMA wires and the evaluation of Braille presentations. J. Biomech. Sci. Eng. **7**(4), 416–432 (2012)
5. Okumoto, S., Zhao, F., Sawada, H.: TactoGlove presenting tactile sensations for intuitive gestural interaction. Proc. IEEE Int. Symp. Ind. Electron. 1680–1685 (2012)
6. Sawada, H., Nakamura, Y., Takeda, Y., Uchida, K.: Micro-vibration array using SMA actuators for the screening of diabetes. Proc. Int. Conf. Hum. Sys. Interact. 620–625 (2013)
7. Jiang, C., Uchida, K., Sawada, H.: Research and development of vision based tactile display system using shape memory alloys. Int. J. Innov. Comput. Inf. Control. **10**(3), 837–850 (2014)

8. Sawada, H., Zhu, G.: Presenting tactile stroking sensations from a touch screen using the cutaneous rabbit illusion. Mecatronics **2014**, 371–376 (2014)
9. Sawada, H., Boonjaipetch, P.: Tactile pad for the presentation of tactile sensation from moving pictures. Proc. Int. Conf. Hum. Sys. Interact. 135–140 (2014)
10. Alles, D.S.: Information transmission by phantom sensations. IEEE Trans. Man-Mach. Syst. **MMS-11**(1), 85–91 (1970)
11. Bekesy, G.V.: Sensations on the skin similar to directional hearing, beats, and harmonics of the ear. J. Acoust. Soc. Am. **29**, 489–502 (1957)
12. Geldard, F.A., Sherrick, C.E.: "The cutaneous rabbit": a perceptual illusion. Science **178**, 178–179 (1972)
13. Takeda, Y., Sawada, H.: Tactile actuators using SMA micro-wires and the generation of texture sensation from images. Proc. IEEE/RSJ Int. Conf. Intell. Robot. Syst. (IROS2013). 2017–2022 (2013)

Chapter 5
Electro-tactile Display: Principle and Hardware

Hiroyuki Kajimoto

Abstract An electro-tactile (electrocutaneous) display is a tactile display that directly activates sensory nerves in the skin by electrical current supplied from an electrode on the skin surface. Compared with a mechanical tactile display that is typically composed of vibrating pins, the electro-tactile display has several merits, such as thinness and mechanical robustness. However, there remain several issues to be solved, such as stabilization. Furthermore, the development of the electro-tactile display requires certain knowledge of electrical circuits. The present paper thus serves as an introduction to research on electro-tactile displays. We start by explaining the principle of electrical nerve stimulation, introducing how the spatial distribution of the electrical current source affects stimulation using the notion of the activating function, and discuss temporal parameters (i.e., pulse width and pulse height) using a simplified nerve model and introducing the strength-duration curve. A typical hardware design, including a voltage-to-current converter circuit and a switching circuit is then introduced, and the electrode fabrication process and necessity of a conductive gel layer are discussed. Finally, the issue of sensation stabilization is treated with possible solutions.

Keywords Activating function • Chronaxie • Electrocutaneous • Electro-tactile display • Rheobase • Strength-duration curve

5.1 Introduction

An electro-tactile (electrocutaneous) display is a tactile display that directly activates sensory nerves in the skin with electrical current from an electrode on the skin surface [1, 2]. Compared with a mechanical tactile display that is typically composed of vibrating pins, the electro-tactile display has several merits such as thinness, deformability, low weight, low energy consumption, and mechanical robustness. Its applications include welfare devices for the visually handicapped, virtual reality and telexistence, and touch panels with haptic feedback (Fig. 5.1) [3–8].

H. Kajimoto (✉)
The University of Electro-Communications, Tokyo, Japan
e-mail: kajimoto@kaji-lab.jp

H. Kajimoto et al. (eds.), *Pervasive Haptics*, DOI 10.1007/978-4-431-55772-2_5

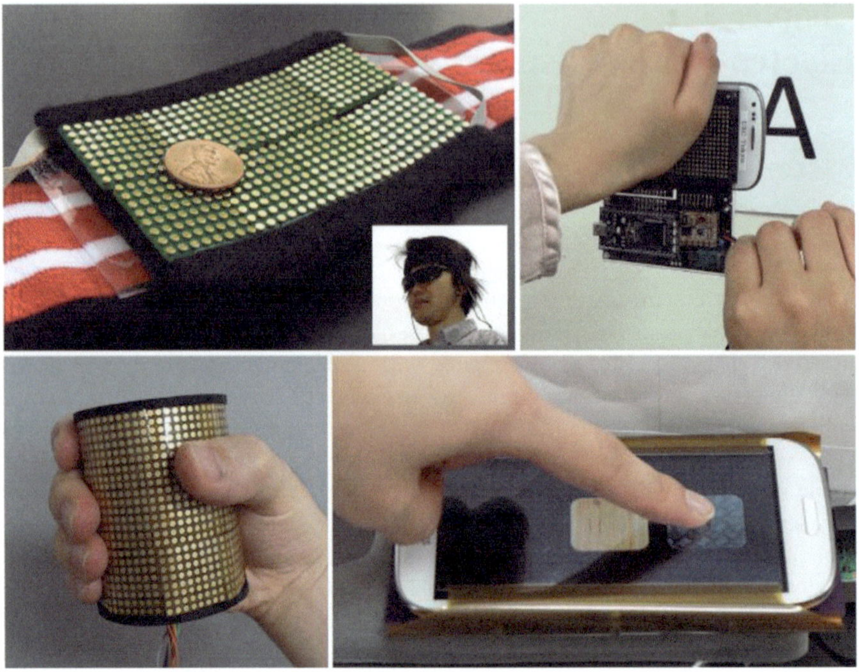

Fig. 5.1 Applications of the electro-tactile display. (*Top left*) Forehead stimulation for a tactile-vision substitution system (TVSS) [1]. (*Top right*) TVSS of palm type [4]. (*Bottom left*) Cylindrical type for virtual-reality application [5]. (*Bottom right*) Touch-panel type using transparent electrodes [6]

Several issues relating to the electro-tactile display remain to be solved; e.g., a method of rendering natural tactile sensation needs to be developed, pain felt by the user needs to be minimized, and the strength of the sensation needs to be stabilized. Furthermore, the development of an electro-tactile display requires certain knowledge of electrical circuits.

Against the background described above, this paper serves as an introduction to research on electro-tactile displays. We first explain the principles of electrical stimulation and typical hardware design. We then go on to discuss current issues relating to the electro-tactile display and their possible solutions.

5.2 Principle of Stimulation

The electro-tactile display activates sensory nerves under the skin with surface electrical current. We first introduce how the spatial distribution of the electrical current source (i.e., electrode placement and polarity) affects stimulation using an activating function, and then discuss temporal parameters (i.e., pulse width and pulse height).

5.2.1 Activating Function

When electric potential is distributed along the outer membrane of a nerve, the nerve is depolarized and activated. This model was first described by McNeal [9], and its use led to the precise modeling and simulation of electrical nerve stimulation. As the model considers the dynamical change of nerve membrane conductance and cannot be used to easily grasp general trends, we introduce Rattay's simple model [10] that sets nerve electric parameters constant and assumes that the nerve is activated when the nerve membrane potential reaches a certain threshold.

Figure 5.2 is a schematic of electrodes, skin, and a nerve and the electrical representation of the system. The nerve is composed of a membrane and internal medium. The system is discretized so that one unit is composed of a membrane with conductance G_m and capacitance C_m and an internal medium with conductance G_a. This discretization matches reality when the nerve is myelinated (i.e., the nerve is insulated by a myelin sheath and is only accessible through small gaps in the sheath).

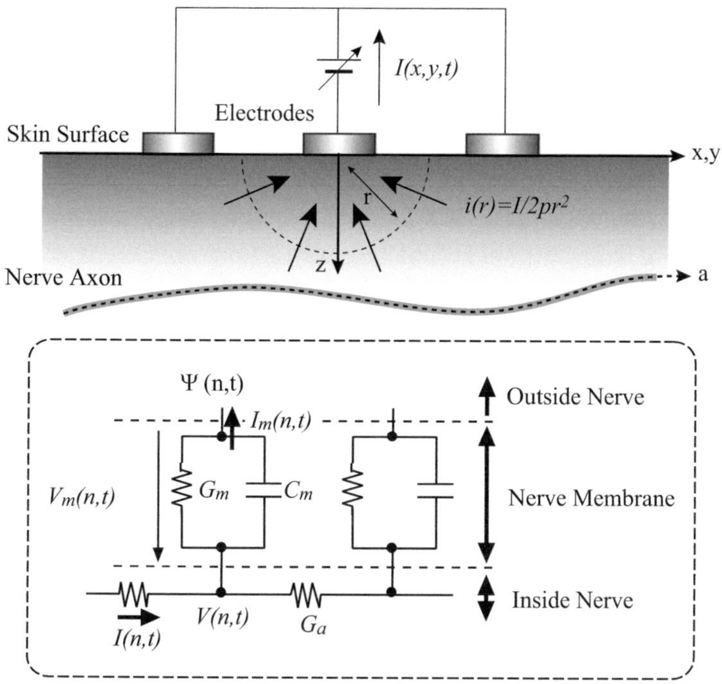

Fig. 5.2 Electrical representation of electrical stimulation. Electrical current I from the skin surface generates a potential distribution. The potential distribution Ψ along the nerve axon generates membrane current I_m, resulting in a membrane voltage difference V_m. G_m: membrane conductance, C_m: membrane capacitance, G_a: internal conductance

Stimulation of the electrical nerve involves changing the membrane voltage difference V_m by changing the external membrane potential Ψ. The stimulation can be represented by a simple control system, with input Ψ and output V_m. Hereafter, we consider the relationship between input and output.

Denoting the membrane current through the nth partition as I_m(n), the internal current as I_a(n), and potentials inside and outside the partition as V_a(n) and Ψ (n), it follows that

$$V_m(n) = V(n) - \Psi(n), \tag{5.1}$$

$$I_m(n) = G_m V_m(n) + C_m \dot{V}_m(n), \tag{5.2}$$

$$I_m(n) = I(n) - I(n+1), \tag{5.3}$$

$$G_a (V(n+1) - V(n)) = -I(n+1). \tag{5.4}$$

From (5.3) and (5.4), it follows that

$$I_m(n) = G_a (V(n+1) - 2V(n) + V(n-1)). \tag{5.5}$$

From (5.2) and (5.5), it follows that

$$G_a (V(n+1) - 2V(n) + V(n-1)) = G_m V_m(n) + C_m \dot{V}_m(n). \tag{5.6}$$

Substituting (5.1) into (5.6) yields

$$\begin{aligned} &C_m \dot{V}_m(n) + G_m V_m(n) - G_a (V_m(n+1) - 2V_m(n) + V_m(n-1)) \\ &= G_a (\Psi(n+1) - 2\Psi(n) + \Psi(n-1)). \end{aligned} \tag{5.7}$$

As the term on the right and the third term on the left are spatial second-order differentials, rearranging the constant values yields

$$-G_a \frac{d^2}{dx^2} V_m + C_m \frac{d}{dt} V_m + G_m V_m = G_a \frac{d^2}{dx^2} \Psi. \tag{5.8}$$

This is a kind of one-dimensional heat transfer equation, where V_m can be regarded as temperature, the right hand side as input heat, and the third term on the left as heat radiation. As the purpose of electrical stimulation is to raise V_m (temperature) above a certain threshold, the right hand side (input heat) directly relates to the strength of the electrical stimulation. This term, a second-order spatial differential of the external membrane potential, is called the activating function [10].

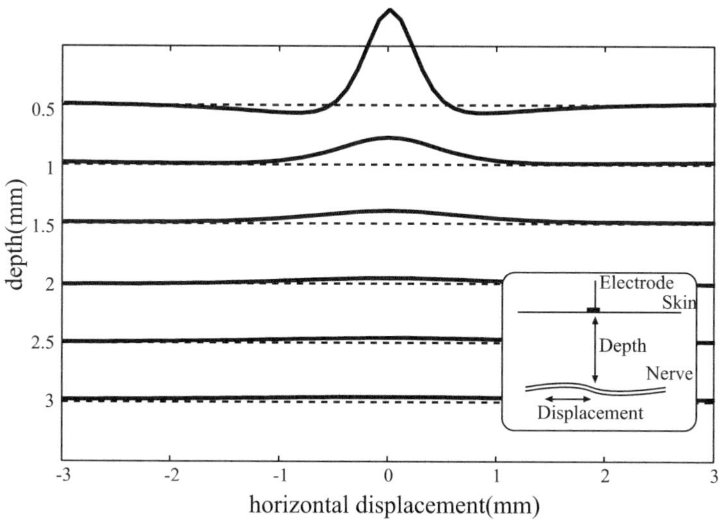

Fig. 5.3 Activating function of cathodic current from a single electrode, assuming the nerve runs parallel to the skin surface

The external membrane potential Ψ is formed by the distribution of electrical current. For example, let us consider the case when the electrical current is generated by a single point current source as in Fig. 5.2, and the medium is uniform. In this case, the equipotential surface is a hemisphere, and the potential distribution can be obtained by integrating current density along distance r from the current source [9]. In the case of multiple electrodes, if all electrodes are current-controlled, we can obtain the potential distribution by the superposition of potentials from each electrode.

Figure 5.3 shows the activating function produced by cathodic (negative) current from a single electrode. Here we assume that the nerve runs parallel to the skin surface, and the second-order differential is taken horizontally. We observe the following characteristics.

- For the nerve that runs parallel to the skin surface, there is a positive activating function in the case of cathodic current. This means that the nerve can be depolarized and activated.
- The activating function decreases as the nerve goes deeper. This means that deeper nerves are more difficult to activate.

The first characteristic is the reason why cathodic current stimulation is normally applied. We note that if the nerve runs perpendicular to the skin surface, the second spatial differential along the nerve takes a negative value, and we choose anodic (positive) stimulation. We address this topic in Sect. 2.3.

The second characteristic holds true not only in the case of a single electrode but also in the case of multiple electrodes. In other words, we cannot form a "local

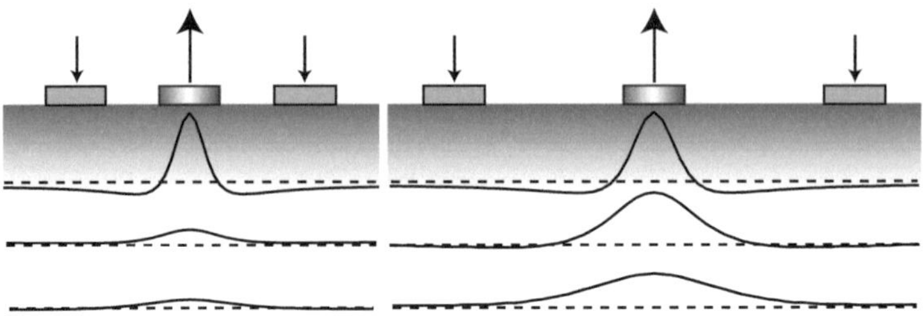

Fig. 5.4 Relationship between the activating function and the diameter of the concentric electrode. A larger inter-electrode distance results in greater and deeper stimulation

peak" of the activating function "inside" the tissue. This interesting limitation can be understood from the previously introduced analogy. No distribution of heaters can generate a local peak in temperature if the medium obeys the heat transfer equation.

However, it does not necessarily follow that selective activation of a deeper nerve is impossible. For example, as we will see in Sect. 2.2, thicker nerves are easier to activate, allowing the selective activation of deeper nerves if the deeper nerves are thicker.

The above discussion is for the case of a single electrode, assuming ground at infinity. In practical situation, concentric electrodes comprising a central stimulating electrode and peripheral ground electrode are used (Fig. 5.4). In this case, the ground is anodic (i.e., electrical current flows from the electrode). Considering the superposition of the activating function for each electrode, the activating function from peripheral ground electrode partly cancels the activating function from the central stimulating electrode. This results in confinement of the stimulation to a small and shallow region.

5.2.2 Strength-Duration Curve

The activating function introduced in Sect. 2.1 provides insight into the spatial design of electrical stimulation, such as the design of the electrode size and polarity. We also need to consider the temporal design; i.e., the waveform. While there have been several studies on the optimal waveform, especially in the field of electrical muscle stimulation. We simply consider a rectangular wave and two parameters: the pulse height and pulse width.

We consider a simplified model of a nerve as in Fig. 5.5. The nerve is represented by internal conductance G_a, membrane conductance G_m, and membrane capacitance C_m. As the electrical current from the surface electrode is converted to a potential distribution, the input to this model is voltage Ψ. The purpose of electrical stimulation is to raise V_m above a certain threshold by operating Ψ.

First, we set the differential equation

$$V_m = V_a - \Psi, \tag{5.9}$$

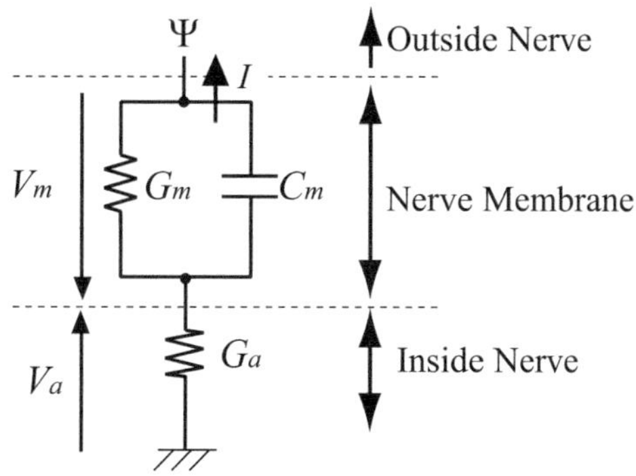

Fig. 5.5 Simplified model of electrical nerve stimulation

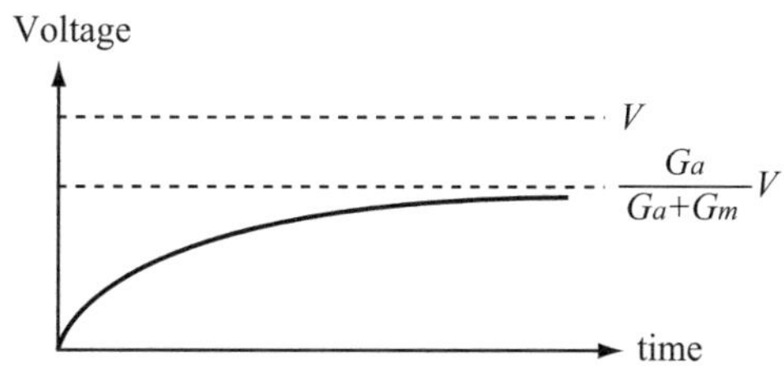

Fig. 5.6 Step response of the membrane voltage difference

$$I = G_m V_m + C_m \dot{V}_m = -G_a V_a. \tag{5.10}$$

Substituting (5.9) into (5.10) yields

$$V_m = -\frac{G_m V_m + C_m \dot{V}_m}{G_a} - \Psi. \tag{5.11}$$

The Laplace transform of (5.11) gives

$$\tilde{V}_m = -\frac{G_a}{C_m}\left(\frac{1}{s + \frac{G_a + G_m}{C_m}}\right)\tilde{\Psi}. \tag{5.12}$$

We now consider a step input $-V$ (cathodic stimulation):

$$\Psi(t) = \begin{cases} 0 & (t < 0) \\ -V & (t \geq 0) \end{cases}. \tag{5.13}$$

Substituting the Laplace transform of (5.13) into (5.12) and taking the inverse Laplace transform gives

$$V_m = \frac{G_a}{G_a + G_m}\left(1 - \exp\left(-\frac{G_a + G_m}{C_m}t\right)\right)V \ (t \geq 0). \tag{5.14}$$

This is the step response of the membrane voltage difference, as in Fig. 5.6.

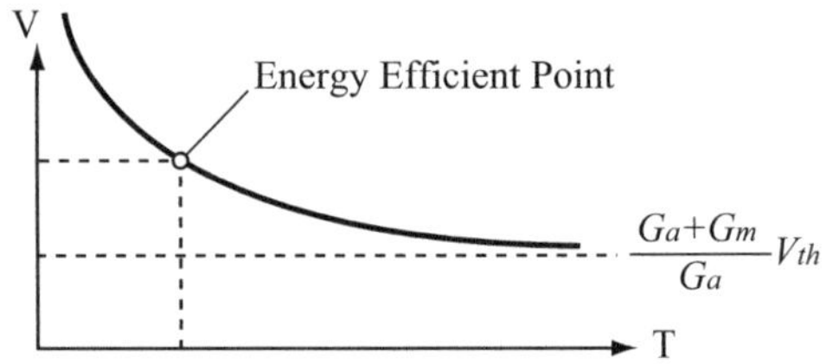

Fig. 5.7 Strength-duration curve showing the relationship between the pulse width and pulse height when the nerve is only just stimulated

Hereafter, we set $b = G_a/(G_a + G_m)$ and $a = (G_a + G_m)/C_m$. The converging value of the step response is written as bV.

Normally, the electrical stimulation is composed of a finite pulse, not an infinite step function. Let the pulse width be denoted T. The condition in which the membrane voltage difference reaches the threshold V_{th} is

$$V_{th} \leq V_m = b\,(1 - \exp(-aT))\,V. \tag{5.15}$$

We thus find the important relationship between pulse width T and pulse height V for which the nerve is only just stimulated:

$$V = \frac{1}{b}\frac{V_{th}}{1 - \exp(-aT)}. \tag{5.16}$$

Figure 5.7 shows this relationship, referred to as the strength-duration curve.

We observe a near inversely proportional relationship between the threshold pulse height and pulse width. This is natural considering that the system is basically supplying an electrical charge to the membrane capacitance, and a larger current results in faster charging.

However, the expression in (5.16) it is not exactly inversely proportional. For example, we require a certain pulse height when the pulse width is infinite. This value, called the rheobase [13], is calculated as

$$\lim_{T\to\infty}\frac{1}{b}\frac{V_{th}}{1 - \exp(-aT)} = \frac{V_{th}}{b}. \tag{5.17}$$

5.2.2.1 Nerve Thickness and Threshold

Let us consider the relationship between the nerve thickness and ease of stimulation. Considering the relationships between the nerve diameter and nerve parameters G_a, C_m and G_m, as the membrane thickness does not change, G_a is proportional to the cross-sectional area (i.e. proportional to the square of the diameter) of the nerve and C_m and G_m are proportional to the perimeter (i.e., proportional to the diameter) of the nerve. Therefore, as the nerve thickens, V_m in (5.14) rises more rapidly. A thick nerve thus has a lower threshold than a thin nerve, especially when the pulse is narrow. This difference diminishes as the pulse width increases.

This is one reason why stimulation with a narrower pulse is generally preferred in electro-tactile display, as pain-related nerves are thinner than mechanoreceptor-related nerves. When we consider myelinated nerves, most of the nerve membrane is covered by insulator and the changes in C_m and G_m are more suppressed, leading to much easier stimulation of thicker nerves [11, 12].

5.2.2.2 Optimal Pulse Width

There might be several criteria used to decide the optimal pulse width, such as lower pain sensation or robustness against variations in skin conditions. Here we consider energy consumption as the most basic criterion for optimal electrical stimulation. This leads not only to lower energy stimulation but also to lower pain stimulation, since one cause of pain during electrical stimulation is Joule heat.

We regard energy E to be proportional to V^2T:

$$E \propto V^2 T \propto \left(\frac{1}{1 - \exp(-aT)} \right)^2 T. \tag{5.18}$$

We set the right-hand side as $F(T)$. Differentiating $F(T)$, we obtain the optimal pulse width T_{optim} that achieves minimum energy, and by substituting T_{optim} into (5.16), we obtain the required pulse height V_{optim}. Numerically obtaining these values, we find the optimal pulse height is 1.4 times the rheobase that we introduced in Sect. 2.2.

The above calculation is derived from the simplified nerve model. However, in the field of electrical stimulation, the relationship between the pulse width and pulse height is generally fitted using an inversely proportional function with an offset:

$$V = c + d/T. \tag{5.19}$$

This modeling does not have a solid physical background. However, with this simple function, the minimum required pulse height is c (rheobase) and the optimal pulse height for minimizing energy V^2T is obtained as

$$E \propto V^2 T \propto (c + d/T)^2 T, \tag{5.20}$$

$$\frac{\partial E}{\partial T} \propto (c + d/T)(c - d/T). \tag{5.21}$$

From (5.21), the optimal pulse width becomes $T = d/c$, and the optimal pulse height is $2c$ (twice the rheobase). The required pulse width for minimizing energy is called chronaxie and is an important parameter for electrical stimulation. The chronaxie differs for different types of nerves, and ranges from 0.35 to 1.17 ms for sensory nerves in arms [13].

The above discussion only dealt with the optimization of energy for a single nerve. However, as mentioned before, there are numerous possible criteria for optimality. As we discussed in Sect. 2.2.1, a thick nerve is easier to stimulate than a thin nerve, and this tendency becomes clearer when the pulse is narrower. To avoid pain sensation from the activities of thin pain nerves, the electro-tactile display uses a shorter pulse, generally ranging from 50 to 200 μs [8]. The pulse height differs for different pulse widths and electrode sizes. When using electrodes with a diameter of 1 mm, the pulse height is approximately 1–10 mA.

5.2.3 Cathodic and Anodic Stimulation

As mentioned in Sect. 2.1, electrical stimulation generally uses cathodic current for stimulation. This means that electrical current enters the skin at the ground electrode, and exits the skin at the stimulating electrode.

There are some cases that anodic (positive) current plays an important role. Kaczmarek discovered that, in the case of fingertip electrical stimulation, the anodic current stimulation has a much lower threshold than the cathodic current [14]. Kajimoto mentioned the possibility of selectively stimulating different types of nerves using different polarities. He further discussed the possibility of achieving any kind of tactile sensation by combining receptor selective stimulations, in what is referred to as the tactile primary color approach [16].

There are two possible reasons why anodic current stimulation is effective in electro-tactile display, although cathodic stimulation is used in general electrical nerve stimulation. The first reason is that, in electro-tactile stimulation, the nerve terminal is close to the electrode, and the simple assumption of an infinitely long nerve fiber does not hold true. The other reason is that the nerves may not run parallel to the skin surface. Physiological studies have revealed that the nerve terminal of Meissner's corpuscle runs perpendicular to the skin surface. This seems to explain the fact that anodic stimulation mostly generates vibratory sensation, which is the role of Meissner's corpuscle.

The anodic stimulation has another merit. In an ordinary cathodic stimulation, the sensation elicited point does not coincide with the stimulation point because the sensation is elicited at the mechanoreceptor (i.e., nerve terminal) whereas the electrical stimulation is conducted at the midpoint of the nerve (Fig. 5.8, left). In

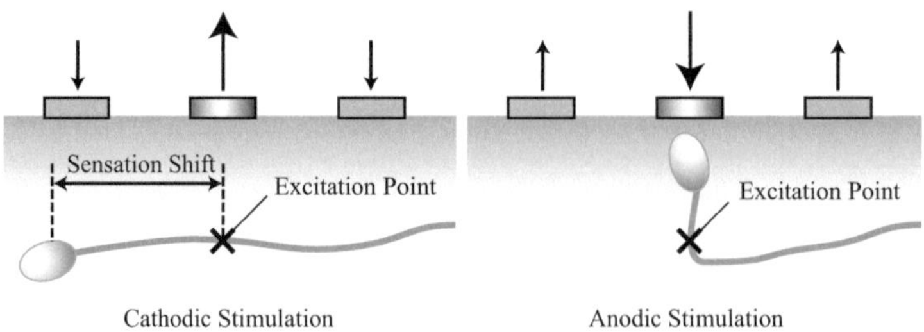

Fig. 5.8 (*Left*) Cathodic stimulation. (*Right*) Anodic stimulation. The anodic stimulation generates a spatially confined sensation because it activates the nerve terminal or vertically oriented nerves

electro-tactile display, this "sensation shift" is a critical issue. Although we stimulate the finger pad, we feel the sensation at the fingertip, and it is thus not possible to realize a high-resolution tactile display. In contrast, anodic stimulation does not generate this kind of spatial gap, since it activates the nerve terminal or vertically oriented nerves as discussed above (Fig. 5.8, right).

5.3 Hardware

The electro-tactile display is composed of electrical components, including a PC (personal computer) for pattern generation, microprocessor for high-speed pulse control, D/A (digital to analog) converter, V/I (voltage-current) converter if the display employs current-controlled stimulation, switching circuit, and electrodes (Fig. 5.9).

5.3.1 *Voltage-Current Conversion*

There are two main designs of the electro-tactile display circuit. One employs voltage regulation [15] and the other current regulation [3, 4, 7, 8]. The voltage-controlled electrical circuit is simpler, but safety issues such as a short circuit between the electrodes and greater electrical current due to sweat must be handled. As the situation of the electro-tactile display is characterized by a frequent change in the contact condition, current control is mainly used.

Figure 5.10 is an example of the voltage-current conversion design. The design is composed of a converter circuit and current mirror circuit that allows high-voltage driving. V_{in} is added to V_{+} of the operation amplifier, the amplifier output voltage,

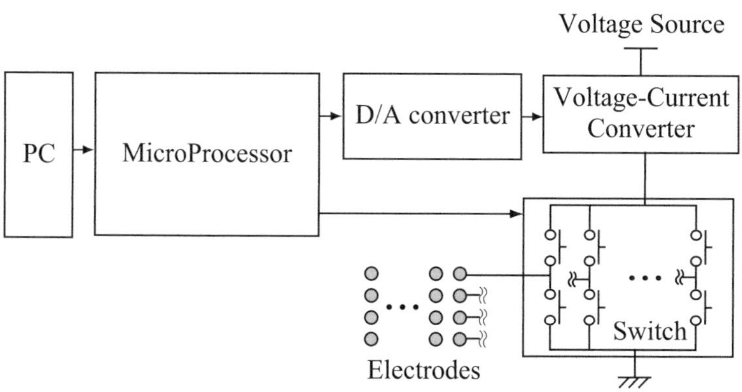

Fig. 5.9 Typical system architecture of an electro-tactile display

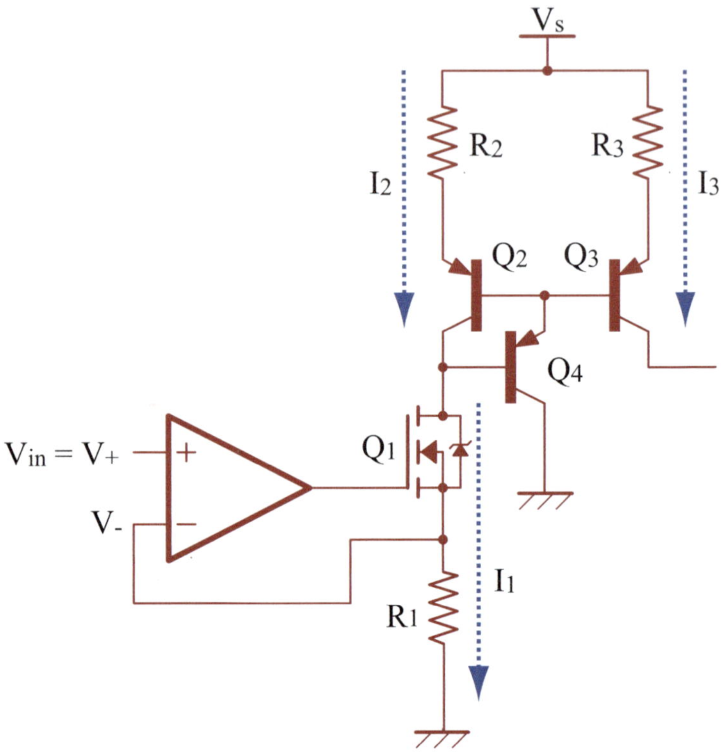

Fig. 5.10 Voltage-current conversion design composed of a converter and current mirror

which is applied to the gate of field-effect transistor Q_1. Electrical current I_1 flows and a voltage is generated at R_1, which is fed back to V_- of the operational amplifier. According to the principle of virtual ground, $V_+ = V_-$, which gives $I = Vin/R_1$. In other words, input voltage is converted to current.

The current mirror circuit works as follows. The electrical current passing through field-effect transistor Q_1 is supplied by PNP transistor Q_1. As the base of PNP transistors Q_2 and Q_3 are connected, the transistors have the same base voltage. As the base-emitter voltage of the transistor is almost constant, the emitter voltages Q_2 and Q_3 are almost the same. Therefore, for electrical current I_2 at R_2 and I_3 at R_3, $I_2R_2 = I_3R_3$ holds true. This means that electrical current is amplified by R_2/R_3. We can also drive the load with high voltage V_s, which is critically important to electro-tactile display.

In an ordinary electro-tactile display, the pulse width is around 50–200 μs, and the pulse height is approximately 1–10 mA. The resistance parameters are thus designed to output maximum current of around 10 mA. The resistance of a fingertip is approximately 50 kΩ for an electrode having a diameter of 1 mm and current pulse of 5 mA, and the voltage source V_s thus needs to be at least 250 V. To compensate for impedance variation of the skin, a voltage of approximately 350 V is desirable.

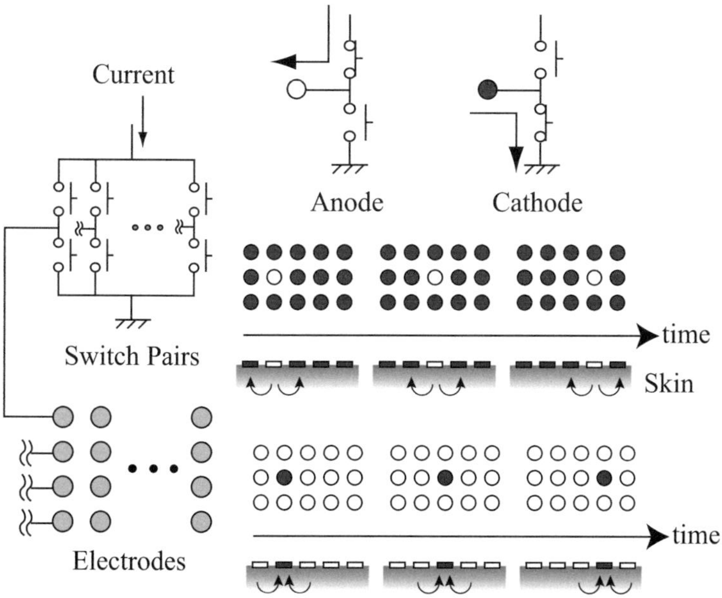

Fig. 5.11 Scanning stimulation using a switching circuit. A pair of switches changes the state of electrodes between anode and cathode. The spatial pattern is obtained by temporal scanning

5.3.2 *Switching Using Half-Bridges*

When driving numerous electrodes, it is impractical to prepare the same number of current sources as electrodes. A switching circuit is thus required to stimulate one electrode at a time and scan the electrodes. Such a method was first presented for cochlea implants [17, 18].

Figure 5.11 shows a schematic circuit diagram. Each electrode is connected to the current source and ground with switches. Such a pair of switches is called a half-bridge, because it is half of the typical bridge circuit used to drive a motor. When the electrode is connected to the current source, it becomes an anode. When it is connected to ground, it becomes a cathode. To achieve anodic stimulation, one electrode is connected to the current source and all other electrodes are connected to ground. This is electrically equivalent to anodic stimulation using a concentric electrode. Conversely, when one electrode is connected to the ground and all others are connected to the current source, cathodic stimulation using a concentric electrode is achieved. In other words, cathodic and anodic stimulation can be achieved using the same circuit.

If the two switches are off, the electrode is insulated, and such an arrangement might be used to realize a concentric electrode with a different diameter. If the two switches are on, there is a short circuit, which might be used to release unnecessary electric charge.

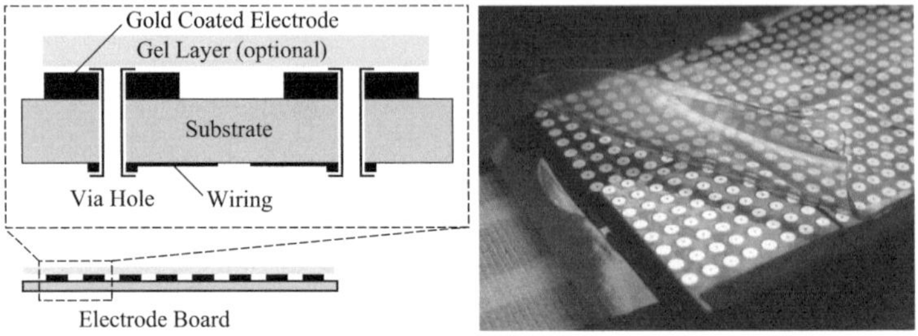

Fig. 5.12 Electrode structure and conductive gel. The electrode surface is gold coated and the wiring is on the other side of the substrate. The conductive gel works as a diffuser

A high-voltage photo coupler is typically used as a switch. A multi-channel switching integrated circuit for driving a microelectromechanical system is also commonly employed.

5.3.3 Electrode and Gel

In the measurement of the myoelectric signal, the material of the electrode directly affects the accuracy of measurement, and an Ag-AgCl electrode is commonly used. In contrast, for an electro-tactile display, the material of the electrode does not directly affect the elicited sensation, and materials that can prevent rust, such as stainless steel, carbon, and gold as a coating, are used. When numerous electrodes are needed, an electronic substrate can be fabricated, including the application of a gold coating as part of the fabrication process. Additionally, a flexible substrate can be used to achieve deformable or curved-surface electrodes [5].

Figure 5.12 (left) shows the structure of the electrode when an electronic substrate is fabricated. The wiring is on the other side of the substrate to prevent the risk of leaking current.

For the measurement of a myoelectric signal or for muscle electrical stimulation, conductive paste is commonly used to stabilize contact. Conductive paste is also effective in the case of the electro-tactile display but only when the electrode is large. When the electrodes are densely distributed and the gap between electrodes is 1 and 2 mm, the electrical current from the stimulating electrode only passes the conductive paste layer and reaches the surrounding electrode, without penetrating the skin. Therefore, for a typical electro-tactile display for the fingertip that uses tens of electrodes spaced at intervals of 2–3 mm, conductive paste is not used and the electrode is in direct contact with the skin.

When the same electrodes are attached to another part of the body such as the forehead or forearm, a strong pain sensation is sometimes elicited, whereas it is not observed in the case of the fingertip. This difference is presumably due to the difference in skin structure. The fingertip has a stratum corneum layer that is approximately 0.6 mm thick and most nerve fibers are below this layer. This thick layer works as a dissipation layer for electrical current. In contrast, other parts of the body have a stratum corneum layer that is a few tens of μm thick [19]. Therefore, local current concentration can sometimes stimulate thin nerves before activating thick nerves, which results in a pain sensation before a tactile sensation.

One way to prevent the issue described above is to use a conductive gel layer that has conductance equivalent to that of skin [1]. In contrast to commonly used conductive paste, this gel layer has a much higher volume resistance, and its thickness allows current dissipation, as in the case of the stratum corneum of the finger pad.

5.4 Stabilization

While there are numerous merits to the electro-tactile display such as its small size, thin dimension, low cost, energy efficiency and durability, the variability of the elicited sensation has hindered its practical use.

This variability is a result of two problems. First, temporal changes occur when the contact conditions are altered by sweat and motion (Fig. 5.13, top). Sweat alters the amount of sensation, while abrupt motion of the skin generates an electric shock sensation, sometimes perceived as an itchy feeling.

Second, spatial effects contribute to variability (Fig. 5.13, bottom). The threshold for tactile sensation (the absolute threshold) is close to the pain threshold [8]. Furthermore, there is substantial spatial variation in the threshold between different

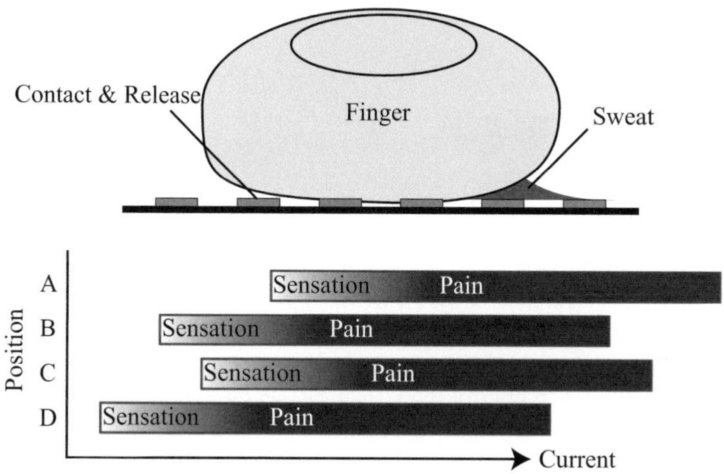

Fig. 5.13 Two factors contributing to the sensation variability of electro-tactile displays. (*Top*) Temporal changes caused by sweat and contact. (*Bottom*) Spatial variation of thresholds

electrodes. Thus, if hundreds of electrodes are used, and the stimulating currents are the same, it is practically impossible to exceed the absolute threshold at all electrodes without also exceeding the pain threshold at some. This problem is specific to the case of numerous electrodes, where the user cannot adjust the volume for each electrode.

Several solutions aimed at stabilizing the sensations generated by electro-tactile displays have been proposed. These can be classified into three categories as follows.

5.4.1 Explicit Feedback by the User

One practical solution is to adjust the stimulation in response to the user's explicit feedback [20]. A pressure sensor is placed under the electrode, and the electrical current is controlled by the sensor value. This simple method is effective for one finger and might be applied to the multi-touch situation if a multi-touch sensor is used, but requires the user to control each finger pressure independently.

5.4.2 Expansion of the Dynamic Range

Solutions in the first category aim to increase the difference between the absolute and pain thresholds. Collins [8] found that using a smaller pulse width (20–50 μs) was effective, while Polleto and van Doren [21] suggested applying a low-level pulse before the main pulse. Kaczmarek et al. [22] optimized several parameters, such as the number of pulses per burst, to obtain the maximal dynamic range. These methods represent important advances, but their effectiveness is limited.

5.4.3 Impedance Feedback

Solutions in the second category measure the electrical impedance of the skin and use the value to control the pulse. The voltage regulation and current regulation that we described in Sect. 3.1 can be regarded as an analog way of achieving impedance feedback in some way. At least, these analog circuit methods provide different strategies for dealing with impedance fluctuations.

To approach the problem in a more general way, it is necessary to construct a measurement-stimulation control loop, such that an electrode can be used for both stimulation and measurement. Tachi et al. [23, 24] regulated the pulse width, on the basis that perceived strength is related to input energy (current × voltage × pulse width). Watanabe et al. [25] reported a correlation between the skin resistance and absolute threshold, and Gregory, Xi and Shen [26] applied a similar technique to fingertips. Additionally, Kajimoto realized impedance feedback with a very fast feedback loop [27].

5.5 Conclusion

Intending to serve as an introduction to research on electro-tactile display, this paper summarized necessary knowledge of the principles of electrical stimulation, hardware design, and current solutions to the problem of stabilization. A search of the literature did not find a derivation of the strength-duration curve from a simple nerve model, and the result presented in Sect. 2.2.2 might therefore have certain novelty.

There are many issues relating to the practical use of electro-tactile displays to be solved, but the electro-tactile display has many merits such as its size, cost, energy efficiency and durability. It is hoped that this paper provides the knowledge required by researchers to participate in the attractive research field of electro-tactile displays.

References

1. Saunders, F.A.: In: Hambrecht, F.T., Reswick, J.B. (eds.) Functional Electrical Stimulation: Applications in Neural Prostheses, pp. 303–309. Marcel Dekker, New York (1977)
2. Saunders, F.A.: Information transmission across the skin: high-resolution tactile sensory aids for the deaf and the blind. Int. J. Neurosci. **19**, 21–28 (1983)
3. Kajimoto, H., Kanno, Y., Tachi, S.: Forehead electro-tactile display for vision substitution. In: Proceedings of EuroHaptics 2006, Paris (2006)
4. Kajimoto, H., Suzuki, M., Kanno, Y.: HamsaTouch: Tactile vision substitution with smartphone and electro-tactile display. In: CHI '14 Extended abstracts on human factors in computing systems, Toronto, 2014, pp 1273–1278 (2014)
5. Kajimoto, H.: Design of Cylindrical Whole-Hand Haptic Interface Using Electrocutaneous Display, Haptics: Perception, Devices, Mobility, and Communication. Lecture Notes in Computer Science, vol. 7283, pp 67–72 (2012)
6. Kajimoto, H.: Skeletouch: Transparent electro-tactile display for mobile surfaces. In: Proceedings of SIGGRAPH ASIA emerging technology session, Singapore (2012)
7. Bach-y-Rita, P., Kaczmarek, K.A., Tyler, M.E., Garcia-Lara, J.: Form perception with a 49-point electrotactile stimulus array on the tongue. J. Rehabil. Res. Dev. **35**, 427–430 (1998)
8. Collins, C.C.: Tactile television: mechanical electrical image projection. IEEE Trans. Man. Mach. **11**, 65–71 (1970)
9. McNeal, D.R.: Analysis of a model for excitation of myelinated nerve. IEEE Trans. Biomed. Eng. **23**(4), 329–337 (1976)
10. Rattay, F.: Modeling axon membranes for functional electrical stimulation. IEEE Trans. Biomed. Eng. **40**(12), 1201–1209 (1993)
11. Rubinstein, J.T., Spelman, F.A.: Analytical theory for extracellular electrical stimulation of nerve with focal electrodes. I. Passive unmyelinated axon. Biophys. J. **54**, 975–981 (1988)
12. Rubinstein, J.T.: Analytical theory for extracellular electrical stimulation of nerve with focal electrodes. II. Passive myelinated axon. Biophys. J. **60**, 538–555 (1991)
13. Geddes, L.A.: Accuracy limitations of chronaxie values. IEEE Trans. Biomed. Eng. **51**(1), 176–181 (2004)
14. Kaczmarek, K.A., Tyler, M.E., Brisben, A.J., Johnson, K.O.: The afferent neural response to electrotactile stimuli: preliminary results. IEEE Trans. Rehabil. Eng. **8**(2), 268–270 (2000)

15. Journée, H.L., Polak, H.E., de Kleuver, M.: Influence of electrode impedance on threshold voltage for transcranial electrical stimulation in motor evoked potential monitoring. Med. Biol. Eng. Comput. **42**(4), 557–561 (2004)
16. Kajimoto, H., Inami, M., Kawakami, N., Tachi, S.: SmartTouch: electric skin to touch the untouchable. IEEE Comput. Graph. Mag. **24**(1), 36–43 (2004)
17. McDermott, H.: An advanced multiple channel cochlear implant. IEEE Trans. Biomed. Eng. **36**(7), 789–797 (1989)
18. Jones, K.E., Normann, R.A.: An advanced demultiplexing system for physiological stimulation. IEEE Trans. Biomed. Eng. **44**(12), 1210–1220 (1997)
19. Montagna, W., Lobitz, W.C.: The Epidermis. Academic, New York (1964)
20. Kajimoto, H., Kawakami, N., Maeda, T., Tachi, S.: Electro-tactile display with force feedback. In: World Multiconf. on Systemics, Cybernetics and Informatics (SCI2001), Orlando, USA (2001)
21. Poletto, C.J., van Doren, C.L.: Elevating pain thresholds in humans using depolarizing prepulses. IEEE Trans. Biomed. Eng. **49**(10), 1221–1224 (2002)
22. Kaczmarek, K.A., Webster, J.G., Radwin, R.G.: Maximal dynamic range electrotactile stimulation waveforms. IEEE Trans. Biomed. Eng. **39**(7), 701–715 (1992)
23. Tachi, S., Tanie, K., Komiyama, K., Abe, M.: Electrocutaneous communication in a guide dog robot (MELDOG). IEEE Trans. Biomed. Eng. **32**(7), 461–469 (1985)
24. Tachi, S., Tanie, K.: U.S. Patent 4,167,189, 1979
25. Watanabe, T., Watanabe, S., Yoshino, K., Futami, R., Hoshimiya, N.: A study of relevance of skin impedance to absolute threshold for stabilization of cutaneous sensation elicited by electric current stimulation. Biomechanisms **16**, 61–73 (2002)
26. Gregory, J., Xi, N., Shen, Y.: Towards on-line fingertip bio-impedance identification for enhancement of electro-tactile rendering. In: Proc. IEEE/RSJ Int. Conf. Intelligent Robots & Systems, pp. 3685–3690 (2009)
27. Kajimoto, H.: Electro-tactile display with real-time impedance feedback using pulse width modulation. IEEE Trans. Haptic **5**(2), 184–188 (2012)

Chapter 6
Solid Ultrasonics Tactile Displays

Masaya Takasaki

Abstract Ultrasonic vibration can generally be excited by piezoelectric materials and can propagate in solids coupled to these materials. Some researchers have applied ultrasonic vibration, when excited on solid surfaces, to the realization of tactile displays. Users touch the display surface with a normal direction vibration. The squeeze film effect, which is induced by ultrasonic vibration, can then modify the surface roughness because of the presence of an air gap. From an alternative viewpoint, the friction between the surface and the user's finger can be reduced. The tactile display can then be configured by controlling this friction reduction process in the time domain. The application of surface acoustic waves (SAWs) to tactile displays has also been proposed. In the case of SAWs, the operating frequency is higher and the friction reduction process can thus be expected to realize presentation of tactile sensation. For future tactile display applications, transparent display technology is expected to enable users to enjoy both visual and tactile information on one screen. Additional topics related to ultrasonic tactile displays are also introduced in this chapter.

Keywords Ultrasonic vibration • Vibration in solid • Friction control • Tactile display

6.1 Introduction

One promising candidate method for presentation of tactile sensation is a type of friction control. If the friction between a finger and a plate can be changed dynamically during a rubbing motion on the plate, the skin of the finger experiences a continuous change in friction. When the change is repeatedly increasing and decreasing, the experience may be considered to be a kind of vibration. If this vibration coincides with or is similar to the vibration that occurs when we rub the surface of an existing object, then friction control can be applied as the principle of a tactile display.

M. Takasaki (✉)
Saitama University, Shimo-Okubo 255, Saitama 3388570, Japan
e-mail: masaya@ieee.org

H. Kajimoto et al. (eds.), *Pervasive Haptics*, DOI 10.1007/978-4-431-55772-2_6

A number of researchers have been focusing on the use of ultrasonic vibration to control friction. In this chapter, some ultrasonic tactile displays based on this type of hardware are introduced. Before these ultrasonic tactile displays are described, a basic overview of ultrasonics is given. For more detailed information on aspects of ultrasonics, such as piezoelectric constitutive relations, the piezoelectric constant or material properties, readers can refer to numerous textbooks. For tactile display applications, transparent tactile displays and related technologies are also introduced.

6.2 Piezoelectric Material and Ultrasonic Transducers

Ultrasonic vibrations occur at frequencies that are higher than the audible range. These vibrations propagate in solids, air and liquids. Today, many applications benefit from the use of ultrasonic waves, including non-destructive testing, medical diagnostics, ocean exploration, mechanical machining, and electronic components for communication devices. In this section, methods for the excitation of ultrasonic vibrations propagating in or on solid materials are introduced.

Material functions such as piezoelectricity, electrostriction, and magnetostriction can be used to excite ultrasonic vibrations. Recently, piezoelectric materials have been widely used as functional materials for ultrasonic wave applications, because various high-quality materials are commercially available, thanks to the findings and efforts of many scientists and engineers over the past few decades. Using these functional materials, we can easily design devices that apply ultrasonic vibrations.

When strain induced by stress is applied to a piezoelectric material, an electric charge appears on its surface (piezoelectric effect). Conversely, when an electric field is applied to the same material, the stress induced inside the material results in the deformation of the solid (inverse piezoelectric effect). These effects are induced by the crystal polarization. The charge and the stress appear in the direction parallel to that of the polarization.

Today, many types of piezoelectric material can be used. These materials can be divided into two groups. The first group is piezoelectric ceramics. Small crystals gather and configure these materials. The polarization direction of each crystal is not unique. Therefore, the polarizations cancel each other out, and a total polarization direction cannot be found. For this kind of material, a polarization process is required before application. Barium titanate ceramics ($BaTiO_3$) and lead zirconate titanate ceramics ($PbZrO_3$-$PbTiO_3$: PZT) are typical piezoelectric ceramic materials. The other type of piezoelectric material is the single crystal material. The volume of the material is filled with a unique crystal structure. In this case, the polarization direction can be defined according to the orientation of the crystal axis. Quartz (SiO_2), lithium tantalate ($LiTaO_3$) and lithium niobate ($LiNbO_3$) are well-known single-crystal piezoelectric materials.

For actual use, piezoelectric materials are shaped into plates. Opposing electrodes are formed on the material surface to apply an electric field, as shown in

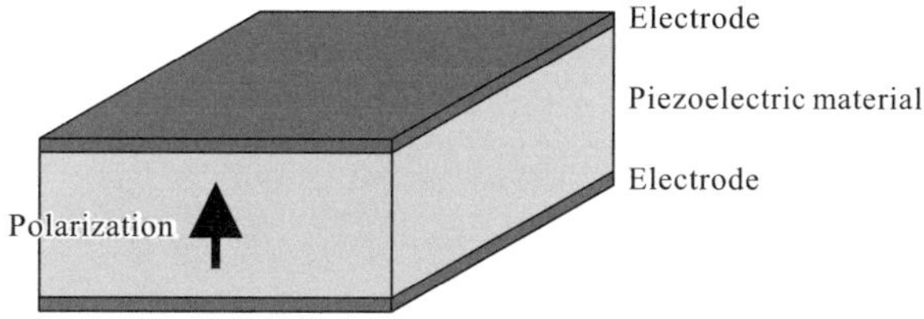

Fig. 6.1 Piezoelectric material with surface electrodes

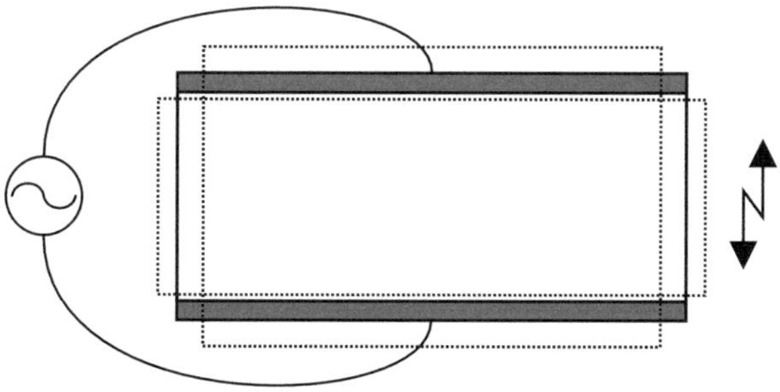

Fig. 6.2 Deformation of piezoelectric material

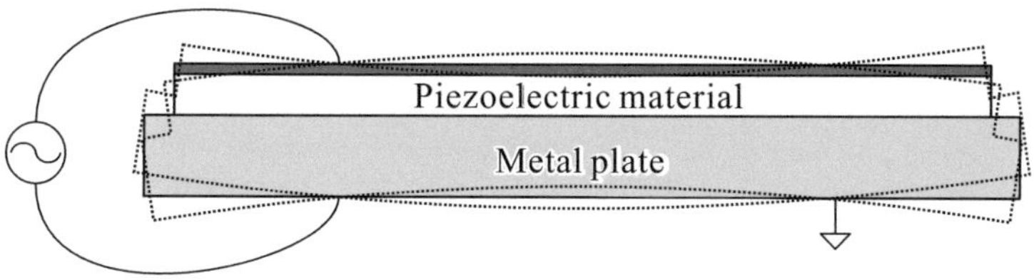

Fig. 6.3 Vibration of unimorph

Fig. 6.1. The polarization direction can be oriented to be perpendicular to the electrodes. When an alternating voltage is applied to these electrodes, the material vibrates along the polarization direction, as shown in Fig. 6.2. The deformation of the piezoelectric material is much smaller than that indicated by the illustration, and is around 1/1000 of the size of the material.

The plate-shaped piezoelectric material can then be bonded on a plate, as shown in Fig. 6.3. This configuration is called a unimorph. The plate material can be selected from a wide range of candidate materials, such as metal or glass. When metal is used, the plate itself can play the role of the electrode and provide the electric field to the piezoelectric material. When an alternating voltage is applied, the plate can be vibrated in a flexural mode, as illustrated in the figure, because the length of the piezoelectric material is changed by the inverse piezoelectric effect while the length of the plate remains constant. This configuration can be used as an ultrasonic transducer.

If the frequency of the applied voltage coincides with the resonance frequency of the vibration mode, then the vibration amplitude can be enlarged. In the resonance

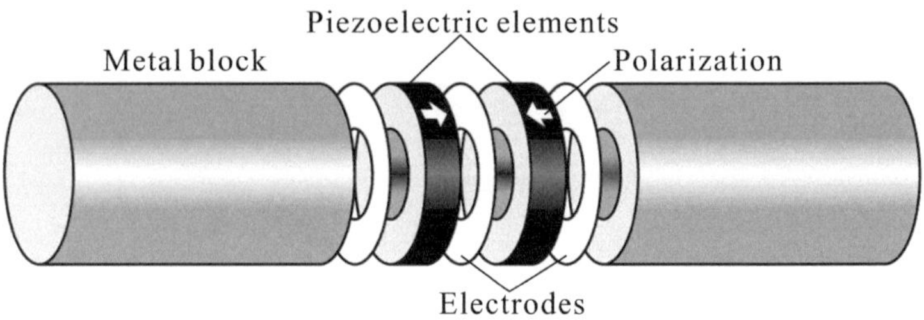

Fig. 6.4 Configuration of bolt-clamped Langevin-type transducer

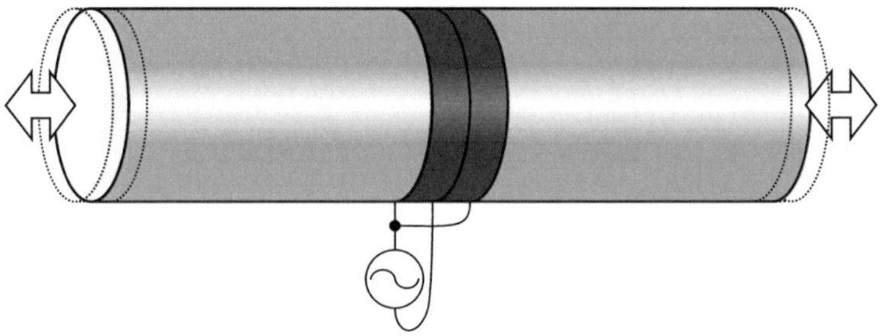

Fig. 6.5 Longitudinal vibration of Langevin-type transducer

state, the applied electric power can be transformed efficiently into mechanical vibration energy, and large currents are observed. This means that the impedance between the electrodes is at a local minimum at this frequency. Therefore, we can search for the resonance frequencies by measuring the impedance during a frequency sweep.

In this configuration, the lower surface of the plate can be grounded and isolated from the high voltage source to drive the piezoelectric material. This surface can then be used as a touching area for tactile displays. The use of larger plate sizes leads to lower bending vibration mode resonance frequencies. To avoid hearing the resulting vibrations, higher order vibration modes can be selected.

Another type of ultrasonic transducer is shown in Fig. 6.4. This transducer consists of metal blocks, ring-shaped piezoelectric elements, and ring electrodes. The elements and the electrodes are clamped together using the metal blocks and a bolt. The polarizations of the two elements oppose each other, as indicated by the white arrows in the figure. The clamped blocks are coupled acoustically and have longitudinal vibration modes. When an alternating voltage with an operating frequency that is equal to the resonance frequency of the first longitudinal vibration mode is applied, a longitudinal vibration is excited and each side vibrates, as illustrated in Fig. 6.5. This configuration is generally used to acquire high-intensity ultrasonic vibrations.

6.3 Control of Surface Roughness

Watanabe et al. proposed a method to control the tactile sensation of surface roughness on an existing object using ultrasonic vibration [1]. Based on their concept, smoothness was added to a rough surface. The ultrasonic vibration helped with the addition of the smoothness. They then developed an experimental apparatus and tested the addition to the surface.

Figure 6.6 illustrates the experimental apparatus used. Ultrasonic vibrations were excited by Langevin-type ultrasonic transducers. A horn was attached to each transducer to enhance the amplitude of the vibration. The top of each horn was then coupled to a plate. By applying an alternating voltage to the transducer, longitudinal vibrations were excited on the top of the horn. These vibrations excited flexural vibrations in the plate, as indicated by the dotted line in the figure. The top surface of the plate, which is designated the display surface, was used to control tactile sensation. The entire ultrasonic transducer system had several resonance frequencies, including 36.5, 67.5, 75.6, and 89.3 kHz.

The display surface was roughened using abrasive paper. This part was a real object. When an alternating voltage at one of the resonance frequencies was applied to one of the Langevin transducers, a standing wave of flexural vibration was excited in the plate. The loop areas of the standing wave were used as the tactile presentation area. The vibration amplitude in the vertical direction in this area was approximately 2 μm. Participants in the experiments touched this area. Under their fingers, an air gap induced by the squeeze film effect was generated, as shown in Fig. 6.7. This gap prevented direct contact between the finger and the rough surface of the plate and helped to add smoothness to the rough sensation provided by the roughened surface.

Short-duration ultrasonic vibrations were also used to present texture sensation with the same experimental apparatus. The vibrations were generated by amplitude modulation of the voltage applied to the transducer. The participants felt "rougher" or "stickier" feelings when they rubbed the plate. It was noted that these feelings

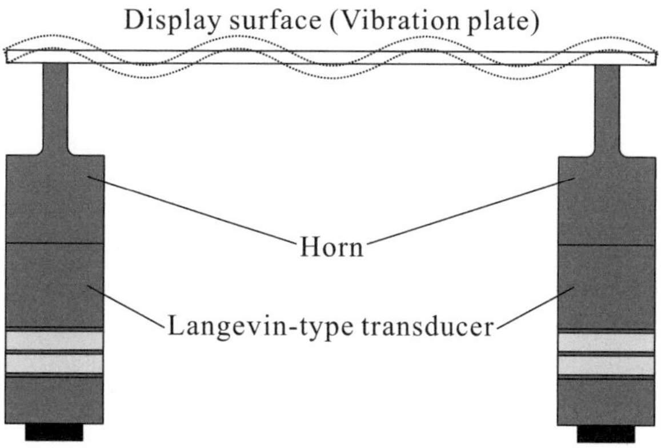

Fig. 6.6 Structure of experimental apparatus [1]

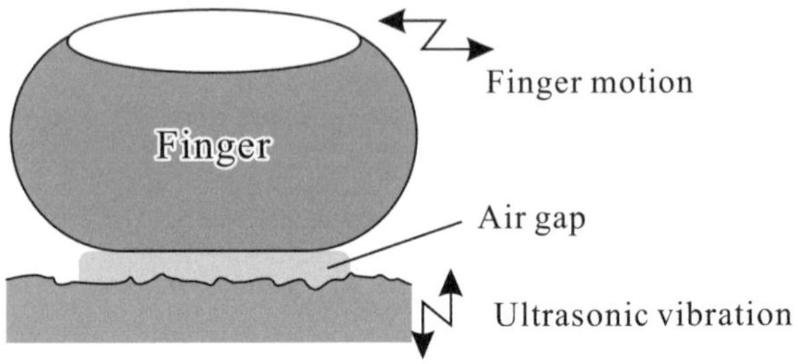

Fig. 6.7 Smoothness by air gap [1]

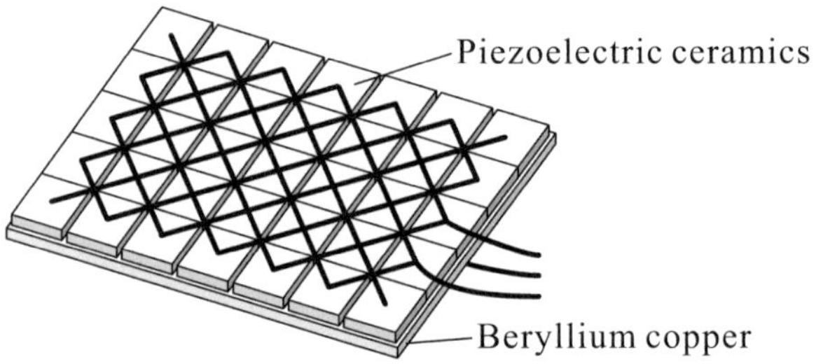

Fig. 6.8 Tactile plate with piezoelectric ceramics [2]

were perceived as resistance in the direction of rubbing motion. The authors of Watanabe and Fukui [1] expected the representation of various textures to be realized through variation of the amplitude modulation.

6.4 Application of Amplitude Modulation

Biet et al. designed a plate ultrasonic transducer [2]. Figure 6.8 illustrates the structure of this transducer. A beryllium copper plate was used as the substrate and the piezoelectric ceramics were glued onto the back side of the substrate in matrix format, as shown in the figure. The dimensions of the ceramics and the substrate were 11 mm × 11 mm × 1 mm, and 83 mm × 49 mm × 2 mm, respectively. The resonance frequency of the vibration mode, which is illustrated in Fig. 6.9, was 30.5 kHz. At the loop point of the mode, a deflection amplitude of 2.3 $\mu m_{p\text{-}p}$ was obtained under an applied voltage of 15 V. Based on this vibration, the squeeze number was 15.9.

Tactile sensation tests were carried out using this tactile plate. Participants were asked to state whether or not the feeling changed while the amplitude of the applied voltage was controlled over four levels. Immediately after a change in the voltage, all participants described a change in sensation. Amplitude modulation of the applied voltage resulted in perception of a change of slipperiness or friction.

Maeno et al. proposed a tactile display for surface texture with amplitude modulation [3]. A stator transducer of an ultrasonic motor was used as an ultrasonic

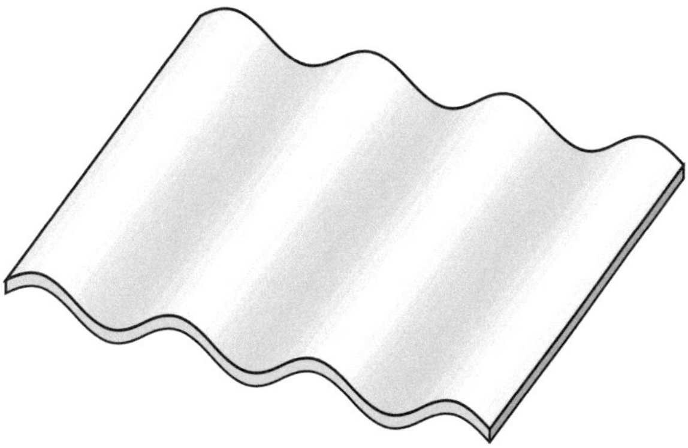

Fig. 6.9 Vibration mode of tactile plate [2]

vibrator. The vibration surface of the transducer contacted the fingertip of the user. When the applied voltage is amplitude-modulated at a lower frequency, then the amplitude of the ultrasonic vibration is also modulated. Only the envelope of the modulated vibration can be detected by the tactile receptors as a strain pattern, because the modulation frequency is within the detection range of the tactile receptors but the frequency of the ultrasonic vibration is outside this range. To display surface textures, the surface geometries of real objects were captured and used as modulation signals. During lateral rubbing motion on the transducer, the amplitude modulation was controlled with reference to the rubbing velocity. More than half of the experimental participants were able to recognize displayed sensations as one of the original textures.

6.5 Application of Surface Acoustic Wave

The surface acoustic wave (SAW) is a type of ultrasonic wave that propagates on a solid surface. The Rayleigh wave is well known as a kind of SAW. When the wave propagates on the solid surface, the surface vibrates along an elliptical locus, as shown in Fig. 6.10. Inside the medium on which the Rayleigh wave propagates, the vibration distribution is seen within two wavelengths of the surface. This means that there is no vibration on the back side of the medium, provided that the thickness of the medium is greater than two wavelengths. We can therefore glue the back side of the medium to fix the vibrator for mechatronics applications of SAW.

Nara et al. applied SAWs to tactile displays [4]. The basic structure of their display is shown in Fig. 6.11. When an alternating voltage is applied to an interdigital transducer (IDT) formed on a piezoelectric material substrate, a Rayleigh wave is generated and propagates on the substrate surface. A $LiNbO_3$ wafer was used as the substrate material in this case. The pitch and the width of the finger electrodes of the IDT were both 100 μm. The frequency of the applied voltage is determined by the IDT electrode dimensions, and was 9.6 MHz in this case. For

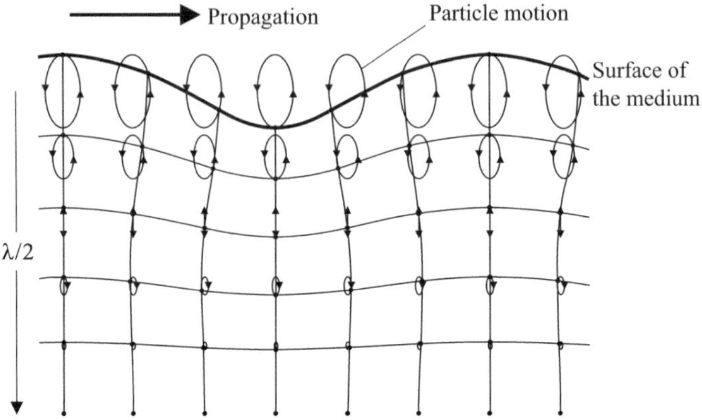

Fig. 6.10 Vibration mode of Rayleigh wave

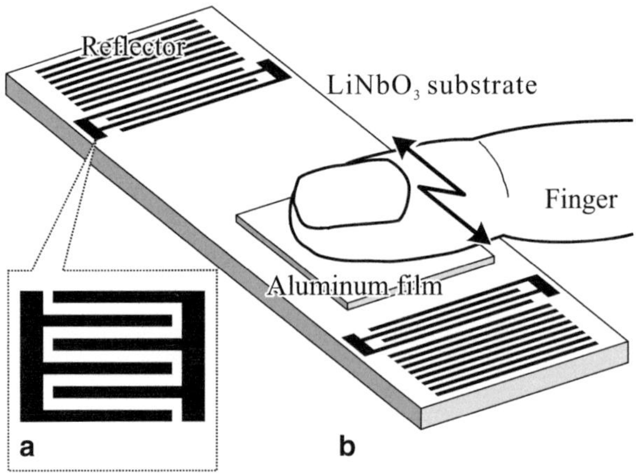

Fig. 6.11 (**a**) IDT and (**b**) basic structure of the tactile display

the SAW tactile display, two IDTs were arranged on the substrate and two sets of reflectors, which were configured using open metal strip arrays, were formed behind each IDT. Vibration energy can be stored between these reflectors as a standing wave. The central area of the substrate was used to present the tactile sensation. Users were able to rub the area through an aluminum film. The film had the dual role of providing finger protection against the high-intensity ultrasonic vibrations and enhancement of the SAW-based friction control.

When the user rubs the tactile sensation presentation area of the SAW substrate with the film, there is kinetic friction between the film and the substrate surface, as shown in Fig. 6.12a. By excitation of a standing Rayleigh wave during the rubbing motion, the friction coefficient is reduced because of intermittent contact induced by 15-MHz vibration of the substrate surface, as shown in Fig. 6.12b. If these two states

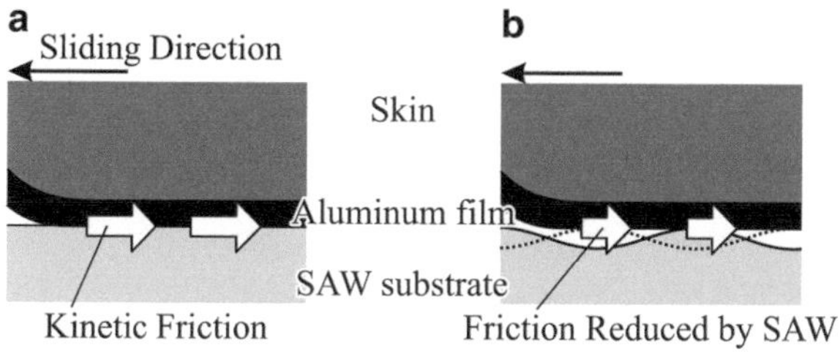

Fig. 6.12 Friction control for tactile display. (**a**) Without SAW. (**b**) With SAW

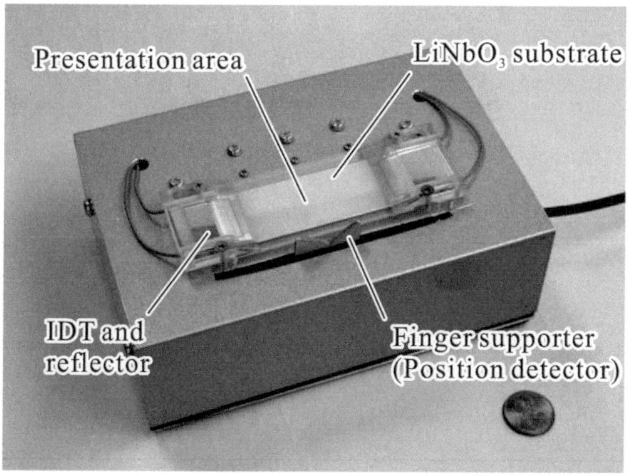

Fig. 6.13 Prototype SAW tactile display

are repeated, then the finger skin in contact with the film experiences fluctuations in the friction and these fluctuations result in the vibration of the skin. The excitation of the wave can be switched easily by an electric circuit. For example, if the switching frequency is 100 Hz, then a vibration with a frequency of 100 Hz can be excited on the skin of the finger. Control of the switching according to the rubbing motion can produce a tactile sensation via the vibration.

One of the developed SAW tactile displays is shown in Fig. 6.13. A 1-mm-thick $LiNbO_3$ 128-Y X-Prop wafer was used as a substrate. An aluminum layer was deposited by vacuum evaporation and was formed into IDTs and reflectors by photolithographic processing and wet etching. The wafer was then cut into 22×98 mm substrates. The IDTs and reflectors were protected by acrylic resin covers. The middle of the substrate was used as the tactile sensation presentation area. The substrate was supported by an acrylic resin base and was fixed on a chassis. To detect the user's finger position, a finger supporter was installed beside the substrate. This supporter was connected to an encoder inside the chassis.

Switching of the SAW driving signal was controlled by a microcomputer (SH2/7045F). The signal flows for control of the display [5] are shown in Fig. 6.14.

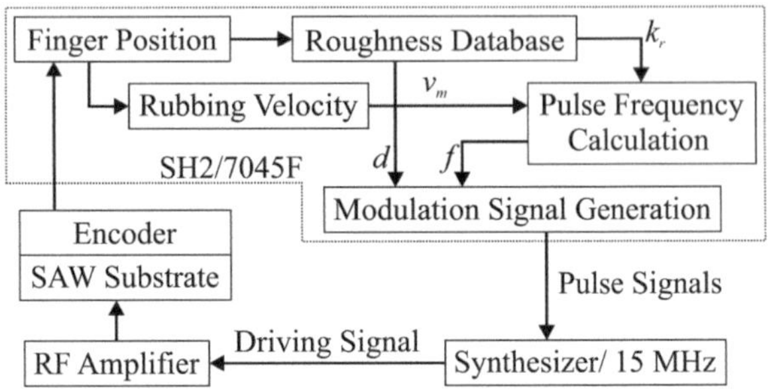

Fig. 6.14 Control process of tactile display [5]

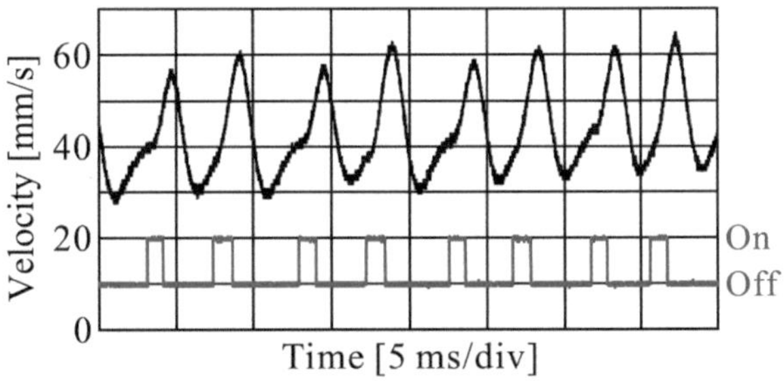

Fig. 6.15 Slider velocity under control ($k_r = 200$ μm)

The finger position was first detected using the encoder connected to the finger supporter and was used to calculate the rubbing velocity v_m. The switching frequency f was then determined using the following equation:

$$f = \frac{v_m}{k_r},$$

where k_r was the roughness factor [m]. k_r was determined by reference to a database inside the controller to enable different sensations to be presented according to the specific finger position. The duty ratio of the pulse wave used to switch the SAW excitation was also determined by reference to the database. Stronger vibrations could be excited behind the film on the SAW transducer by using higher numbers for the ratio. Based on the frequency and the duty ratio, the microcomputer generated a pulse wave. This pulse wave switched the SAW excitation signal in a signal synthesizer. The synthesized driving signal was then amplified by a radio-frequency (RF) amplifier and applied to the SAW substrate.

The velocity of the slider was measured by a laser Doppler velocity meter under the control process illustrated in Fig. 6.14. The measurement results are plotted in Fig. 6.15. The black line is the lateral velocity of the slider. The gray line is the control signal that was used to switch the SAW driving signal. The velocity increased after the edges of the control signal rose and decreased after the edges fell. The average value of the velocity, i.e., the DC component of the plotted signal, indicated the velocity of the rubbing motion. In contrast, the fluctuations in the velocity, which can be assumed to be the AC component, indicated the vibrations generated on the skin of the finger. With higher duty ratios and higher roughness factors, the vibrations generated can be perceived as rougher tactile sensations [5].

6.6 Transparent Tactile Display

The combination of visual information with tactile information is one potentially interesting application of tactile displays. To realize this type of application, one proposed solution is the development of a transparent tactile display that can be installed on the surface of a visual display such as a liquid crystal display (LCD). Users would then see images projected on the LCD and also touch these images to perceive tactile information. To realize such a transparent ultrasonic tactile display, the material of the ultrasonic medium must be transparent. One possible solution is glass.

Marchuk et al. glued a 16-mm-diameter piezoelectric ceramic disk on the surface of a $7.6 \times 7.6 \times 0.3$ mm borosilicate glass plate, as shown in Fig. 6.16 [6]. The gray solid lines indicate the nodal lines when the driving frequency was 36.1 kHz. In addition, this ultrasonic transducer had three more vibration modes, and each of these modes had its own nodal line pattern. On these lines, we cannot expect friction reduction, because there is a small vibration amplitude on each line. To solve this problem, mode switching was proposed. Every point on the surface of the plate may be a loop area for one of four vibration modes.

Giraud et al. designed a tactile stimulator [7], which is illustrated in Fig. 6.17. A $93 \times 65 \times 0.9$ mm glass plate was selected as a transparent ultrasonic medium to cover a 4.3-in. LCD screen. Two beryllium copper plates with piezoelectric ceramics were prepared as exciters and were glued on both edges of the glass plate, as shown in the figure. With this configuration, the stimulator had a flexural mode similar to

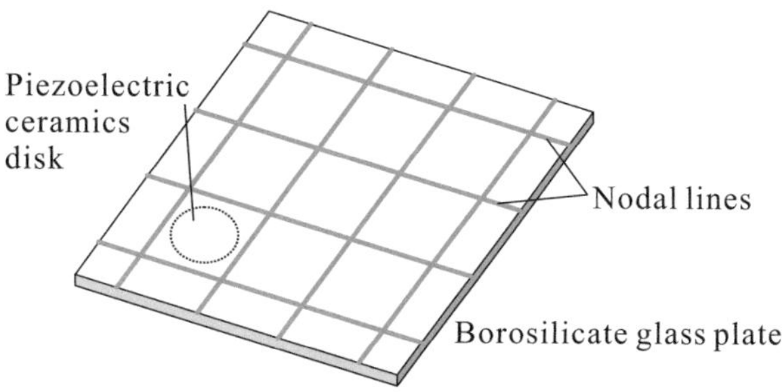

Fig. 6.16 Schematic view of large area TPaD [6]

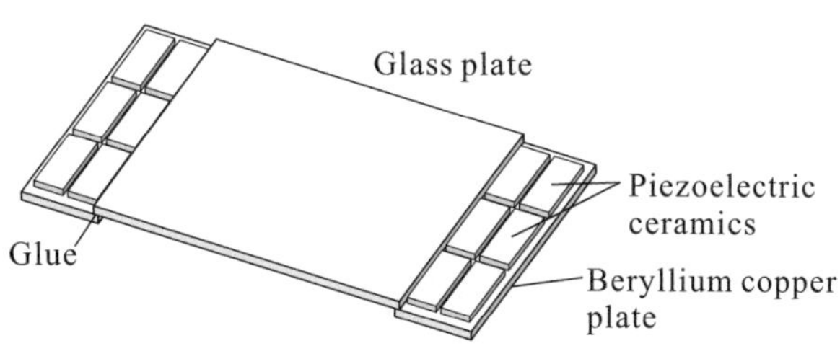

Fig. 6.17 Schematic view of tactile stimulator with glass plate and piezoelectric ceramics [7]

that shown in Fig. 6.9 at 31.2 kHz. A 1-$\mu m_{p\text{-}p}$ vibration amplitude was acquired on the loop area with 400-mW input electric power. Four force sensors were placed behind the stimulator to detect the finger position. The friction to be rendered on the stimulator was calculated based on the measured position. By modulating the voltage across the piezoelectric ceramics, the friction was modified dynamically as a lateral force [8].

Lithium niobate is originally a transparent material. If both sides of a $LiNbO_3$ wafer are polished, then the wafer appears to be transparent. Using such a wafer, a transparent SAW transducer can be realized as shown in Fig. 6.18. In the $LiNbO_3$ wafer case, the maximum dimension of the transducer is limited by the 4-in. wafer, which is the largest commercially available size. To overcome this limitation, SAW excitation on a glass substrate was proposed [9]. In this proposal, a piece of $LiNbO_3$ was coupled on the substrate to excite SAW, as shown in Fig. 6.19. The substrate thickness was 3 mm. The glass substrate was not suitable for combination with a touch screen, because the touch sensors on the screen cannot sense the position of a finger 3 mm away. The challenge of ultrasonic vibration excitation on a 0.3-mm-thick glass plate was also addressed [10].

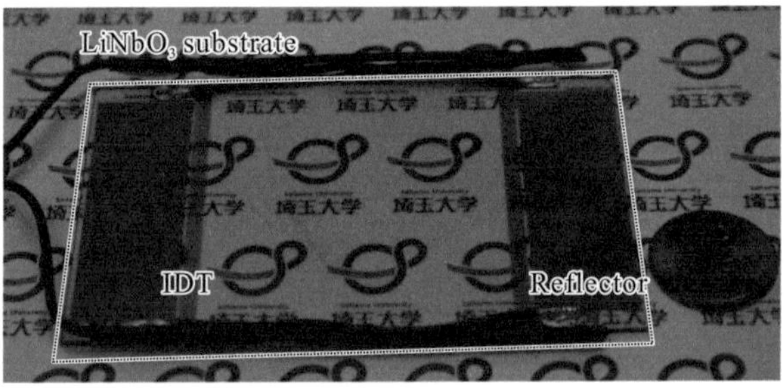

Fig. 6.18 Transparent SAW transducer

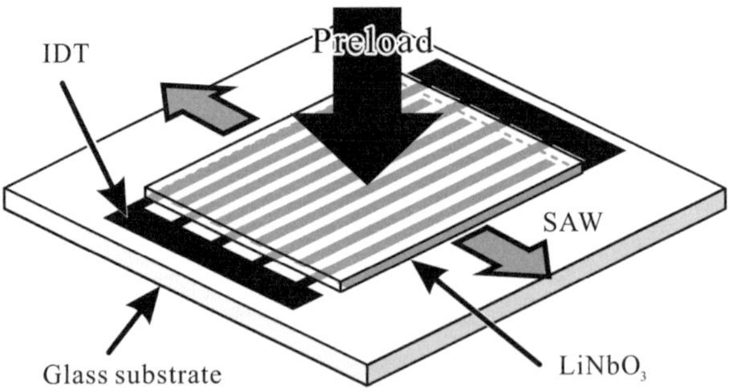

Fig. 6.19 Indirect SAW excitation on glass substrate

6.7 Related Technologies

In the case of the ultrasonic tactile display, friction can be reduced but not increased. Giraud et al. combined friction reduction by the squeeze film effect and friction enhancement by electrostatic force to expand the lateral force modification range [11].

Takasaki et al. reported pen tablet-type tactile displays [12]. Using the same principles and controls as the SAW tactile display, the friction between the pen top and the tablet was controlled and the vibration of the pen was presented as a drawing sensation. Under the controls described above, the pen vibration frequency was constant if the drawing speed was constant and resulted in an unnatural sensation. By referencing m-sequence random numbers to add frequency fluctuations to the signal used to switch the SAW excitation, the drawing sensation presented to the user became more natural, and similar to the sensation when drawing a line on a canvas [13].

By considering the transfer functions of the pen on a real object and the pen on the tactile display, the vibration of the pen when exploring real objects can be reproduced in the pen on the tactile display [14]. This concept is called "tele-touch," where real drawing sensations from a remote site can be reproduced on the tactile display at a local site.

Shiokawa et al. proposed a method to convey roughness, softness and friction sensations simultaneously. In this method, softness presentation by the squeeze film effect induced by high-intensity ultrasonic vibrations at 28.5 kHz and roughness sensation presentation by amplitude modulation was realized. A friction sensation was also rendered using a force feedback device [15].

6.8 Conclusion

In this chapter, an introduction to the basics of ultrasonic vibration was provided. Some of the technologies used to realize tactile displays were then described by categorizing their designs. Further information was added to expand upon the ultrasonic tactile display applications. The author of this chapter expects that this content will be helpful to readers who are interested in the development of ultrasonic tactile displays.

References

1. Watanabe, T., Fukui, S.: A method for controlling tactile sensation of surface roughness unsing ultrasonic vibration. In: Proceedings of IEEE International Conference on Robotics and Automation, pp. 1134–1139 (1995)

2. Biet, M., Giraud, F., Lemaire-Semail, B.: Squeeze film effect for the design of an ultrasonic tactile plate. IEEE Trans. Ultrason. Ferroelectr. Freq. Control **54**(12), 2678–2688 (2007)
3. Maeno, T., Otokawa K., Konyo, M.: Tactile display of surface texture by use of amplitude modulation of ultrasonic vibration. In: Proceedings of 2006 IEEE Untrasonics Symposium, pp. 62–65 (2006)
4. Nara, T., Takasaki, M., Maeda, T., Higuchi, T., Ando, S., Tachi, S.: Surface acoustic wave tactile display. IEEE Comput. Graph. Appl. **21**(6), 56–63 (2001)
5. Takasaki, M., Nara, T., Mizuno, T.: Control parameters for an active type SAW tactile display. In: Proceedings of 2004 IEEE/RSJ International Conference on Intelligent Robots and Systems, pp. 4044–4049 (2004)
6. Marchuk, N.D., Colgate, J.E., Peshkin, M.A.: Friction measurement on a large area TPaD. Proc. IEEE Haptics Symp. **2010**, 317–320 (2010)
7. Giraud, F., Amberg, M., Lemaire-Semail, B., Casiez, G.: Design of a transparent tactile stimulator. Proc. IEEE Haptics Symp. **2012**, 485–489 (2012)
8. Giraud, F., Amberg, M., Lemaire-Semail, B., Giraud-Audine, C.: Using an ultrasonic transducer to produce tactile rendering on a touchscreen. In: Proceedings of 2014 Joint IEEE International Symposium on the Applications of Ferroelectrics, International Workshop on Acoustic Transduction Materials and Devices & Workshop on Piezoresponse Force Microscopy (ISAF/IWATMD/PFM), pp. 1–4 (2014)
9. Kotani, H., Takasaki, M., Nara, T., Mizuno, T.: Glass substrate surface acoustic wave tactile display with visual information. In: Proceedings of IEEE International Conference on Mechatronics and Automation, pp. 1–6 (2006)
10. Takasaki, M., Suzaki, M., Takada, H., Mizuno, T.: Sheet-like ultrasonic transducer and its application for tactile display. In: Proceedings of 32nd Symposium Ultrasonic Electronics, pp. 21–22 (2011)
11. Giraud, F., Amberg, M., Lemaire-Semail, B.: Merging two tactile stimulation principles: electrovibration and squeeze film effect. In: Proceedings of IEEE World Haptics Conference 2013, pp. 199–203 (2013)
12. Takasaki, M., HongXu, J., Kotani, H., Mizuno, T.: Application of surface acoustic wave tactile display for pen tablet interface with visual information. In: Proceedings of IEEE International Conference on Robotics and Biomimetics 2007, pp. 1024–1028 (2007)
13. Tamon, R., Takasaki, M., Mizuno, T.: Generation of drawing sensation by surface acoustic wave tactile display on graphics tablet. SICE J Control Meas. Syst. Integr. **5**(4), 242–248 (2012)
14. Takada, H., Tamon, R., Takasaki, M., Mizuno, T.: Stylus-based tele-touch system using a surface acoustic wave tactile display. Int. J. Intel. Mechatron. Robot. **2**(4), 41–57 (2012)
15. Shiokawa, Y., Tazo, A., Konyo, M., Maeno, T.: Hybrid display of roughness, softness and friction senses of haptics. In: Proceedings of International Conference on Artifitial Reality and Telexistance 2008, pp. 72–79 (2008)

Chapter 7
Lateral-Force-Based Tactile Display

Satoshi Saga

Abstract With the expansion of touchscreen interface, expectation for tactile/haptic technologies increases. Many intuitive access have been realized, however, responses to virtual objects is still not providing any intuitive interactivity. In this chapter I introduce lateral-force-based tactile display. By employing haptic illusion of human, the lateral force display realize displaying bump and texture information simultaneously. The rendering method realize simple and intuitive interaction with virtual objects through touchscreen.

Keywords Tactile display • Touchscreen • Lateral force • Haptic illusion

7.1 Introduction

With the recent popularity of touchscreen interfaces and large distribution of smartphones, expectation for tactile/haptic technologies increases. Touchscreens and other position sensors provide intuitive access to digital content such as augmented-reality content or computer graphics games. However, interaction with virtual objects still requires more intuitive responses, which give us further haptic information and dextrous manipulation ability. This is partly because of the low capability of output devices compared with input devices. Thus our objective is to construct a general-purpose haptic display and develop the simultaneous display of bump and texture information.

In the current research field, many texture displays [1–4] have been developed employing high-frequency mechanical vibrations. However, these devices only display small textural information. Here we propose a method that allows the user to simultaneously feel both large bumps and small textures via a screen. Our method employs lateral forces and direction-controlled mechanical vibration (around 0–400 Hz). The technology allows not only geometrical shapes but also textures to be felt. Our experiments demonstrate that with our proposed method

S. Saga (✉)
University of Tsukuba, 1-1-1, Tennodai, Tsukuba, Japan
e-mail: saga@saga-lab.org

H. Kajimoto et al. (eds.), *Pervasive Haptics*, DOI 10.1007/978-4-431-55772-2_7

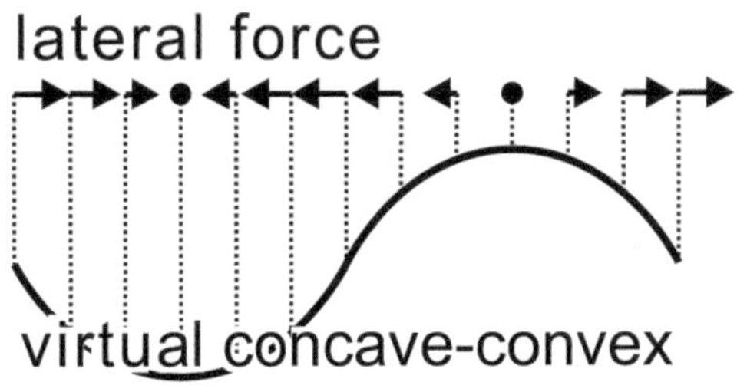

Fig. 7.1 Lateral-force-based haptic illusion: If the distribution of force as shown in upside is displayed to the user when he/she rubs a surface with his finger, he/she feels virtual bumps shown in downside. This is a well-known haptic illusion [5]

most people can simultaneously detect geometric and textural information on a touchscreen. In addition, providing adequate direction for the vibration enhances the sensation of softness.

7.1.1 Haptic-Illusion-Based Bump Display

This section introduces a rendering method based on lateral force for displaying bump information. The lateral-force-based haptic illusion displaying geometrical shapes is a well-known phenomenon [5–8]. This illusion arises when distributed lateral forces (shown in upside of Fig. 7.1) are displayed to the user, he feels virtual bumps (shown in downside of Fig. 7.1).

By combining touchscreen and haptic illusion, we describe a proposal that produces virtual bump illusions [9]. The hardware system comprises four motors, strings, a touchscreen, and a controller. The touchscreen detects the position of a finger and calculates the displayed force. We control four motors placed at the corners of the touchscreen that pull on strings connected to a finger pad. The finger pad applies two-dimensional forces to the users finger. By controlling the pulling torques of the four strings, an appropriately regulated force is applied to the users finger. Our previous work revealed that if one wants to display same height information from many waves that have several spatial wavelengths, we have to control the magnitude of the displayed lateral force according to the dominant spatial wavelengths. To measure the wavelengths, we employed multi-resolution analysis based on Haar wavelets.

7.1.2 Acquiring Gain Values from Depth Maps

First we have to acquire the spatial parameters from the displayed 3D shape. To measure the distribution of the spatial frequency of the shape, we employ multi-resolution analysis using Haar wavelets. The required distribution is around two centimeters squared, which covers the area of the fingertip. Hence, we crop a 128-pixel-squared area around the fingertip and analyze it (Fig. 7.2).

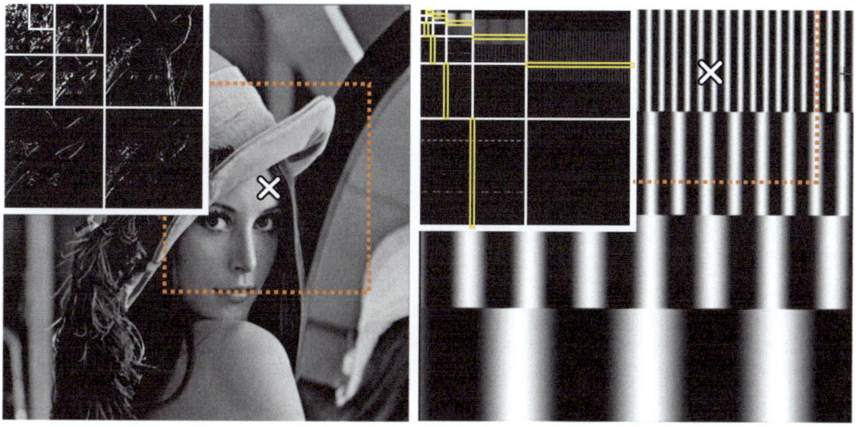

Fig. 7.2 Example of wavelet translation. *Left*: An example result of Lenna, *Right*: An example result of an image which has several spatial frequencies

Figure 7.2 shows the result of multi-resolution analysis applied to the cropped area marked by the dotted rectangle. A 128-pixel-squared area of the cropped image around the cursor was used in the calculation. Each rectangle shows results of each level. We calculated the mean values of intensities along the center lines on the images, and expressed them as logarithms representing gains of each spatial frequency. A previous study clarified that one has to control the displayed gain in accordance with the spatial frequency. Hence, we employ a mediating function, G(x), and calculate the displaying force, F(x), as

$$G(x) = C \exp |2.0 - \log_2 A_x|, \tag{7.1}$$

$$F(x) = G(x) \cdot k \cdot Sobel(I(x)) \tag{7.2}$$

where C and k are appropriate constant values. Here we employ the Sobel filter to calculate gradients of the 3D shape.

7.1.3 Displaying Experiment

To evaluate the proposed displaying method and to compare the method with conventional one, we performed an experiment displaying several virtual stimulations. Participants were asked to rub a virtual semicylinder shape, which under Haar analysis involves several spatial frequencies, and after rubbing to freely decide the shape perceived. Figure 7.3 shows the results. From the results, we found that with the conventional method participants were unable to feel the correct shape; the hemisphere felt trapezoidal in shape. In contrast, with our proposed method, each felt a hemi-spherical shape. That is, our proposed method can correctly display 3D bump shapes.

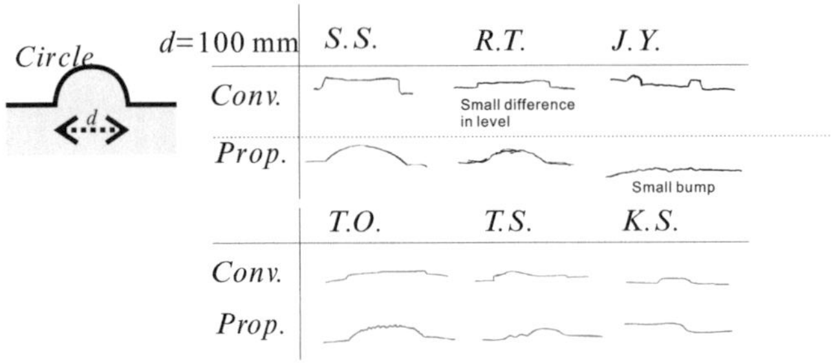

Fig. 7.3 Example results of rubbing experiment: In conventional method, semicylinder stimuli are perceived as center-concaved step-like shapes. In proposed method, on the other hand, semicylinder stimuli are perceived as correct shapes

7.2 Vibration-Based Texture Display on Flat Screen

In this section we describe how to display high-definition texture information by employing high-frequency mechanical vibrations. Lateral-force-based vibrational force can convey high-definition texture information.

7.2.1 Recording Method

To record the texture information, we used an accelerometer (KXM52-1050, Kionix, its bandwidth is 0 to 3 kHz, and has 3 axes) and a touchscreen (XE700T1A, Samsung). The accelerometer was attached by adhesive tape to the side of a model of a fingertip made of silicone rubber. The accelerometer measured the mechanical vibration induced by the rubbing movement between finger and object, and the touchscreen measured the velocity of the finger. The average vibration on each axis was recorded by an audio input at 44.1 kHz and the velocity was recorded at a frequency around 30 Hz, which was based on the sampling rate of the touchscreen.

7.2.2 Display Method

The display has a resolution of 1366 pixels per 256.8 mm. If the touchscreen has the same resolution, the spatial resolution is 5.3 pixels/mm. Hence, it can display some fine textures such as sandpaper. Furthermore, our recorded data have a sampling rate of 44.1 kHz. However, the temporal resolution of the touchscreen (around 30 Hz) is not capable of displaying them. Moreover, the control frequency of the four motors is around 400 Hz. To compensate for the different sampling rates and different velocities between the recording and playing phases, we employed a precision

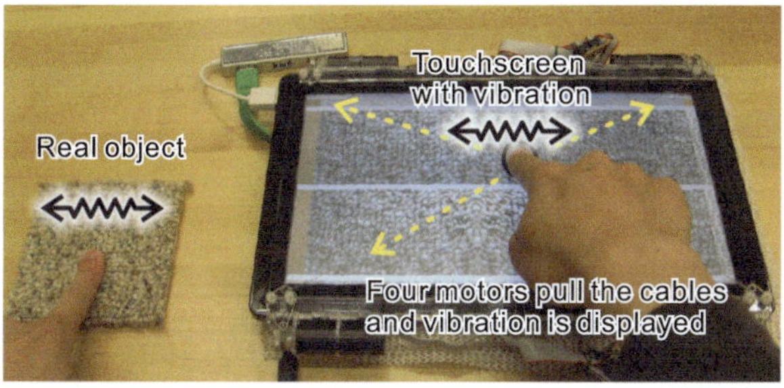

Fig. 7.4 Overview of the texture display system

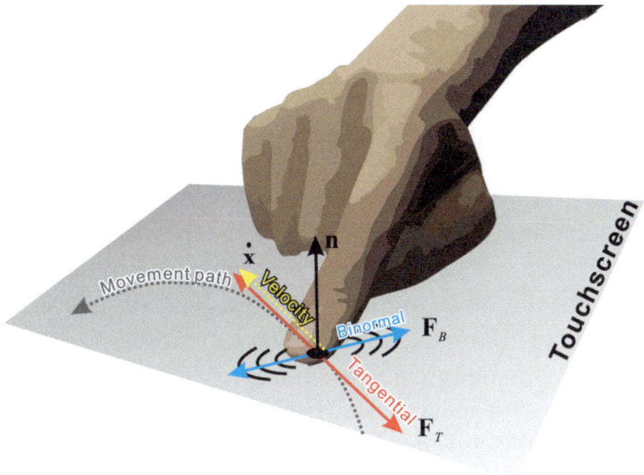

Fig. 7.5 Direction of vibration: $\boldsymbol{F}_T$ exerts lateral force tangential to movement of finger, and $\boldsymbol{F}_B$ is displayed as a binormal force to movement of finger

timer and recorded the velocities of the finger movement, and calculated virtual movements of the finger and resampled the vibration from the recorded one. With this method, we can display high-definition textural information.

7.2.3 Two Methods for Displaying Vibration

Most of the conventional research on vibratory texture display employ voice coil motors. Hence the direction of vibration is limited to one degree of freedom. Therefore, they cannot validate the effect of the direction of vibration. In our system, in which we use a string-based two-dimensional force displaying system (Fig. 7.4), the direction of vibration is also controllable (Fig. 7.5). Here we propose two methods for displaying the recorded vibration by selecting the direction, and examine which vibration is more suitable for textural display through a psychophysical experiment.

The characteristics of these display methods are as follows;

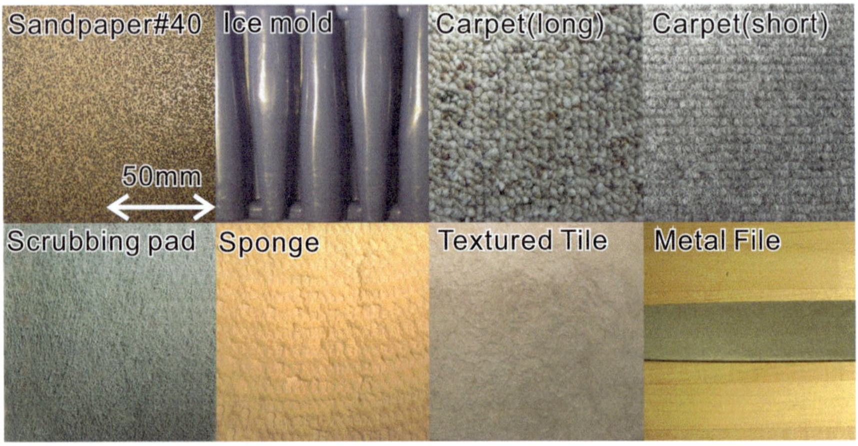

Fig. 7.6 Several recorded textures

- Vibration tangential to the movement direction of the finger ($\boldsymbol{F}_T$) The method realizes similar algorithm with geometrical display. Thus it exerts lateral force tangential to movement of finger. Both of geometrical virtual displacement and texture information are always displayed as resistive force.
- Vibration binormal to the movement direction of the finger ($\boldsymbol{F}_B$) The method realizes independent display between geometry information and texture information. The displacement of the geometrical shape is represented as a resistive force, and the vibration of the texture is displayed as a binormal force that does not interfere with the geometrical shape information.

7.2.4 Comparison of Each Method

Based on the aforementioned methods, we recorded vibration data for eight textures (Fig. 7.6). Each data is recorded for three seconds. Under the playback vibration of the recorded acceleration data, participants were requested to rub with bare fingers a real texture and two virtual textures (vibrations) rendered by the ($\boldsymbol{F}_T$) and ($\boldsymbol{F}_B$) methods, and to decide which of the virtual textures was more similar to the real one. The recorded data were repeated many times during rubbing. We repeated this experiment three times per texture using eight textures (sandpaper, ice mold, two carpets, two sponges, textured tile, and metal file). These experiments took less than 30 min per participant. The results are shown in Fig. 7.7.

Most objects were reported to have higher similarity with the textures rendered by the ($\boldsymbol{F}_B$) method. For the scrubbing pad, metal file, and both carpets, statistically significant results were seen. These results show that, in softer textures (such as carpets or a scrubbing pad), ($\boldsymbol{F}_B$) produces a better rendering than ($\boldsymbol{F}_T$) of the textures. The result for sponge shows no significant result; however, the result is

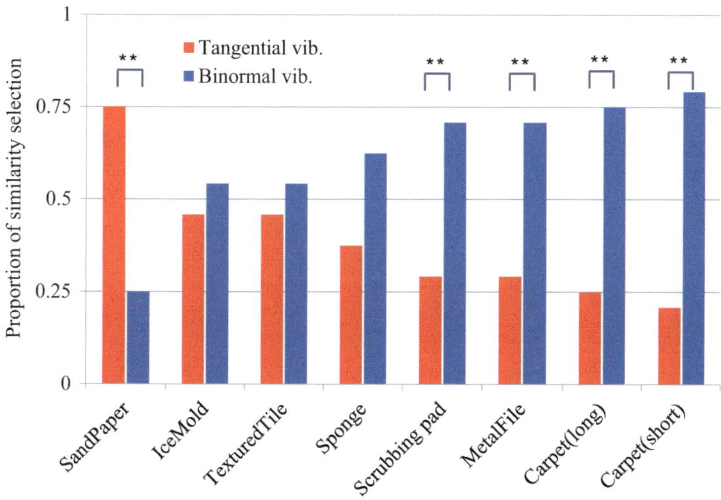

Fig. 7.7 Rating result of reality between tangential and binormal. Each bar shows mean and standard deviation. Subjects are asked to rate each virtual texture with free number between 0.0 and 1.0. Each ** shows $p < 0.01$ which is significance test results of t-test

similar to other softer textures. In contrast, for harder textures (such as textured tile or ice mold), no significant difference can be found between the two methods. Thus, in displaying harder textures, either method can be employed. The result for sandpaper and metal file show exceptions compared with the other textures. This is probably because the feel of the metal file is different to other harder textures. In contrast, sandpaper has a randomized distribution of spatial frequencies. The randomness is similar to other textures such as carpets. Nevertheless the material of the texture is harder than others. This fact probably implies that the resistive forces produced by the ($\boldsymbol{F}_T$) method renders adequate stimuli. That is, in those textures that are hard and have random-type surfaces, the ($\boldsymbol{F}_T$) method renders a more realistic feeling than ($\boldsymbol{F}_B$). In summary, these results show the following:

- In softer textures (carpets, scrubbing pad), $\boldsymbol{F}_B$ is the better method
- In harder textures (textured tile or ice mold), significant difference between the methods cannot be found
- In harder textures, which has random frequency (sandpaper), $\boldsymbol{F}_T$ is the better method

7.3 Combination Display with Geometrical Information

With the same participants, we performed a display test of geometric and textural information. In the combination display, the displayed force was a summation of the static-force-based geometric information and the vibratory-force-based textural information. The separation of both types of information appears difficult. Subjects

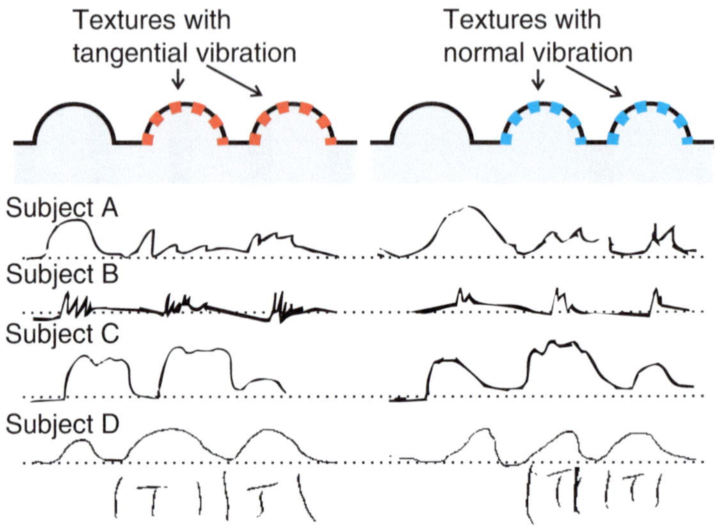

Fig. 7.8 Example results of rubbing experiment (with texture): They experienced the stimuli only with tactile and haptic feelings, without visual information, and were requested to freely write down the shape of the displayed geometry and texture. The subjects were novice to such kind of stimulation. In spite of such condition, these results show many subjects could feel not only the geometry but also the texture information simultaneously

were requested to rub six different virtual stimuli and describe the shapes and textures they felt. The displayed stimuli consisted of *shape-only* stimuli and *shape + texture* stimuli. In each experiment, three bumps were displayed. Some of the bumps were displayed with recorded vibration of metal file. They experienced the stimuli only through tactile and haptic sensing, without visual information, and were requested to freely describe the shape of the displayed object and its texture.

Figure 7.8 shows representative examples of the results. In instances when the displayed information and the written figure included the same information, the response was correct. Thus, the following two cases are the ones when the displayed information and the drawing were both *shape-only*, or the displayed information is *shape + texture* and the drawing is also *shape + texture*. The participants had no prior acquaintance with this sort of stimulation. Despite this, these results show many participants could feel not only the geometry but also the textural information simultaneously [10].

Several studies on haptic-object processing in the brain support these results [11]. In the human brain, the microgeometric properties are processed in a hierarchical pathway that is at least partially distinct from the hierarchical pathway that processes macrogeometric properties. Taking this into consideration, the input signals of the geometric shape and the texture should be displayed independently.

7.4 Conclusions

In this chapter, we employed lateral force and mechanical direction-controlled vibration. This technology provides tactile geometry as well as textural information. By employing recorded mechanical vibration, our system displays textural information. Furthermore, the direction of the vibration can be controlled. Here we proposed two rendering methods, the tangential ($\boldsymbol{F}_T$) and binormal ($\boldsymbol{F}_B$) method. In addition, the extracting display method easily displays many kinds of texture information without requiring a library of vibration signals.

The psychophysical experiment results show that ($\boldsymbol{F}_B$) is better at displaying softer materials, whereas no significant difference between the two was evident for the display of harder materials. In addition, under challenging textures such as carpets and sponges, participants rate the realism at over 50 % for displays based on the ($\boldsymbol{F}_B$) method. Furthermore, in combination display experiments, several results for tangential vibrations show a flattened geometrical shape when compared to the result for binormal vibrations. Five of the eight participants felt not only the geometric but also the textural information simultaneously. Our proposed method enables virtual touch of a displayed object on a screen. Technology based on this method will enable a more tactile interactivity enhancing experiences during internet shopping, screenings of television programs, and telecommunications, allowing users to feel clothing fabrics, teddy bears, and even pets via the screen.

In the near future, many kinds of remote or virtual communications will have tactile media, too. Eventually, the method proposed here will become a simple, feasible, and fundamental technology taking its place alongside other touchscreen technologies.

References

1. Chubb, E.C., Colgate, J.E., Peshkin, M.A.: Shiverpad: a glass haptic surface that produces shear force on a bare finger. IEEE Trans. Haptics **3**(3), 189–198 (2010)
2. Wang, D., Tuer, K., Rossi, M., Shu, J.: Haptic overlay device for flat panel touch displays. In: Proceedings of 12th International Symposium on HAPTICS'04, Chicago, p. 290. IEEE (2004)
3. Bau, O., Poupyrev, I., Israr, A., Harrison, C.: TeslaTouch: electrovibration for touch surfaces. In: Proceedings of the 23nd Annual ACM Symposium on User Interface Software and Technology, New York, pp. 283–292. ACM (2010)
4. Romano, J.M., Kuchenbecker, K.J.: Creating realistic virtual textures from contact acceleration data. IEEE Trans. Haptics **5**(2), 109–119 (2012)
5. Minsky, M.R., Ouh-young, M., Steele, O., Brooks, F.P. Jr., Behensky, M.: Feeling and seeing: issues in force display. Comput. Graphics **24**(2), 235–243 (1990)
6. Levesque, V., Hayward, V.: Experimental evidence of lateral skin strain during tactile exploration. In: Proceedings of Eurohaptics, Dublin (2003)
7. Smith, A.M., Chapman, C.E., Donati, F., Fortier-Poisson, P., Hayward, V.: Perception of simulated local shapes using active and passive touch. J. Neurophys. **102**(6), 3519 (2009)
8. Robles-De-La-Torre, G., Hayward, V.: Force can overcome object geometry in the perception of shape through active touch. Nature **412**(6845), 445–448 (2001)

9. Saga, S., Deguchi, K.: Lateral-force-based 2.5-dimensional tactile display for touchscreen. In: Proceedings of IEEE Haptics Symposium 2012, Vancouver, pp. 15–22 (2012)
10. Saga, S., Raskar, R.: Simultaneous geometry and texture display based on lateral force for touchscreen. In: Proceedings of IEEE World Haptics 2013, Daejeon, pp. 437–442 (2013)
11. Kim, S., James, T.W.: The Handbook of Touch, chapter 6, pp. 143–159. Springer, New York (2011)

Chapter 8
Airborne Ultrasound Tactile Display

Takayuki Hoshi and Hiroyuki Shinoda

Abstract The authors and colleagues invented an ultrasound-based noncontact tactile display in 2008 and have been developing this technology since then. It is suitable for gesture input systems and aerial imaging systems because no physical contact is required to provide haptic feedback. An ultrasonic phased array generates a focal point of airborne ultrasound to press the skin surface. The amplitude modulation of ultrasound provides vibrotactile stimulation covering the entire frequency range of human tactile perception. The position of the focal point is computationally controlled to follow users' hands and/or provide a trajectory of stimulation. While this technology was originally invented as a tactile display, a wide variety of other applications has been recently reported that exploit noncontact force generated at a distance. Examples include noncontact measurement by pressing or vibrating objects, levitation and manipulation of small objects, and actuation of fragile or soft materials. The present chapter describes the background, principles, systems, and applications of this ultrasonic technology.

Keywords Tactile display • Aerial interface • Airborne ultrasound • Acoustic radiation pressure • Phased array

8.1 Introduction

Aerial touch panels are being tested as next-generation touch-panel interfaces. When users reach toward floating images in front of them, contact between the users' hands and the images is detected, and the images react accordingly. Such interactive floating images [1–3] provide clean, dirt-free touch panels that can be used in medical operation rooms, hospital waiting rooms, food processing plants, restaurants, entertainment, games, exhibitions, and digital signage.

T. Hoshi (✉)
Nagoya Institute of Technology, Nagoya, Japan
e-mail: star@nitech.ac.jp

H. Shinoda
The University of Tokyo, Tokyo, Japan

H. Kajimoto et al. (eds.), *Pervasive Haptics*, DOI 10.1007/978-4-431-55772-2_8

In light of the foregoing, demands for noncontact and mid-air haptic feedback are emerging. People who experience aerial touch panels feel a strong need for haptic feedback. Human vision, despite its high lateral resolution, is poor at determining the distance to an object. Therefore, haptic feedback is helpful for a user to perceive and confirm that his/her finger contacts the aerial touch panel. The absence of haptic feedback yields uncertainty regarding the contact instant and input operation. Moreover, haptic feedback is indispensable in manipulating three-dimensional (3D) volumetric images.

There are two known methods for realizing haptic feedback in air. The first is to wear wearable devices such as a glove with small vibrators [4], motor-driven belts [5], or pin arrays [6]. One problem of this method is that users must wear the devices in advance. The second method involves transmitting tactile stimulation via air. For example, air jets [7] and air vortices [8, 9] were used for providing noncontact haptic feedback to bare hands.

Our research group first demonstrated that focused ultrasound produces perceivable force on a bare hand in air [10]. Our ultrasonic device (airborne ultrasound tactile display, AUTD) has hundreds of ultrasonic transducers and creates an ultrasonic focal point using phase control. High-amplitude ultrasound within the focal point presses objects in the propagation direction. This phenomenon is known as acoustic radiation pressure. The output force is sufficient to stimulate human tactile sensation. The focal point is generated at an arbitrary position in a 3D space. One of the greatest strengths of the AUTD is that both the spatial and temporal resolutions are high, and hence, various tactile patterns can be reproduced. It can produce not only active-touch sensation for aerial touch panels but also passive-touch sensation for alerts, communication, guidance, and other purposes [11]. Additionally, the AUTD offers precise movement of the focal point, which can be used to reproduce trajectories of hand-written letters and/or illustrations [12].

Other researchers have tried to use the AUTD for their own purposes. Ciglar proposes the AUTD as a musical instrument with tactile feedback [13]. Researchers at Bristol Interaction and Graphics combined the AUTD with a mobile TV [14], demonstrated an interactive table on which tangible objects were freely moved [15], developed an aerial interaction system comprising the AUTD and an acoustically transparent screen [16], and reported the simultaneous generation of multiple focal points [17].

In this chapter, first, the theoretical background of AUTD is explained. There are two main principles: acoustic radiation pressure and phased array focusing. Second, an example device is introduced. Last, developed systems based on the AUTD are shown. The latest application trials other than tactile displays are also introduced.

8.2 Ultrasound-Based Tactile Stimulation

Production of tactile stimulation by ultrasound has been attempted underwater [18–21], and there are two methods for this. The first involves direct stimulation of nerve fibers underneath the skin by ultrasound [18, 19]. In this method, ultrasonic waves penetrate the skin because their characteristic acoustic impedances are comparable. The second method is to press the skin surface by acoustic radiation pressure [20, 21]. In this method, a reflective membrane or plate is used to reflect ultrasonic waves.

In contrast, the proposed method is conducted in air and requires no reflective material. Users can receive tactile stimulation with their bare hands because total reflection occurs at the skin surface owing to the large difference in acoustic impedance between air and skin. This feature reduces the complexity of the AUTD.

In this section, we review the theoretical background of the AUTD. There are two main principles: acoustic radiation pressure and phased array focusing. The control methods and characteristics of ultrasound are also discussed.

8.2.1 Acoustic Radiation Pressure

The acoustic radiation pressure [22, 23] is a nonlinear phenomenon of ultrasound. When an ultrasound beam is reflected vertically at an object's surface, the surface is subjected to a constant vertical force in the direction of the incident beam (Fig. 8.1). Assuming a plane wave, the acoustic radiation pressure P [Pa] acting on the object's surface is written as

$$P = \alpha E = \alpha \frac{I}{c} = \alpha \frac{p^2}{\rho c^2} \tag{8.1}$$

where E [J/m^3] is the energy density of the incident wave, I [W/m^2] is the sound intensity, c [m/s] is the sound speed, p [Pa] is the RMS sound pressure of ultrasound, ρ [kg/m^3] is the density of the medium. α is a constant depending on the amplitude reflection coefficient R and the amplitude transmittance coefficient T at the object's surface ($\alpha = 1 + R^2 - T^2$). In the case of air and human skin, α is 2 because of total reflection: $R = 1$ and $T = 0$.

Note that p^2 is typically assumed to be negligible (approximately zero) in the wave theory for linearization. The radiation pressure P becomes noticeable when the sound pressure p is relatively large such that p^2 is not negligible. Additionally, (8.1)

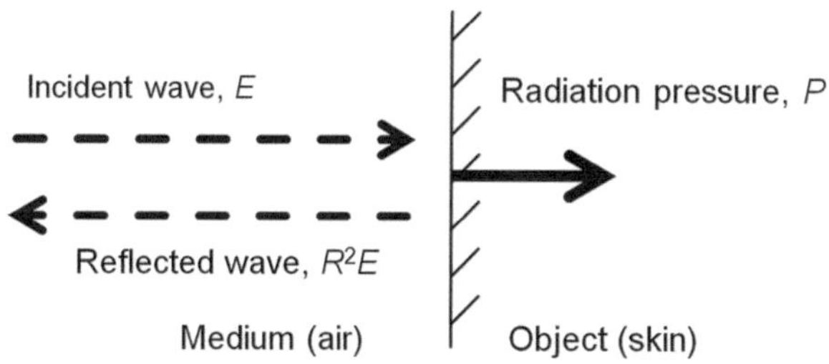

Fig. 8.1 Radiation pressure on object's surface

suggests that the spatial distribution of the radiation pressure P can be controlled by synthesizing the spatial distribution of the sound pressure p.

8.2.2 Phased Array Focusing

The phased array focusing technique is used to achieve sufficiently high-amplitude ultrasound. An ultrasonic focal point is generated by setting adequate phases. The phase θ_i [rad] for the ith transducer is written as

$$\theta_i = 2\pi \frac{d_i - d_0}{\lambda} \tag{8.2}$$

where d_i [m] and d_0 [m] are the distances from the focal point to the ith and 0th transducers, respectively (Fig. 8.2). λ [m] is the wavelength. Theoretically, the peak value of the sound pressure in the center of the focal point is proportional to the number of ultrasonic transducers. The focal point can be generated at an arbitrary position by tuning the phase delays.

There is a trade-off between the spatial resolution (size of focal point) and array size. It is theoretically derived that the spatial distribution of ultrasound generated from a square transducer array is approximately sinc-shaped [24]. The width of the main lobe (w [m]) parallel to a side of the square is written as

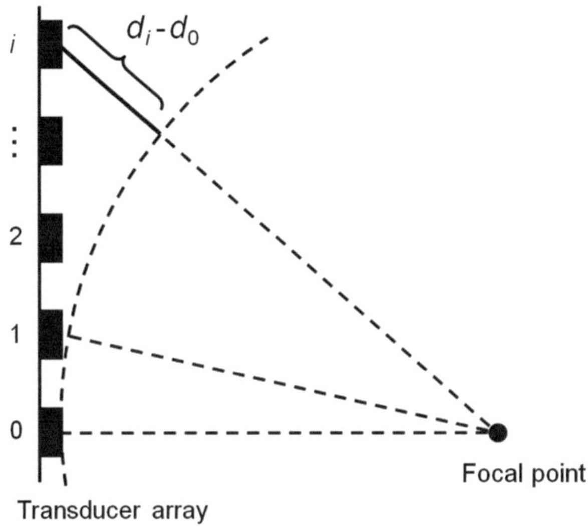

Fig. 8.2 Difference in distance from focal point between ith and 0th transducers

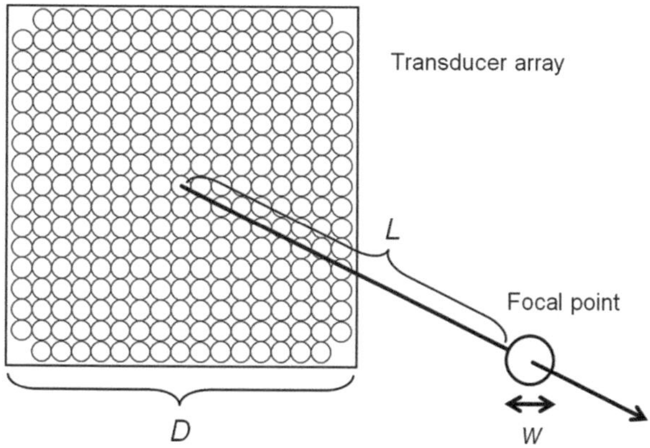

Fig. 8.3 Device size, focal length, and size of focal point

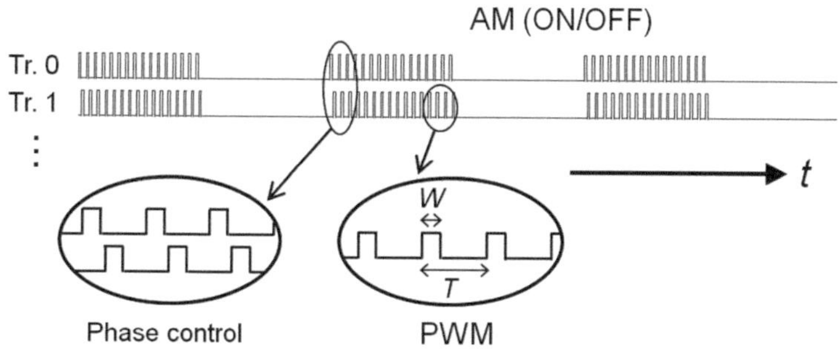

Fig. 8.4 Overview of driving signals

$$w = 2\lambda \frac{L}{D} \tag{8.3}$$

where L [m] is the focal length, and D [m] is the side length of the square (Fig. 8.3).

Note that (8.3) is obtained using an approximation of the distance. It is no longer valid when the focal length L is far shorter than the array size D. Additionally, the directivity of ultrasonic transducers become noticeable at such a distance.

8.2.3 *Vibrotactile Stimulation*

Vibrotactile stimulation can be provided by the amplitude modulation (AM) of ultrasound. The acoustic radiation pressure is controlled between zero and the maximum value. The simplest method of control is turning the ultrasound ON/OFF at a constant frequency (Fig. 8.4). Although conventional vibrators have the possibility of loss of contact because of the viscosity of the skin, the AUTD can apply the intended vibration to the skin because it requires no physical contact.

Regarding human tactile properties, a bandwidth of 1 kHz is sufficient for tactile displays [25]. The AUTD can produce various tactile feelings because the carrier frequency, 40 kHz, is far higher than the required modulation bandwidth, 1 kHz. The frequency characteristics of the transducer are important here.

8.2.4 Intensity Control

Pulse-width modulation (PWM) can be used to easily solve electrical circuits, rather than modulating the amplitude of driving signals. The driving signal $V(t)$ [V] is a rectangular wave written as

$$V(t) = \begin{cases} V_0 & (nT \leq t < nT + W) \\ 0 & (nT + W \leq t < nT + T) \end{cases} \quad (8.4)$$

where W [s] and T [s] are the pulse width and cycle time of ultrasound, respectively (Fig. 8.4), and n is an integer. We control the pulse width W. The amplitude of the fundamental frequency component a_1 [V] of $V(t)$ is written as

$$a_1 = \frac{2}{\pi} V_0 \left| \sin \pi \frac{W}{T} \right| \quad (8.5)$$

According to (8.5), a 50 % duty ratio (i.e. $W = T/2$) yields the maximum carrier amplitude. The output sound pressure is proportional to a_1 because the driven device is a resonator, and the resulting acoustic radiation pressure is accordingly proportional to $|\sin(\pi W/T)|^2$.

8.2.5 Absorption in Air

Air is a lossy medium, and its attenuation coefficient β [Np/m] for a plane sound wave varies according to the wave frequency. The energy density E at the distance z [m] is written as

$$E = E_0 e^{-2\beta z}, \quad (8.6)$$

where E_0 is the energy density at the transducer surface ($z = 0$ mm). According to [26], the attenuation coefficient β at 40 kHz is 1.15×10^{-1} Np/m (i.e., 1 dB/m), and β is proportional to the square of the frequency. Figure 8.5 shows the relationship between the frequency and the energy loss rate at $z = 200$ mm. The energy loss is 4 % when the frequency is 40 kHz but over 50 % when the frequency is 200 kHz. We use 40-kHz ultrasound because the attenuation is relatively small and 40-kHz

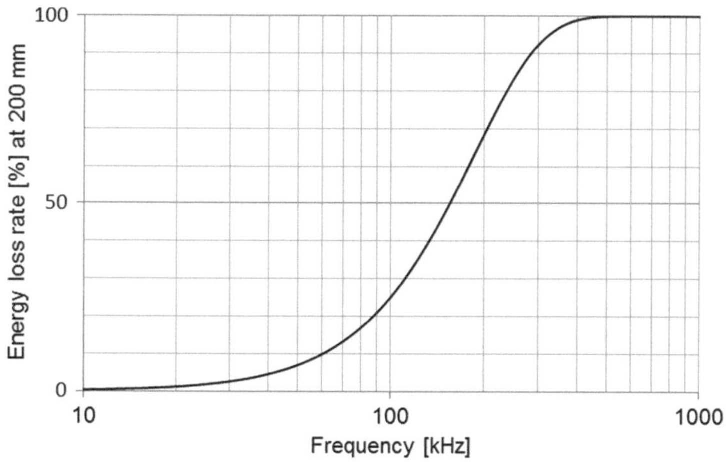

Fig. 8.5 Energy loss rate with respect to frequency

ultrasound transducers are commercially available. If we limit the range to 400 mm, the energy loss is less than 10 % everywhere in the workspace.

As shown the previous section, the wavelength also determines the spatial resolution. Thus, there is a trade-off between the spatial resolution and effective distance.

8.2.6 *Safety Limits*

Regarding human safety, there are two ultrasound-intensity limits. The first concerns heat damage to tissue under the skin; therefore, we estimate the reflection coefficient of the skin surface. The characteristic acoustic impedance of skin (soft tissue) Z_s and that of air Z_a are 1.63×10^6 N•s/m^3 and 0.0004×10^6 N•s/m^3, respectively [27]. The reflection coefficient R is determined as

$$R = \left| \frac{Z_s - Z_a}{Z_s + Z_a} \right| \approx 0.9995 \tag{8.7}$$

In this case, 99.9 % ($= R^2 \times 100$) of the incident acoustic energy is reflected at the skin surface, i.e., an approximately total reflection. In the field of ultrasonography, a safe level of tissue exposure to ultrasound is defined as 100 mW/cm^2 [26]. Because 0.1 % of ultrasound power is absorbed by skin, ultrasound up to 100 W/cm^2 can be applied to the skin surface. The sound intensity I gives the acoustic radiation pressure: $P = 2I/c = 5882$ Pa.

The second safety limit concerns the impingement on the human ear. The limit recommended in [28] is 110 dB SPL. The sound pressure at a distance of 300 mm from a single transducer is 121.5 dB SPL, which is larger than the recommended limit. Furthermore, hundreds of transducers are used in the AUTD. Until the safety of the AUTD is confirmed, users are advised to wear headphones or earmuffs to protect their ears.

8.3 Implementation

Various AUTD prototypes and systems were developed. Their performance differs because of the number and arrangement of ultrasonic transducers and other design parameters. In this section, one of the prototypes [29] (Fig. 8.6) is considered as a typical example. It has 285 ultrasonic transducers [30] (Nippon Ceramic, T4010A1, 40 kHz, 10 mm in diameter) arranged within a 170×170 mm^2 square area. The power-supply voltage is 24 V. The resolution of the position control of the focal point is 0.5 mm, which is determined by the digitization of the phases in the control circuit.

Firstly, the spatial distribution of the sound pressure of a 40-kHz ultrasound was measured by a microphone mounted on an XYZ stage. The array size D was 170 mm, the focal length L was 200 mm, and the wavelength λ was 8.5 mm. Then, the diameter of the focal point w was estimated as 20 mm. The measurement results on the focal plane show that the generated focal point has an equal diameter (Fig. 8.7). The measurement results along the acoustic axis are shown in Fig. 8.8. The peak value was 2,585 Pa RMS (162 dB SPL), which is 100 times larger than that for a single transducer, although 285 transducers were used. This decrease may originate from the directivity (100° half-amplitude full angle), the incident angle of each ultrasonic wave into the microphone, and/or other problems arising in practice.

Fig. 8.6 Focused ultrasound flips up paper strips

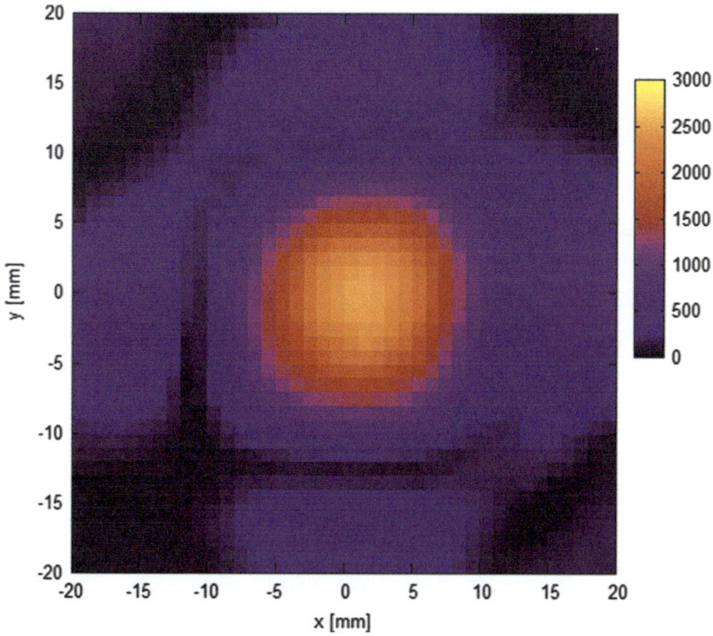

Fig. 8.7 Spatial distribution of sound pressure on focal plane (x-y)

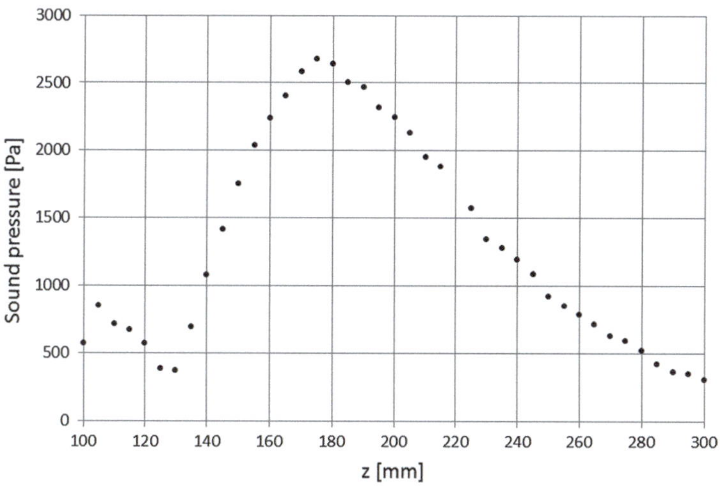

Fig. 8.8 Spatial distribution of sound pressure along acoustic axis (z)

Secondly, the time-response characteristics were tested. The focal length was 200 mm, and the microphone was fixed at the focal point. The sound pressure and radiation pressure are shown in Fig. 8.9. The 40-kHz driving signal was modulated by a 100-Hz rectangular wave. CH1 is the sound pressure that is the normalized

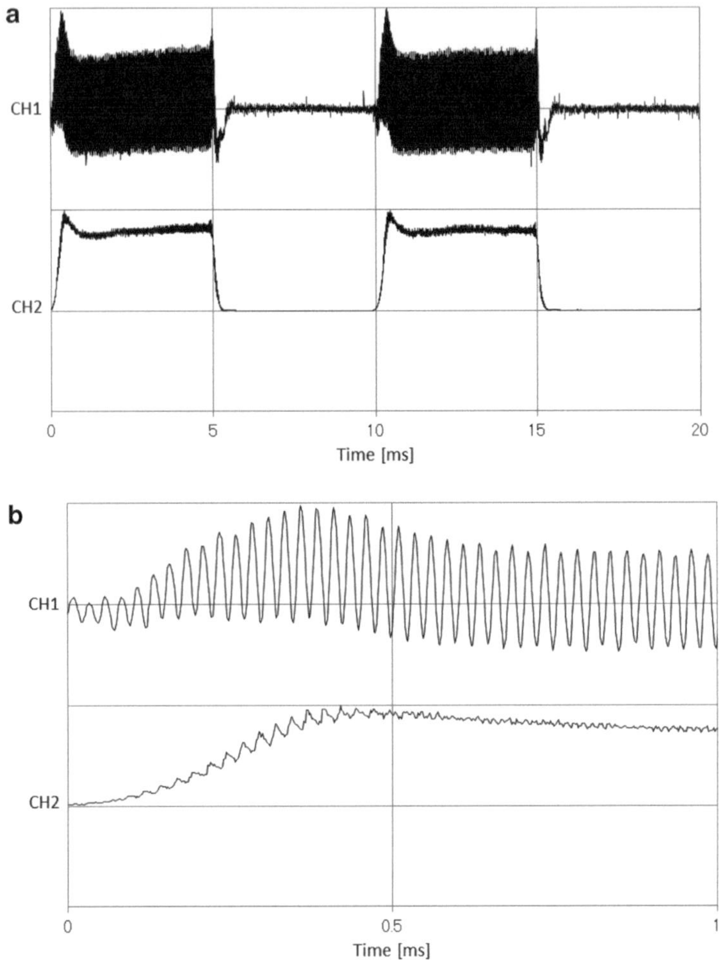

Fig. 8.9 Normalized waveforms of ultrasound (CH1) and radiation pressure (CH2). (**a**) 100-Hz modulated waveforms. (**b**) Close-up of (**a**)

waveform of the raw data, and CH2 is the radiation pressure that is the squared and normalized waveform of the waveform extracted by a lock-in amplifier. The rising time is shorter than 1 ms, indicating that the refresh rate is higher than 1 kHz. Figure 8.10 shows the frequency characteristics of the radiation pressure from 1 to 1,000 Hz. The radiation pressure at 1,000 Hz is nearly 5 dB less than that at 400 Hz.

Thirdly, the output force was measured by a digital scale. The prototype vertically pressed down the center of the scale as a result of focused ultrasound from a 200-mm height. The maximum measured force was 16 mN, which is sufficient to produce vibrotactile stimulation on a palm.

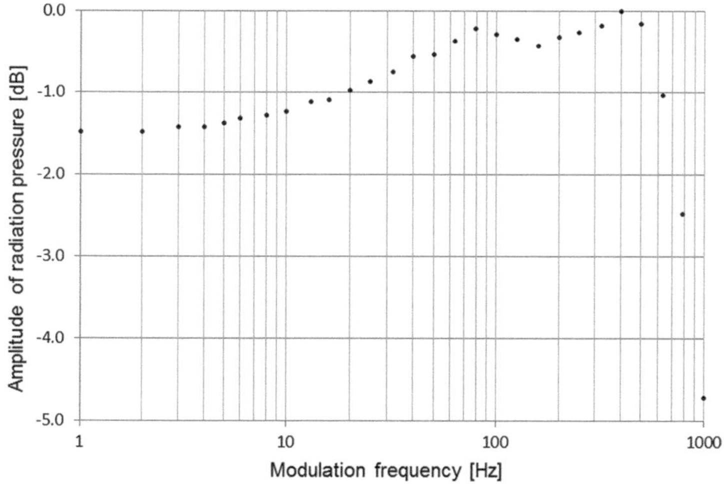

Fig. 8.10 Frequency characteristics of radiation pressure

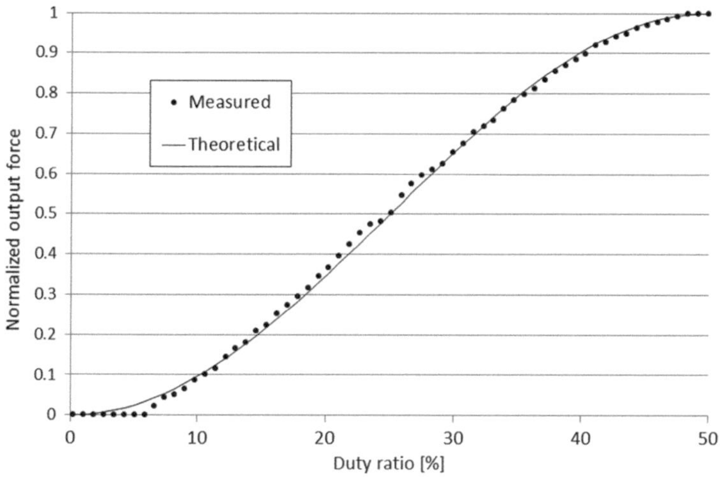

Fig. 8.11 Normalized output force vs. duty ratio

Lastly, the PWM-based intensity control was confirmed. Figure 8.11 shows the normalized force, which was measured by a digital scale. This force is a squared sine function of the duty ratio, as theoretically predicted.

8.4 Applications

We developed application systems using AUTDs, as discussed in this section. Applications other than a tactile display are also introduced.

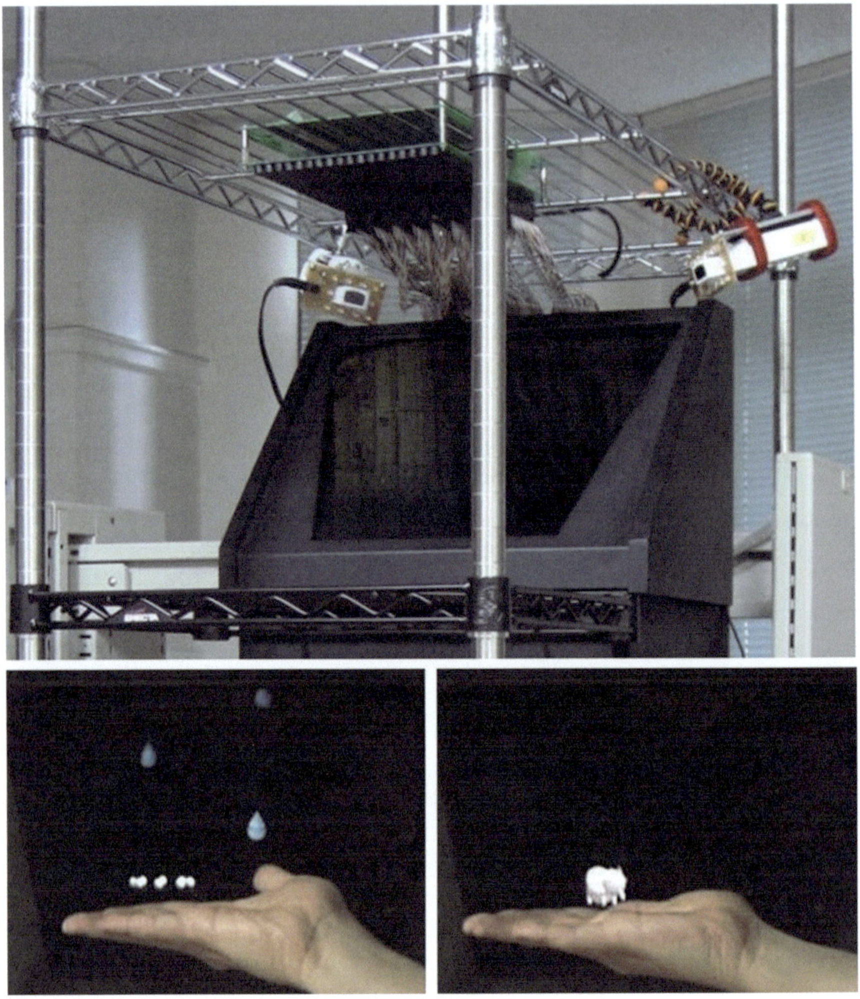

Fig. 8.12 Touchable holography; (*top*) system overview; (*bottom-left*) raindrops; (*bottom-right*) small animal wandering on a palm

8.4.1 Touchable Holography

A system that adds tactile feedback to floating images was developed [31] (Fig. 8.12). The images are projected by HoloVision (Holo17T, Provision Interactive Technologies, Inc.), which provides floating images from a liquid-crystal display (LCD) by utilizing a concave mirror. The projected images float 300 mm away from the HoloVision. The position of a user's hand is estimated by triangulation using two infrared (IR) cameras (Wii Remote, Nintendo Co., Ltd.). A retroreflective marker is attached on the user's finger and illuminated by IR LEDs. According to interactions between the user's hand and floating images, noncontact tactile feedback is adequately provided by the AUTD.

8.4.2 Tactile Projector

A noticeable feature of the AUTD is the freedom of stimulation-point location. The AUTD can generate passive haptic sensations at various parts of the human skin for alert and guidance, e.g., in a driver's seat. To demonstrate this application, the large-aperture AUTD shown in the top photos in Fig. 8.13 was developed [11]. It comprises 3×3 units that are designed to be connected to each other and collaborate together. By connecting more units, the workspace can be extended to realize a larger-aperture AUTD, under which people can receive haptic information even while walking or working.

An example of a combination of the AUTD with visual projection is shown in the bottom photos of Fig. 8.13. A moving image projected on skin, together with its haptic stimulation, evokes the realistic sensation of a small animal or bug. This system can also be used for communication. Suppose that you are under a large-

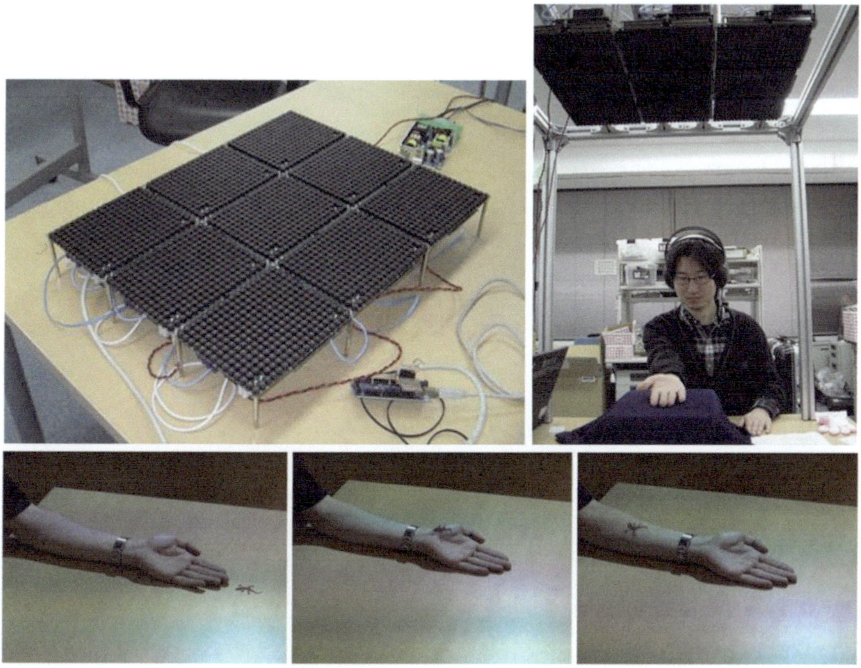

Fig. 8.13 Haptic projector; (*top-left*) large-aperture AUTD using 3×3 units; (*top-right*) system overview; (*bottom*) combination with visual projection

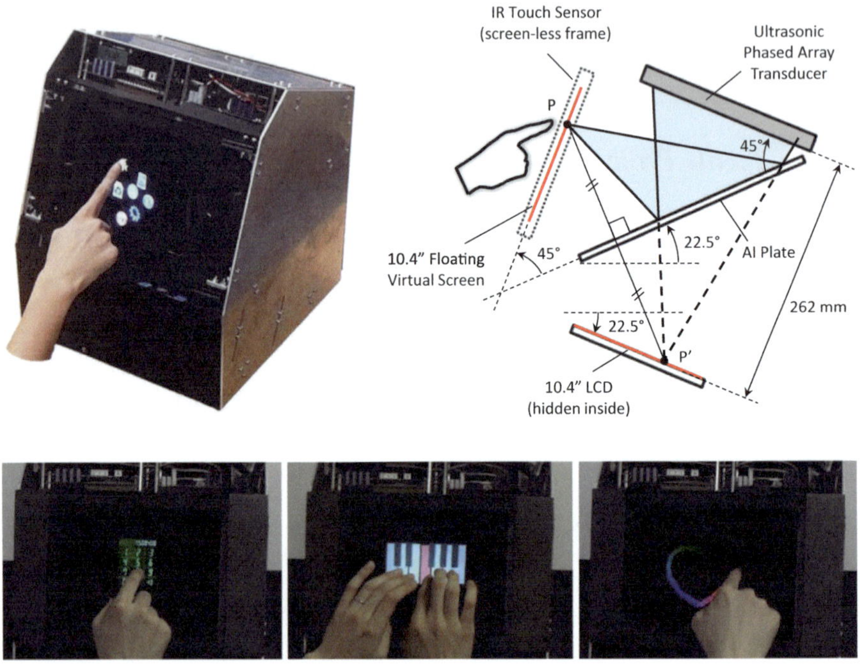

Fig. 8.14 HaptoMime; (*top-left*) system appearance; (*top-right*) system structure using reflected ultrasound; (*bottom*) floating-touch-panel demos of numeric pad, keyboard, and free drawing with haptic feedback

aperture AUTD and observe a visual image of another person. If tactile stimulation on your skin is synchronized with the person's image touching you, you feel as if touched by the person.

8.4.3 HaptoMime

This system demonstrates an aerial touch panel with haptic feedback [32]. As shown in the upper-right image of Fig. 8.14, an aerial-imaging (AI) plate manufactured by Asukanet Co. Ltd. produces an aerial image as a copy of the LCD image [3]. The ultrasonic wave radiated from the AUTD is reflected by the AI-plate surface and forms a focus at the finger position detected by an IR touch sensor. Because the visual response and haptic feedback are well synthesized, users can operate the aerial touch panel comfortably. A drag operation for aerial icons is shown in the top-left photo of Fig. 8.14. A numeric pad, a keyboard, and free drawing are shown in the bottom photos. For the number input, the finger is stimulated in a short time around the touch moment. In the drag operation and free drawing, the tactile stimulation is produced while the finger is detected in the virtual plane.

8.4.4 HORN (3D Haptic Hologram)

The aforementioned examples involve traveling ultrasonic waves synchronized with finger motion. The HORN (hapt-optic reconstruction) system produces a floating image with haptic feedback using a stationary ultrasonic wave [33]. The energy density of the stationary wave is spatially modified to be consistent with the visual image; thus, multiple users can touch the image at different positions simultaneously. For a stationary virtual object, people feel tactile sensation even without temporal amplitude modulation; hence, silent (noiseless) tactile feedback is realized. The top figures of Fig. 8.15 show HORN, which creates haptic holograms in a space surrounded by an octagonal layout of 16 AUTD units with visual floating images. The bottom-left photo shows stationary objects (letters) with rough textures. This photo does not suffer from delay or error due to a motion sensor, because the ultrasonic wave is stationary and the haptic sensation is produced by the user's hand motion. The bottom-right photo shows a dynamically deformable object according to the finger motion sensed by a Leap Motion Controller (Leap Motion, Inc.).

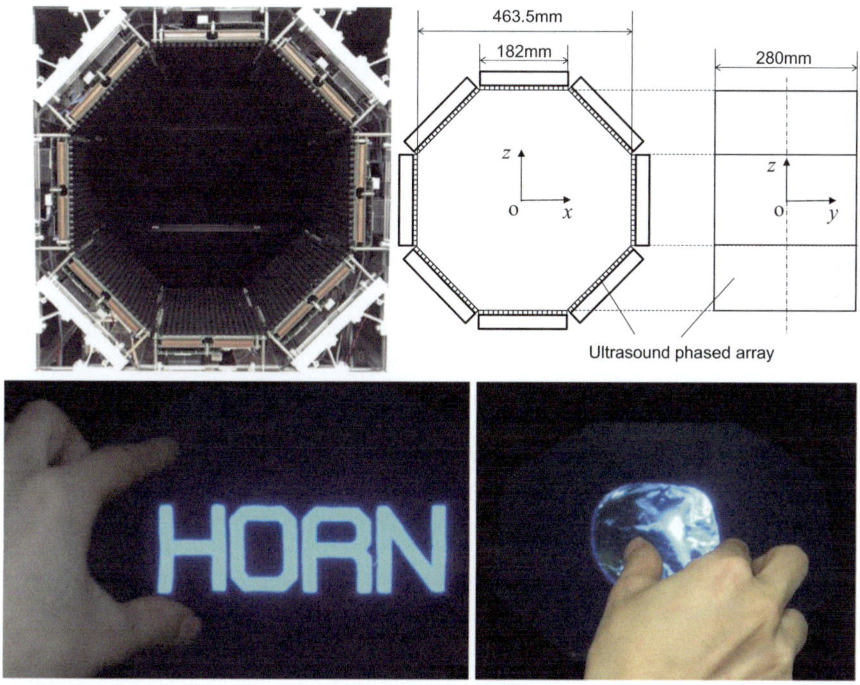

Fig. 8.15 HORN; (*top-left*) system appearance; (*top-right*) dimensions of system; (*bottom*) demos of 3D volumetric image with haptic feedback

8.4.5 Other Applications

Applications other than a noncontact tactile display have been explored. The advantages of AUTDs are the noncontact feature and spatial control. On the other hand, the drawback is that the output force is weak and in only one direction. Our investigation revealed that many applications benefit from the advantages of AUTDs in spite of the drawbacks.

The following are the authors' ongoing studies. Here, we call our device an airborne ultrasound focusing device (AUFD) instead of the AUTD because it is not restricted to a tactile display.

(a) The surface-compliance distribution of a soft material is measured [34]. The surface is locally pressed by an AUFD and observed using a laser ranger. The compliance is estimated according to the surface deformation. The distribution is obtained by scanning the measured point.
(b) The static-electricity distribution on a thin film of polyimide or other material is measured [35]. A low-frequency radio wave is radiated while the film is locally vibrated by an AUFD. The static electricity is estimated according to the amplitude of the radio wave detected by an antenna. Scanning yields a distribution result.
(c) A medical training system is developed using an AUFD [36]. A thin membrane representing skin is deformed by an AUFD to reproduce the pulse/thrill sensation for palpation training. The softness of the membrane is not lost, because there is no hard structure.
(d) Artwork is created that exhibits creature-like motion by standing-wave acoustic levitation [37]. An AUFD generates a movable acoustic standing wave. A small, black particle representing a fruit fly is levitated in the standing wave and moved above a plate containing leftovers.
(e) The reflection property of a soap film is controlled [38]. The reflection state of the film vibrated by AUFD is diffuse, whereas the original reflection state is mirror. By altering these two states at a high speed, the reflectance of various materials is represented owing to the persistence of vision.
(f) Three-dimensional noncontact manipulation of small particles is demonstrated using multiple AUFDs [39]. This technology is further extended to real-world computer graphics [40], in which the levitated objects play the role of physical pixels.
(g) A fur carpet turns into a computer display when graphics are drawn on it and erased [41]. Two AUFDs control the distribution of flattened and raised fibers in a noncontact manner.
(h) Artificial pollination is conducted in plant factories [42]. Flowers are vibrated by an AUFD mounted on a monitoring robot moving in the room. Thus, pollination is automatically achieved, without bees and humans.

8.5 Conclusion

The airborne ultrasound tactile display (AUTD) produces noncontact tactile stimulation. Its theoretical background, including acoustic radiation pressure, phased array focusing, vibrotactile stimulation, intensity control, absorption in air, and safety limits, was discussed. Next, a prototype device was introduced as an implementation example. The focal point was 20 mm in diameter, the maximum output force was 16 mN, and the refresh rate was 1 kHz. Applications of the AUTD were demonstrated: floating images with haptic feedback, a tactile projector, a floating touch display, and all-directional tactile feedback. Additionally, applications other than a noncontact tactile display were introduced.

References

1. Rodriguez, T., de Leon, A.C., Uzzan, B., Livet, N., Boyer, E., Geffray, F., Balogh, T., Megyesi, Z., Barsi, A.: Holographic and action capture techniques. Proc. ACM SIGGRAPH 2007. Emerging Technologies, article no. 11 (2007)
2. Floating Touch Display, http://www.nict.go.jp/en/press/2009/04/15-1.html. Last accessed on 30 Apr 2015
3. Aerial Imaging Plate, http://aerialimaging.tv/ (in Japanese). Last accessed on 30 Apr 2015
4. CyberTouch, http://www.est-kl.com/products/data-gloves/cyberglove-systems/cybertouch.html. Last accessed on 30 Apr 2015
5. Yoshida, T., Shimizu, K., Kurogi, T., Kamuro, S., Minamizawa, K., Nii, H., Tachi, S.: RePro3D: full-parallax 3D display with haptic feedback using retro-reflective projection technology. Proc. IEEE ISVRI **2011**, 49–54 (2011)
6. Kim, S.-C., Kim, C.-H., Yang, T.-H., Yang, G.-H., Kang, S.-C., Kwon, D.-S.: SaLT: small and lightweight tactile display using ultrasonic actuators. Proc. IEEE RO-MAN **2008**, 430–435 (2008)
7. Suzuki, Y., Kobayashi, M.: Air jet driven force feedback in virtual reality. IEEE Comput. Graph. Appl. **25**, 44–47 (2005)
8. Sodhi, R., Poupyrev, I., Glisson, M., Israr, A.: AIREAL: interactive tactile experiences in free air. ACM Trans. Graphics **32**, article no. 134 (2013)
9. Gupta, S., Morris, D., Patel, S.N., Tan, D.: AirWave: non-contact haptic feedback using air vortex rings. Proc. ACM Ubiquit. Comput. **2013**, 419–428 (2013)
10. Iwamoto, T., Tatezono, M., Shinoda, H.: Non-contact method for producing tactile sensation using airborne ultrasound. Proc. Eur. Haptics **2008**, 504–513 (2008)
11. Hasegawa, K., Shinoda, H.: Aerial display of vibrotactile sensation with high spatial-temporal resolution using large-aperture airborne ultrasound phased array. Proc. IEEE World Haptics Conf. **2013**, 31–36 (2013)
12. Hoshi, T.: Handwriting transmission system using noncontact tactile display. Proc. IEEE Haptics Symp. **2012**, 399–401 (2012)
13. Ciglar, M.: An ultrasound based instrument generating audible and tactile sound. Proc. NIME **2010**, 19–22 (2010)
14. Alexander, J., Marshall, M.T., Subramanian, S.: Adding haptic feedback to mobile TV. CHI Extended Abstr. **2011**, 1975–1980 (2011)
15. Marshall, M.T., Carter, T., Alexander, J., Subramanian, S.: Ultra-tangibles: creating movable tangible objects on interactive tables. Proc. CHI **2012**, 2185–2188 (2012)
16. Carter, T., Seah, S.A., Long, B., Drinkwater, B., Subramanian, S.: UltraHaptics: multi-point mid-air haptic feedback for touch surfaces. Proc. UIST **2013**, 505–514 (2013)

17. Long, B., Seah, S.A., Carter, T., Subramanian, S.: Rendering volumetric haptic shapes in mid-air using ultrasound. ACM Trans. Graph. **33**, article no. 181 (2014)
18. Gavrilov, L.R., Tsirulnikov, E.M., Davies, I.a.I.: Application of focused ultrasound for the stimulation of neural structures. Ultrasound Med. Biol. **22**(2), 179–192 (1996)
19. Iwamoto, T., Maeda, T., Shinoda, H.: Focused ultrasound for tactile feeling display. Proc. ICAT **2001**, 121–126 (2001)
20. Dalecki, D., Child, S.Z., Raeman, C.H., Carstensen, E.L.: Tactile perception of ultrasound. J. Acoust. Soc. Am. **97**, 3165–3170 (1995)
21. Iwamoto, T., Shinoda, H.: Two-dimensional scanning tactile display using ultrasound radiation pressure. Proc. IEEE Haptics Symp. **2006**, 57–61 (2006)
22. Awatani, J.: Studies on acoustic radiation pressure. I. (General considerations). J. Acoust. Soc. Am. **27**, 278–281 (1955)
23. Hasegawa, T., Kido, T., Iizuka, T., Matsuoka, C.: A general theory of Rayleigh and Langevin radiation pressures. Acoust. Sci. Technol. **21**(3), 145–152 (2000)
24. Hoshi, T., Takahashi, M., Iwamoto, T., Shinoda, H.: Noncontact tactile display based on radiation pressure of airborne ultrasound. IEEE Trans. Haptics **3**(3), 155–165 (2010)
25. Vallbo, Å.B., Johansson, R.S.: Properties of cutaneous mechanoreceptors in the human hand related to touch sensation. Hum. Neurobiol. **3**, 3–14 (1984)
26. Bass, H.E., Sutherland, L.C., Zuckerwar, A.J., Blackstock, D.T., Hester, D.M.: Atmospheric absorption of sound: further developments. J. Acoust. Soc. Am. **97**, 680–683 (1995)
27. Togawa, T., Tamura, T., Öberg, P.Å.: Biomedical Transducers and Instruments. CRC Press, Boca Raton (1997)
28. Howard, C.Q., Hansen, C.H., Zander, A.C.: A review of current ultrasound exposure limits. J. Occup. Health Saf. Aust. N. Z. **21**(3), 253–257 (2005)
29. Hoshi, T.: Development of portable device of airborne ultrasound tactile display. Proc. SICE Annu. Conf. **2012**, 290–292 (2012)
30. Ultrasound Transducer, http://www.nicera.co.jp/pro/ut/pdf/T4010A1(ENG).pdf. Last accessed on 30 Apr 2015
31. Hoshi, T., Takahashi, M., Nakatsuma, K., Shinoda, H.: Touchable holography. Proc. ACM SIGGRAPH 2009, Emerging Technologies, article no. 23 (2009)
32. Monnai, Y., Hasegawa, K., Fujiwara, M., Inoue, S., Shinoda, H.: HaptoMime: mid-air haptic interactions with a floating virtual screen. Proc. ACM UIST **2014**, 663–667 (2014)
33. Inoue, K.K., Kirschvinkand, Y., Monnai, K., Hasegawa, Y., Makino, Shinoda, H.: HORN: the hapt-optic reconstruction. Proc. ACM SIGGRAPH 2014, Emerging Technologies, article no. 11 (2014)
34. Fujiwara, M., Nakatsuma, K., Takahashi, M., Shinoda, H.: Remote measurement of surface compliance distribution using ultrasound radiation pressure. Proc. World Haptics Conf. **2011**, 43–47 (2011)
35. Kikunaga, K., Hoshi, T., Yamashita, H., Fujii, Y., Nonaka, K.: Measuring technique for static electricity using focused sound. J. Electrost. **71**(3), 554–557 (2012)
36. Hung, G.M.Y., John, N.W., Hancock, C., Gould, D.A., Hoshi, T.: UltraPulse – simulating arterial pulse with focused airborne ultrasound. Proc. EMBC **2013**, 2511–2514 (2013)
37. Kono, M., Hoshi, T., Kakehi, Y.: Lapillus bug: creature-like behaving particles based on interactive mid-air acoustic manipulation. Proc. ACE 2014, article no. 34 (2014)
38. Ochiai, Y., Oyama, A., Hoshi, T., Rekimoto, J.: The colloidal metamorphosis: time division multiplexing of the reflectance state. IEEE Comput. Graph. Appl. **34**(4), 42–51 (2014)
39. Hoshi, Ochiai, Y., Rekimoto, J.: Three-dimensional noncontact manipulation by opposite ultrasonic phased arrays. Jpn. J. Appl. Phys. **53**, 07KE07 (2014)
40. Ochiai, Y., Hoshi, T., Rekimoto, J.: Pixie dust: graphics generated by levitated and animated objects in computational acoustic-potential field. ACM Trans. Graph. **33**, article no. 85 (2014)
41. Sugiura, Y., Toda, K., Hoshi, T., Kamiyama, Y., Inami, M., Igarashi, T.: Graffiti Fur: turning your carpet into a computer display. Proc. ACM UIST **2014**, 149–156 (2014)
42. Shimizu, H., Nakamura, K., Hoshi, T., Nakashima, H., Miyasaka, J., Ohdoi, K.: Development of a non-contact ultrasonic pollination system. Proc. CIOSTA 2015, paper ID 43 (2015)

Chapter 9
Tactile Sensing Techniques That Use Intrinsic Force Sensors

Toshiaki Tsuji

Abstract Many types of tactile sensing techniques have been studied by now. This chapter describes the techniques that use intrinsic force sensors. First, classification of tactile sensing is described and the merits and demerits of intrinsic tactile sensing are discussed based on comparisons with other methods. Then, a simple algorithm of intrinsic tactile sensing is described and some extended methods based on the algorithm are discussed.

Keywords Intrinsic force sensing • Tactile sensing • Haptic sensing • Haptic signal processing

9.1 Introduction

Based on the location of the tactile sensors, tactile sensing can be categorized as intrinsic or extrinsic tactile sensing [2]. Intrinsic sensing is based on a force/torque sensor that is placed within the mechanical structure of the sensing system. The intrinsic sensing system derives contact information such as the magnitude of force or the contact location by using the force/torque sensor. In contrast, extrinsic sensing is based on sensors that are often arranged in arrays that are mounted at or near the contact interface.

Extrinsic sensing is widely studied, especially in the field of robotics [6, 7, 9] where applying sensor arrays is a good solution to gather rich tactile sensing information. Although array sensors on the surface of a tactile sensing system are useful, it is also often that array sensors on the surface of a tactile sensing system often present restrictions because of their high cost, large number of wires, and difficulty in measuring the resultant force. This section focuses on intrinsic tactile sensing as a simple solution in practical use.

The basic principle of intrinsic tactile sensing was first studied in the 1980s. Salisbury proposed a tactile sensing method that identifies the contact point and the force vector on an insensitive end effector by using a six-axis force sensor [8]. Bicchi

T. Tsuji (✉)
Saitama University, Saitama, Japan
e-mail: tsuji@ees.saitama-u.ac.jp

H. Kajimoto et al. (eds.), *Pervasive Haptics*, DOI 10.1007/978-4-431-55772-2_9

proposed the concept of intrinsic contact sensing that involves the identification of the contact point and force vector [1]. Iwata and Sugano developed an interface for human symbiotic robots. They realized whole-body tactile sensation by molding the end effector of the force sensor into a shell shape with touch sensors [4]. Tsuji et al. have also proposed "haptic armor," a simple intrinsic contact sensing calculation technique in which the spin resisting torque is ignored [11]. The mechanism has the advantage of reduced cost and less wiring, while allowing for six-axis resultant force detection.

It can be argued that intrinsic tactile sensing is a kind of a signal processing, which integrates new information from force/torque response. Although intrinsic tactile sensing is a simple solution for tactile sensing systems, the methods have not been widely applied and studied largely because such methods are sensitive to the accuracy of force/torque sensor calibration and can provide erroneous information because of unmodeled dynamic forces. Further, the compliance and inertia of the manipulator may also interfere in such cases. Some recent studies show that such problems can be solved in most cases, and some applications have been developed, too. Therefore, the following part describes the technique of "haptic armor" as a simple style of intrinsic tactile sensing, with some additional technical extension.

9.2 Intrinsic Tactile Sensing Algorithm

Figure 9.1 shows a diagram of making a force and moment measurement for a tactile sensing system with a cover supported by a force sensor. All external forces are detected by the sensor device because the force is transmitted through the cover, which is the end effector of the force sensor.

First, we describe the simplest intrinsic tactile sensing algorithm. This simple algorithm needs assumptions that the end effector is rigid and that its shape is a convex hull. The equilibrium of forces on the end effector leads to the following:

$$\mathbf{F}^e + \mathbf{F}^o = \mathbf{0}, \tag{9.1}$$

where $\mathbf{F}$ denotes the triaxial force, superscript e denotes the external force, and superscript o denotes the reaction force generated at the fixing point.

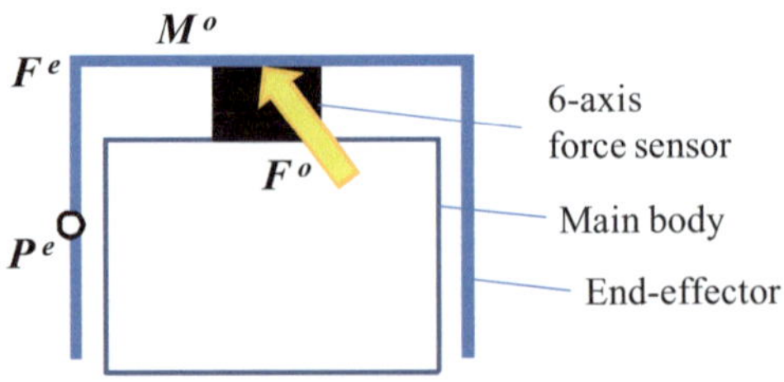

Fig. 9.1 Diagram for making a force and torque measurement on haptic armor

Although the mechanism does not have an array structure, it is possible to calculate the contact point only if a single contact point is assumed. The contact point calculation is based on the equilibrium of torque acting on the end effector:

$$\mathbf{F}^e \times (\mathbf{P}^e - \mathbf{P}^o) + \mathbf{M}^o = \mathbf{0}, \tag{9.2}$$

where $\mathbf{P}$ denotes the triaxial position of a contact point, $\mathbf{P}^e$ denotes the contact point, $\mathbf{P}^o$ denotes the point at which the end effector is fixed to the sensor device, and $\mathbf{M}^o$ is the triaxial reaction torque acting at the fixed point.

Equation (9.2) can be rewritten as follows:

$$\begin{aligned} P_z^e &= \frac{-M_x^o - F_z^e P_y^o}{F_y^e} + P_z^o + \frac{F_z^e}{F_y^e} P_y^e \\ &= \frac{M_y^o - F_z^e P_x^o}{F_x^e} + P_z^o + \frac{F_z^e}{F_x^e} P_x^e. \end{aligned} \tag{9.3}$$

Here, the subscripts x, y, and z refer to the axes of the Cartesian coordinates. M_x^o, M_y^o, F_x^e, F_y^e, and F_z^e are measured by the force sensor, whereas P_x^o, P_y^o, and P_z^o are derived by direct kinematics. As is evident from the red broken line in Fig. 9.1, (9.3) yields a straight line action of force, which is along $\mathbf{F}^e$. The straight line constrains $\mathbf{P}^e$. Since a pushing force is applied from outside the end effector, the contact point position $\mathbf{P}^e$ can be estimated.

Suppose that the shape in this case is given by the following equation:

$$f_k(\mathbf{P}^e) = 0 \qquad (k = 1, 2, \ldots, p), \tag{9.4}$$

where p denotes the number of surfaces. f_k is a function that calculates the distance from the surface to the input point.

The abovementioned method can be verified by a mobile robot. The overview of the experimental setup is shown in Fig. 9.2. The main body with a sensor device in the left figure is covered with a cubic end effector constructed of 3-mm acrylic boards.

We next examine the accuracy of the calculation of the contact location. We determine 9 points on each surface as sampling points. The total number of sampling points was 45 because the outer shell has five surfaces.

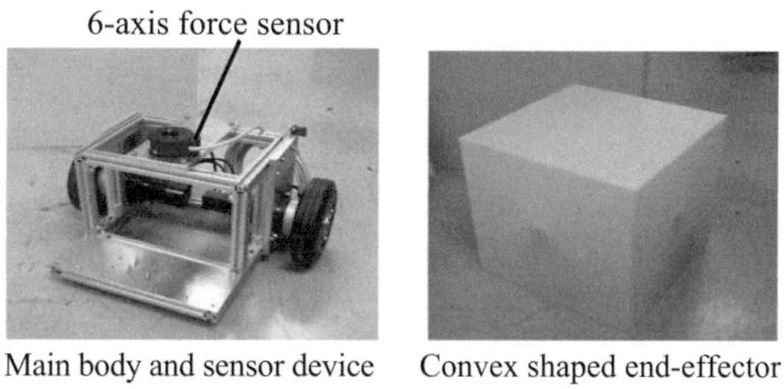

Fig. 9.2 Experimental system

Fig. 9.3 Results of contact location calculation. (a) Calculated contact location. (b) Contact location displayed on screen

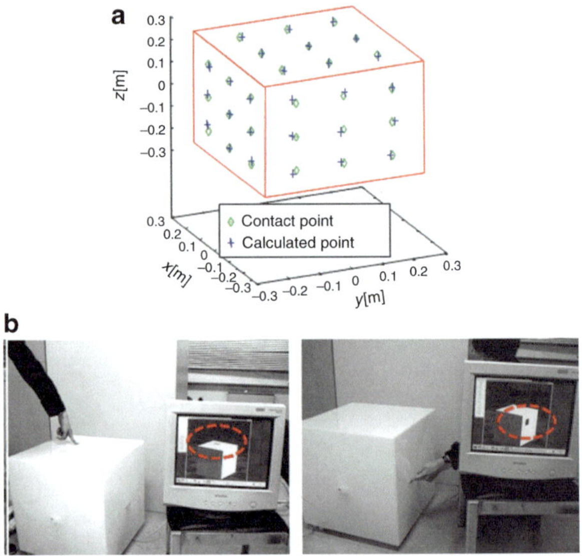

An external force was applied sequentially at each sampling point. Fig. 9.3 shows the calculated contact location corresponding to the respective sampling points. The average error was 2.64 % (1.32 cm), whereas the maximum error was 6.60 % (3.30 cm). The position measurement at the top surface had a higher accuracy, with an average error of 1.57 % (0.79 cm) and a maximum error of 2.13 % (1.06 cm). The main sources of these errors are the nonlinearity and hysteresis of the force sensor. Although this method cannot deal with multiple contacts, the gravity force of the load can be canceled with a simple calculation [11].

9.3 Compensation of Inertia Force

Force and torque equilibria in (9.1) and (9.2) may produce a certain amount of errors during acceleration because these equations do not take into account the inertia force. A large mass end effector or an end effector with a fixed load may also lead to a similar result. Hence, these equations are extended as follows:

$$\mathbf{F}^e + \mathbf{F}^o = w^e \ddot{\boldsymbol{P}}^o - w^e \mathbf{r}^g \times \ddot{\boldsymbol{\phi}}, \tag{9.5}$$

$$\mathbf{F}^e \times (\mathbf{P}^e - \mathbf{P}^o) + \mathbf{M}^o = w^e \mathbf{r}^g \times \ddot{\boldsymbol{P}}^o + \mathbf{I}^e \ddot{\boldsymbol{\phi}}, \tag{9.6}$$

$$\mathbf{r}^g = \mathbf{P}^g - \mathbf{P}^o.$$

Here, w^e and $\mathbf{I}^e$ are the weight and moment of inertia (MOI) of the end effector, respectively, $\ddot{\boldsymbol{\phi}}$ is the angular acceleration about the vertical axis, and $\mathbf{P}^g$ is the center of mass (CoM) of the end effector.

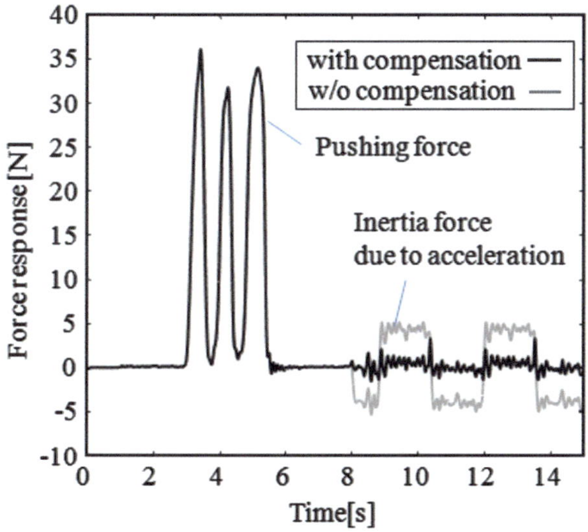

Fig. 9.4 Experimental result showing cancellation of inertia force

Equations (9.5) and (9.6) realize higher accuracy while they require additional acceleration information. The easiest way to detect acceleration is to add an acceleration sensor, but adding a sensor is undesirable in some cases. For tactile sensing for robots, it is also possible to measure acceleration from optical encoders on motors [11].

The experimental result in Fig. 9.4 shows how the inertia force of the end effector affects the force measurement. The weight of the end effector was 4.4 kg in total. First, an external force was applied three times by a hand while the vehicle was paused. After 8 s, a command velocity derived by $v_{cmd} = a \sin \omega(t - 8)$ was given to the vehicle. Here, $a = 1.0$ [m/s] and $\omega = 2.0$ [rad/s].

The gray line in Fig. 9.4 shows the force response derived by using (9.2), a result without a measurement of acceleration. In contrast, the black line shows the force response derived by using (9.6), a result with inertia force cancellation. Almost all of the inertia force was canceled, while some error still remained.

The root mean square (RMS) errors of force sensation are compared in Fig. 9.5. Some error remains along the y axis, the axis of vehicle movement, while the differences along the other axes are relatively small.

9.4 Compensation for Soft Material Deformation

Tactile sensing systems often need soft material on their surfaces. "FuwaFuwa" [10] is a good example of such a system. Although soft material transfers static force, in the common method of intrinsic tactile sensing one assumes a rigid end effector. The reason is to avoid the error produced by the deformation of the material. This

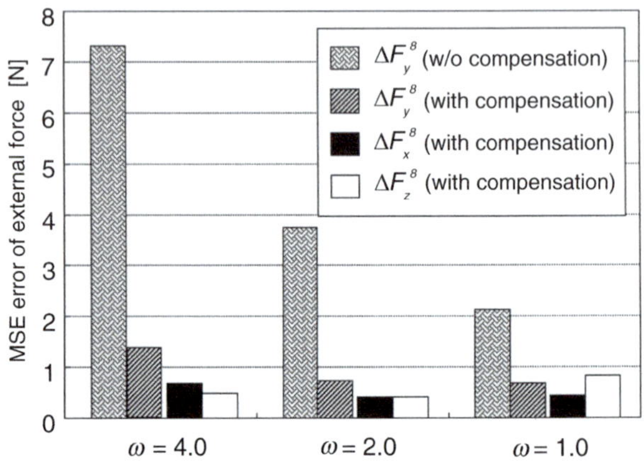

Fig. 9.5 Comparison of RMS errors of force sensation

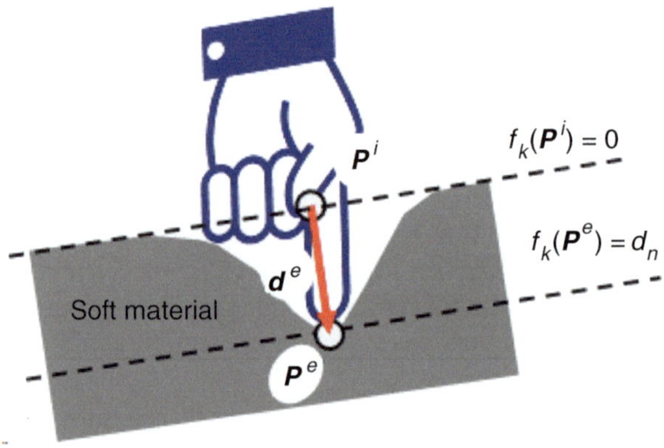

Fig. 9.6 Model of deformation

section discusses a method to realize intrinsic tactile sensing by compensating for the deformation of the soft material [5].

An end effector made of a soft material is deformed under an external force as shown in Fig. 9.6. Here, $\mathbf{P}^i$ denotes the point at which the external object initially contacts the surface. $\mathbf{P}^e$ denotes the contact point on the surface deformed by the external force. Deformation $\mathbf{d}$ is derived as follows:

$$\mathbf{d} = \mathbf{P}^e - \mathbf{P}^i. \tag{9.7}$$

Then, based on (9.4), equations representing rigid surfaces can be developed as follows:

$$f_k(\mathbf{P}^i) = 0, \tag{9.8}$$

$$f_k(\mathbf{P}^e) = d_n. \tag{9.9}$$

Here, d_n denotes the component of **d**, which is vertical to the surface.

Equations (9.8) and (9.9) show that the method described in the previous section contains error owing to the deformation of the soft material. Hence, we propose a method to compensate for the error based on an estimation of the deformation. In this study we introduce a simplified model of soft material deformation, which is estimated via a linear approximation of the external force and the displacement characteristics of the soft material under an applied external force. By considering deformation only in the nominal direction, a linear approximation for estimating the displacement of the soft material can be described as follows:

$$d_n = KF_n + b. \tag{9.10}$$

Here, F_n denotes the normal-direction force on the soft material surface, and K and b denote linear approximation parameters, which are provided by a preliminary test recording the external force acting on and the displacement characteristics of the soft material.

It is possible to estimate soft material deformation using (9.10), while the equation calculates the deformation from the force along the nominal axis. Absolute coordinates can be transformed to contact point coordinates, relative coordinates that have a z axis in the nominal direction of the surface, by using

$$\mathbf{d}^r = \mathbf{R}_k \mathbf{d}^e, \tag{9.11}$$

$$\mathbf{F}^r = \mathbf{R}_k \mathbf{F}^e, \tag{9.12}$$

where $\mathbf{d}^r$ and $\mathbf{F}^r$ are the external force and deformation vectors in contact point coordinates, respectively, and $\mathbf{R}_k$ is a rotational matrix that transforms absolute coordinates to contact point coordinates of the kth surface. Since F_n is the normal direction component of the force, $F_n = F^r_z$. By substituting these equations into (9.10), the deformation d_n of the soft material can be estimated.

Calculation of the compensated position of the contact point is shown below. By developing (9.11), we obtain

$$\mathbf{d}^e = \mathbf{R}_k^{-1} \mathbf{d}^r. \tag{9.13}$$

Substituting (9.13) into (9.7), we acquire the compensated contact position $\mathbf{P}^e$:

$$\mathbf{P}^e = \mathbf{P}^i + \mathbf{d}^e \tag{9.14}$$

$$= \mathbf{P}^i + \mathbf{R}_k^{-1} \mathbf{d}^r. \tag{9.15}$$

In many cases, the soft material is deformed in the normal direction because the external force is applied in the normal direction. Although there are small deformations in other directions, it is assumed that these are small enough. Then, from (9.10) and (9.15), we have

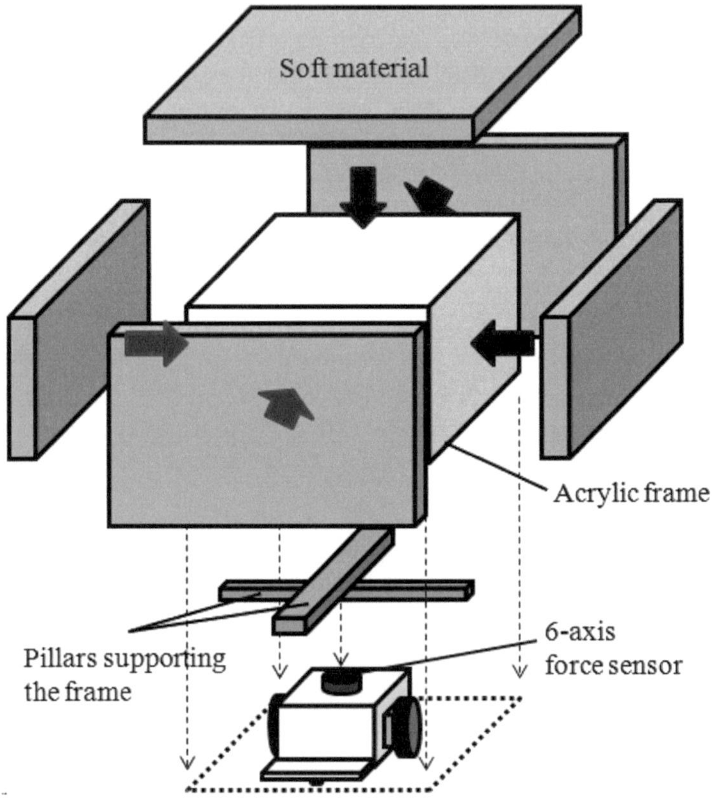

Fig. 9.7 Structural diagram of haptic armor with soft material

$$\begin{aligned}\mathbf{P}^e &= \mathbf{P}^i + \mathbf{R}_k^{-1}\,[0 \quad 0 \quad d_n]^T \\ &= \mathbf{P}^i + \mathbf{R}_k^{-1}\,[0 \quad 0 \quad KF_n + b]^T\,. \end{aligned} \tag{9.16}$$

Equation (9.16) shows that $\mathbf{P}^e$ can be calculated from the force information if the shape of the end effector and contact parameters K and b are known.

Figure 9.7 shows a structural diagram of haptic armor composed of soft material. An end effector made of soft material is supported by an acrylic frame. Figure 9.8a, b show diagrams of mechanisms for fixing soft material on a force sensor. In the following two cases, this fixation method may produce errors:

- For an external force applied close to the edge of the end effector, the correct contact point is not determined because the external force and displacement characteristics are different.
- In the case of a large external force, the end effector may come into contact with the robot body. This degrades the performance because (9.1) and (9.2) hold under the assumption that the end effector is supported only on the force sensor.

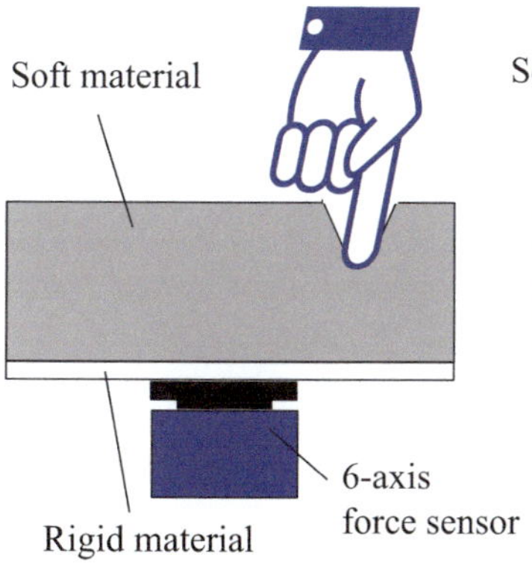

Fig. 9.8 Mechanisms of fixing soft material on the force sensor

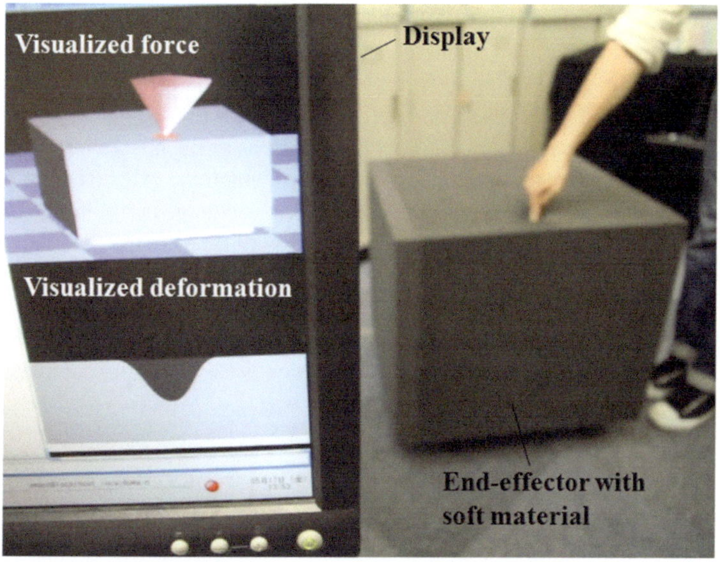

Fig. 9.9 Demonstration of intrinsic contact sensing with soft material

Therefore, this method should be applied to a mechanism in Fig. 9.8a. If the end effector has a rigid framework surrounded by soft material with even thickness, an accurate compensation is maintained. It is quite common that some areas have different thickness owing to irregularity of the shape. The performance of the error compensation will be degraded in such areas, while the accuracy is still better than the result without any compensation. Figure 9.9 shows an experimental view of demonstration with soft material. The display in the left side of the photo visualizes both the external force vector and the deformation of the soft material.

9.5 Estimation of Contact Location in Arbitrarily Shaped end Effectors

The abovementioned method needs an assumption that the end effector is a convex hull, because the contact point calculation may find multiple solutions with a nonconvex shape. As many instruments and robots have nonconvex shape, it is quite obvious that this assumption strongly limits its applicability. Hence, the following part describes a method to estimate the contact location on nonconvex-shaped end effectors.

In determining a contact point, the change in an external force vector upon contact, with respect to time, is utilized as important information. In most cases in practice, the force vector does not have a fixed direction. Figure 9.10 shows the measurement results for an actual external force. The outer shell of a robot was pressed with an index finger, and a straight line derived from the starting time of contact using (9.3) was visualized. In the figure, we can verify that the slope of the line changes with the contact point as its center.

If the direction of the external force vector changes with time, the slope of the straight line derived from (9.3) changes; this is shown in Fig. 9.11, as the time changes from t_1 to t_2 to t_3, etc. Although the positions of the possible contact points do not change with respect to time for the point at which the external force is being applied, the positions of other possible contact points change owing to the direction of the external force vector that is also changing with time. Thus, the contact point can be calculated by treating the variability of the candidate contact points on each surface as the standard deviation.

To use the standard deviation, $\mathbf{F}^o$ and $\mathbf{M}^o$ are acquired from the sensor within a certain period of time, and the candidate contact points are calculated. The

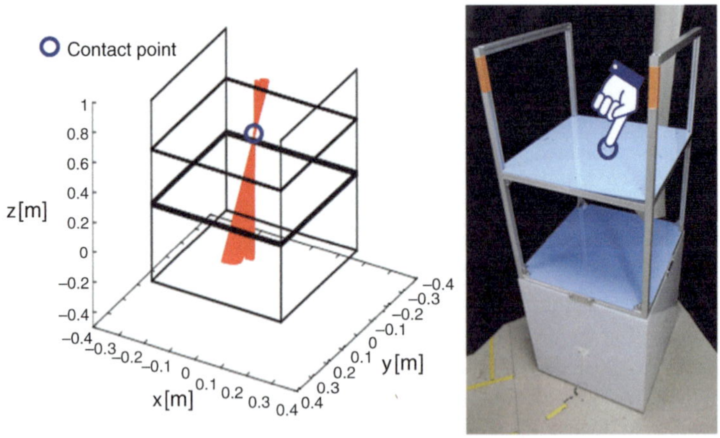

Fig. 9.10 Force and torque measurement on haptic armor

Fig. 9.11 Variation in external force when the end effector has contact

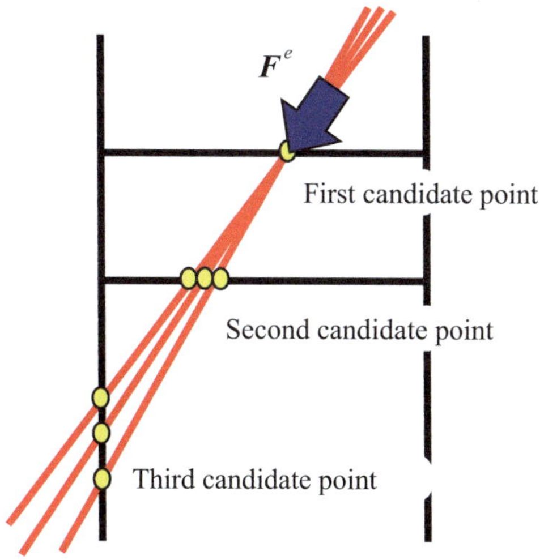

following equation shows the conditions that are necessary for sensor devices to begin acquiring information:

$$|\mathbf{F}^o| \geq F_{th}. \tag{9.17}$$

The calculation of possible contact points begins when the size of F^o detected by the sensor exceeds the threshold F_{th}. T_{sd} is the constant time period during which the points are continuously calculated. If the sensor device acquires information in sampling period T_s, then the total number of single-component data n is calculated as $n = T_{sd}/T_s$. Here, T_{sd} is assumed to be a value that is divisible by T_s. Here, let us assume that the time at which (9.17) is satisfied is t_1. From t_1 acquired within T_{sd}, each piece of sensor information within each component $t_1 + T_{sd}$ is assigned a number from 1 to n. The number of end-effector surfaces is p, and the possible contact point at t_1 on the k_{th} surface is

$$\mathbf{P}^{cand}_{kt_1} = (x_{k1}, y_{k1}, z_{k1}). \tag{9.18}$$

Possible contact points are obtained for p surfaces using (9.3) and (9.4), based upon each surface of the end effector. At any given time, a number of contact points equal to the number of surfaces can be obtained, but the proposed approach also takes into account changes in time. Thus, possible contact points are treated as a matrix. The following equation shows the set $\mathbf{X}^{cand}$ of candidate contact points along the x axis in a time period T_{sd}:

$$\mathbf{X}^{cand} = \begin{bmatrix} \mathbf{x}_1 \\ \vdots \\ \mathbf{x}_p \end{bmatrix} = \begin{bmatrix} x_{11} & \cdots & x_{1n} \\ \vdots & \ddots & \vdots \\ x_{p1} & \cdots & x_{pn} \end{bmatrix}. \tag{9.19}$$

$\mathbf{X}^{cand}$ is a $p \times n$ matrix consisting of row vectors $\mathbf{x}_1, \cdots, \mathbf{x}_p$. Row vectors are the possible contact points derived in time T_{sd} along the x axis of each surface. Likewise, column vectors are the possible contact points at each point in time along the x axis for p surfaces. The standard deviation σ_{kx} of the possible contact points along the x axis for surface k, using components in vector $\mathbf{x}_k$, is determined by using

$$\sigma_{kx} = \sqrt{\frac{\sum_{i=1}^{n}(x_{ki} - \bar{x}_k)^2}{n}} \quad \bar{x}_k = \frac{1}{n}\sum_{i=1}^{n} x_{ki}. \tag{9.20}$$

Similarly, the sets of possible contact points Y^{cand} and Z^{cand} in a given time period T_{sd} along the y and z axes, respectively, can be expressed using (9.19). The standard deviations σ_{ky} and σ_{kz} of the possible contact points on k faces along the y and z axes, respectively, can also be determined. The standard deviation of the position of possible contact points, σ_{kx}, σ_{ky}, and σ_{kz}, for surface k can be determined using

$$\sigma_k = \sqrt{\sigma_{kx}^2 + \sigma_{ky}^2 + \sigma_{kz}^2}. \tag{9.21}$$

Because the standard deviation is smallest on the surface on which the external force is applied, the contact point is calculated by determining σ_{min} from the range of σ_1 to σ_p and the surface that is equivalent to σ_{min}:

$$\sigma_{min} = \min[\sigma_1, \sigma_2, \ldots \sigma_p]. \tag{9.22}$$

This approach calculates the surface on which an external force is being applied using the standard deviation of the position of possible contact points, and it determines the possible point on that surface as the contact point.

Figure 9.12 shows a scene of contact point estimation with a nonconvex-shaped end effector and Fig. 9.13 shows the result. The figures show that the method can estimate the proper contact point from multiple candidates. Its detection rate was over 95 % with 400 force/torque data ($n = 400$) gathered by a 1-ms sampling time system.

9.6 Configuration of Haptic Desk

Application of the abovementioned theory makes it possible to detect external force application points even without installing tactile sensors on the surface skin of objects. The configuration of a system capable of converting a variety of objects into interfaces by mounting the required mechanisms on commonly found machines, furniture, and consumer electronics is described below. Force sensors are installed on the supports of objects subject to conversion into interfaces. A mechanism with

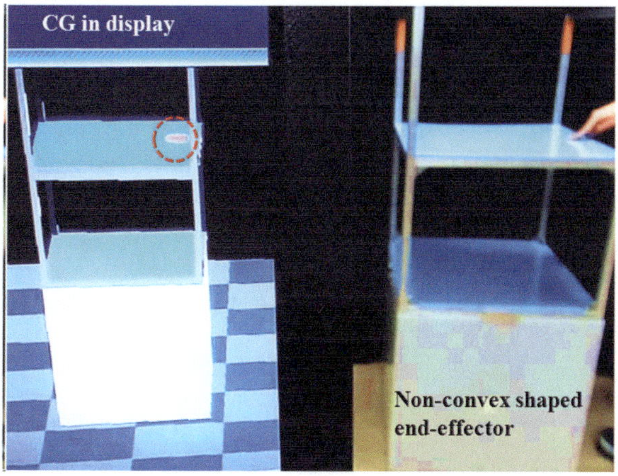

Fig. 9.12 Contact point estimation with a nonconvex-shaped end-effector

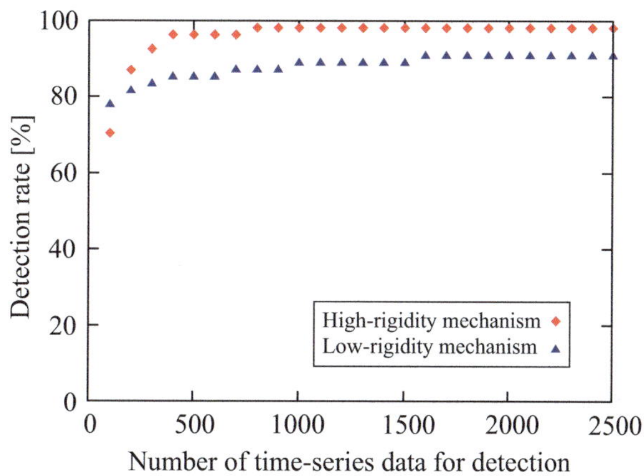

Fig. 9.13 Result of contact point estimation with a nonconvex-shaped end-effector

a fixture attached to a force sensor was developed [12]. The setup is shown in Fig. 9.14. The end effector of the force sensor was secured to a leg by using a clamp mechanism. Since the desk has four legs, four force sensors are needed. Equations (9.1) and (9.2), the force and torque equilibria using one force sensor, should be extended to equilibria for multiple force sensors as follows [3]:

$$\mathbf{F}^e + \sum_{i=1}^{n} \mathbf{F}_i^s = \mathbf{0}, \tag{9.23}$$

$$\mathbf{F}^e \times (\mathbf{P}^e - \mathbf{P}^o) + \sum_{i=1}^{n} (\mathbf{F}_i^s \times (\mathbf{P}_i^s - \mathbf{P}^o)) = \mathbf{0}, \tag{9.24}$$

where $\mathbf{F}_i^s$ denotes the force measured by the ith sensor and $\mathbf{P}_i^s$ denotes the location of the force sensors.

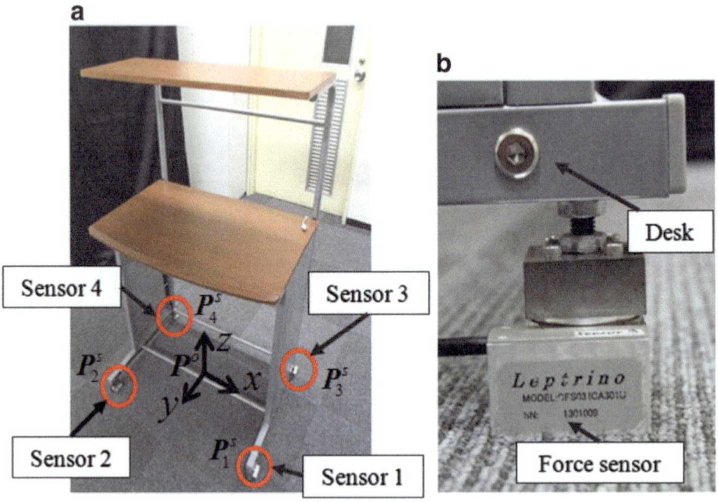

Fig. 9.14 Photographs of the force-sensing device with the mounting mechanism. (**a**) Photo of haptic desk. (**b**) Sensor fixation

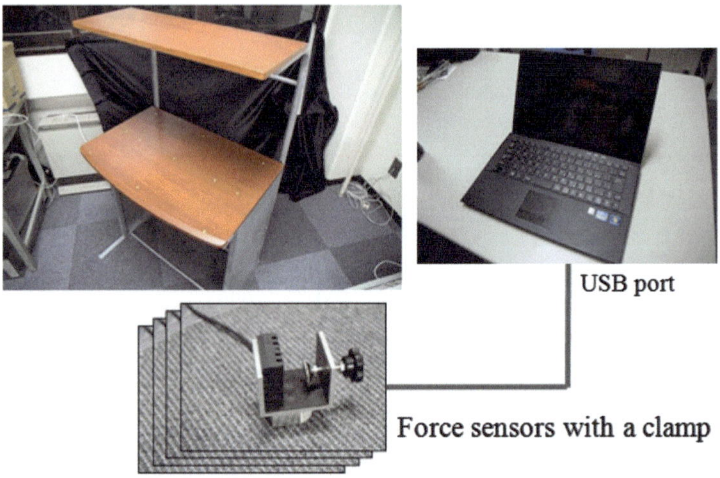

Fig. 9.15 Photographs of the system configuration

The six-axis force sensor CFS034CA301U from Leptrino was used as the force sensor. These sensors could be replaced by other three-axis force sensors, since force information of only three axes was used to calculate the contact position.

The actual system configuration is shown in Fig. 9.15. The objects that were converted into interfaces were not included in the basic configuration, since in

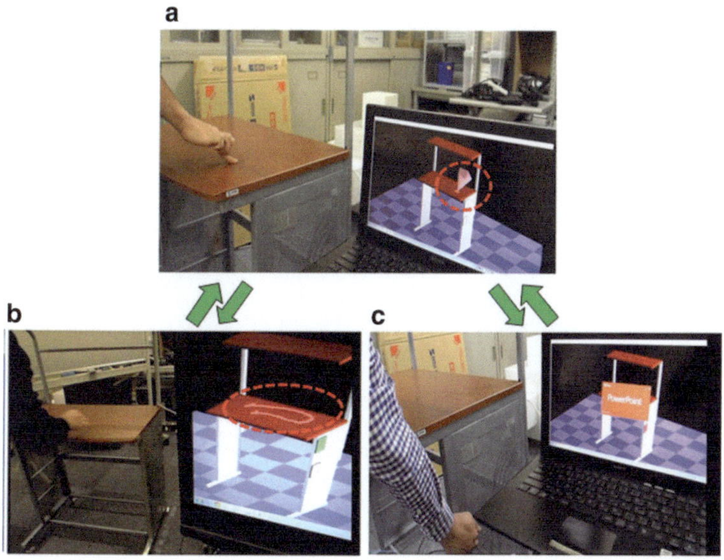

Fig. 9.16 Photographs showing the respective modes of the haptic desk. (**a**) Basic mode. (**b**) Drawing mode. (**c**) Power point mode

this research we only attempted to convert a regular desk into an interface. This system comprised force sensors installed on the legs of objects as well as a USB hub and a laptop PC. The theory presented here can be applied provided the force and the resultant force of moments of the six axes can be derived, even if the number of sensors is decreased. Information from the force sensors is sent to the laptop PC, and the external forces as well as their application points are identified by using software running on the PC. The objects can be used as interfaces for the PC, by sending the information to the mouse driver or software. As shown in Fig. 9.16, the haptic desk features three modes with different applications: (a) a basic mode, (b) a drawing mode, and (c) a PowerPoint mode. Furthermore, object recognition based on loads can be performed, regardless of the currently executed mode. (a) The basic mode displays the contact points and external force vectors. (b) The drawing mode allows the continuous drawing of contact points and the retention of these drawing strokes. The thickness of the drawing strokes can be controlled according to the sizes of the external force vectors. (c) In the PowerPoint mode, PowerPoint is manipulated by pressing virtual switches. Figure 9.17 shows the results from the drawing mode. The letter on the display is “Ki,” (or wood in Japanese), and it shows that writing in brush-like strokes can realized by adjusting the thickness of the drawing strokes according to the writing pressure.

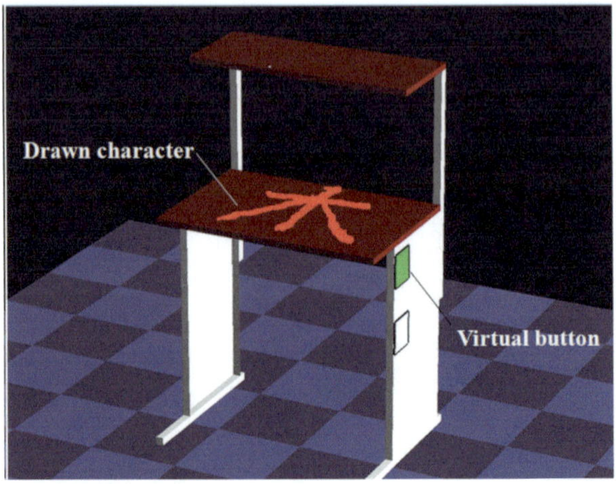

Fig. 9.17 Results of the drawing mode

9.7 Conclusion

Intrinsic tactile sensing is a technique, which requires no sensor devices on the surface of the end-effector. Hence, it is a good solution for tactile sensing systems in some cases. Several practical issues for intrinsic tactile sensing exist: error due to inertia force, deformation of soft material, and extension to an end-effector with a non-convex shape. The solutions to these issues were described and such techniques were introduced to a novel application of a desk with tactile sensation. These studies suggest that signal processing on force information has a potential to be extended for further applications.

References

1. Bicchi, A.: Intrinsic contact sensing for soft fingers. In: Proceedings of the IEEE International Conference on Robotics and Automation, Tsukuba, pp. 968–973 (1990)
2. Dahiya, R.S., Metta, G., Valle, M., Sandini, G.: Tactile sensing-from humans to humanoids. IEEE Trans. Robot. **26**(1), 1–20 (2010)
3. Hanyu, R., Tsuji, T.: Development of a whole-body haptic sensor with multiple supporting points and its application to a manipulator. IEEJ Trans. Ind. Appl. **131**, 1128–1134 (2011)
4. Iwata, H., Sugano, S.: Whole-body covering tactile interface for human robot coordination. In: Proceedings of the IEEE International Conference on Robotics and Automation, Washington, pp. 3818–3824 (2002)
5. Kurita, N., Hasunuma, H., Sakaino, S., Tsuji, T.: Simplified whole-body tactile sensing system using soft material at contact areas. In: Proceedings of the 39th Annual International Conference on IEEE Industrial Electronics Society, Vienna, pp. 4264–4269 (2013)
6. Mukai, T., Onishi, M., Odashima, T., Hirano, S., Luo, Z.: Development of the tactile sensor system of a human-interactive robot RI-MAN. IEEE Trans. Robot. **24**(2), 505–512 (2008)

7. Ohmura, Y., Nagakubo, A., Kuniyoshi, Y.: Conformable and scalable tactile sensor skin for curved surface. In: Proceedings of the IEEE International Conference on Robotics and Automation, Orlando, pp. 1348–1353 (2006)
8. Salisbury, J.: Interpretation of contact geometries from force measurements. In: Proceedings of the IEEE International Conference on Robotics and Automation, Atlanta, pp. 240–247 (1984)
9. Schmitz, A., Maiolino, P., Maggiali, M., Natale, L., Cannata, G., Metta, G.: Methods and technologies for the implementation of large-scale robot tactile sensors. IEEE Trans. Robot. **27**(3), 389–400 (2011)
10. Sugiura, Y., Kakehi, G., Withana, A., Lee, C., Sakamoto, D., Sugimoto, M., Inami, M., Igarashi, T.: Detecting shape deformation of soft objects using directional photoreflectivity measurement. In: Proceedings of the Annual ACM Symposium on User Interface Software and Technology, Scottsdale, pp. 509–516 (2011)
11. Tsuji, T., Kaneko, Y., Abe, S.: Whole-body force sensation by force sensor with shell-shaped End-effector. IEEE Trans. Ind. Electron. **56**(5), 1375–1382 (2009)
12. Tsuji, T., Seki, T., Sakaino, S.: A Mounting foot type force sensing device for a desk with haptic sensing capability. In: Proceedings of the Asia Haptics, Tsukuba, (2014)

Chapter 10
Reflection-Image-Based Tactile Sensor

Satoshi Saga

Abstract In this chapter I introduce a novel precise tactile sensor, reflection-image-based tactile sensor. Today many touchscreen interfaces have become popular worldwide, however, they cannot sense forces or vibrations. For richer interfaces, precise and robust tactile sensors are required. Then we propose a reflection-image-based tactile sensor. The sensor employs deformable transparent silicone rubber and distortion of a reflection image. The construction of the sensor solves wiring issues and degradation problems. In addition, full tactile information can be acquired from the reflection images. Furthermore, active sensing enhance the robustness of the proposed sensor.

Keywords Tactile sensor • Total internal reflection • Optical lever • Active sensing

10.1 Introduction

Touchscreen interfaces have recently become popular worldwide. Moreover, popular consumer electronics devices such as the iPad use a dedicated touchscreen as an interface. These touchscreens require for input multiple finger-point contact or sliding finger movements, but they cannot sense forces or vibrations transmitted by the fingers. To achieve richer input methods, such as more forceful twisting or soft pinching, sensing technologies that acquire rich tactile information are necessary. Furthermore, industrial robots are required that perform in a work-cell capacity in a small-lot-size production. Such robots need to manipulate objects of various shapes and sizes with accuracy. To achieve precise sensing of position and shape of an object as humans do is indispensable. Therefore we aimed for a robust and precise tactile sensor that has characteristics similar to humans and have proposed a reflection-image-based tactile sensor.

To create novel sensing devices for richer tactile information, we propose a precise real-time reconstruction method that employs an optimization method.

S. Saga (✉)
University of Tsukuba, 1-1-1, Tennodai, Tsukuba, Japan
e-mail: saga@saga-lab.org

H. Kajimoto et al. (eds.), *Pervasive Haptics*, DOI 10.1007/978-4-431-55772-2_10

Furthermore, we propose using active patterns. Distortions in the reflected image caused by deformations in the surface in a reflection-based tactile sensor causes reconstruction errors. If the image pattern can be controlled, the sensor can reconstruct the surface deformation more precisely.

10.2 Principle

In several recent studies, computer vision has been employed to measure the distribution of force field [1–7]. These sensing methods use cameras to circumvent wiring issues and degradation problems caused by repetitive measurements. Furthermore, full tactile information can be acquired from the captured images. In several researches optical information is employed for precise measurement [8–10], however these researches require repetitive measurements and calculations. In our research on novel tactile sensing [11], we employ not only computer vision but also the reflected images.

10.2.1 Detection of Feature Points from Reflection Image

Our sensor is composed of transparent silicone rubber, an image pattern, and video camera. The difference between the refractive indices of air and silicone rubber produces total-internal-reflection at the boundary (Fig. 10.1). By this means, the

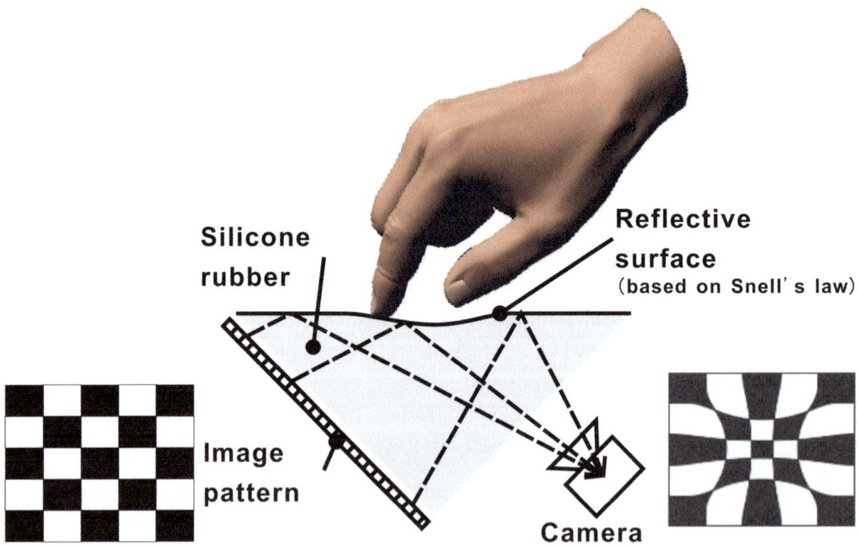

Fig. 10.1 Cross section of the sensor: *Grey triangle* shows transparent silicone rubber and the *left side* of image pattern is reflected on the *upper side* of the silicone rubber by internal reflection. The reflection image is captured by a camera

reflection of the image pattern is captured by the camera. With the boundary as the sensor surface, any deformation of it shows up as a distortion in the reflected image of the image pattern. By exploiting the distortion of the image, we can reconstruct the deformation of the sensor surface. Because the reflection amplifies deformations in the surface as described by the principle of the optical lever, the sensor is able to measure small deformations precisely.

10.2.2 Reconstruction of Deformed Surface from Tracked Feature Points

In the proposed reconstruction method, we implement the following procedures. First, we extract and track feature points from the reflected image and obtain the corresponding incident optical vectors (Fig. 10.2). Second, by considering the geometric optical constraints of the sensor, we formulate the reflection points on the sensor surface and their partial differential coefficients. Third, by Taylor expansion, we formulate a system of equations relating adjacent points on the sensor surface. Finally, we perform an optimization by solving the system of equations using Newton's method to reconstruct the shape of the sensor surface. A typical result of a reconstruction is shown in Fig. 10.3: the left panel shows the original image, the center is the reconstructed result, and the right shows the errors between them. This result highlights the precision in the reconstruction. Moreover, in tracking feature points, several errors are included in the tracking results. Figure 10.4 presents a reconstruction result in which error has been included. The left panel shows reconstruction result and the right gives the errors. Even when error is included, our

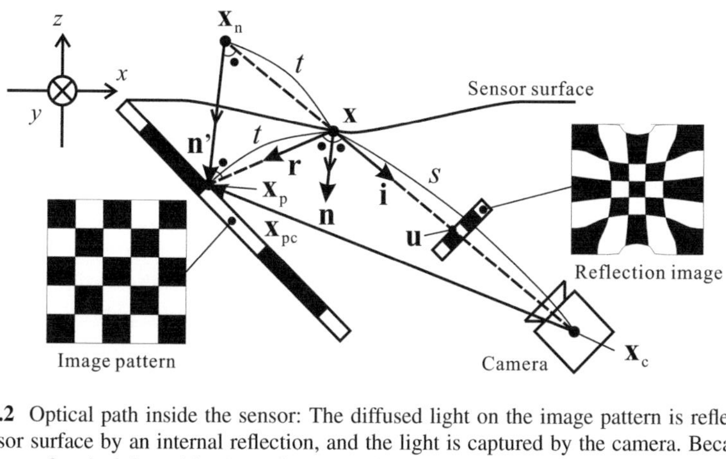

Fig. 10.2 Optical path inside the sensor: The diffused light on the image pattern is reflected on the sensor surface by an internal reflection, and the light is captured by the camera. Because the reflective surface is deformable, the angle and position of normal vector ***n*** is variable. We estimate the angle from the acquired reflection image

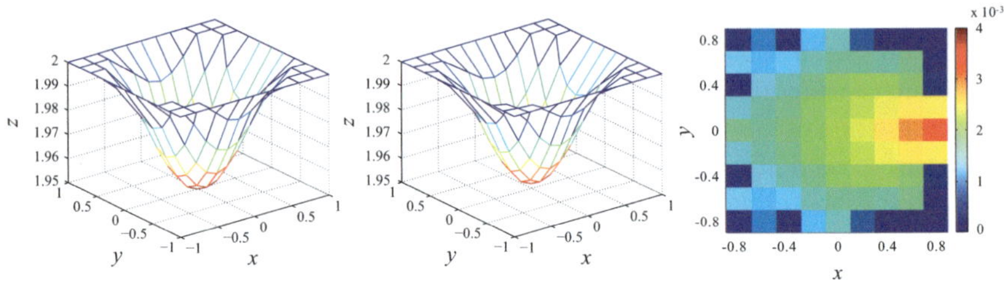

Fig. 10.3 Simulated result of reconstruction: (*left*) true image, (*center*) reconstructed result, (*right*) distribution of errors

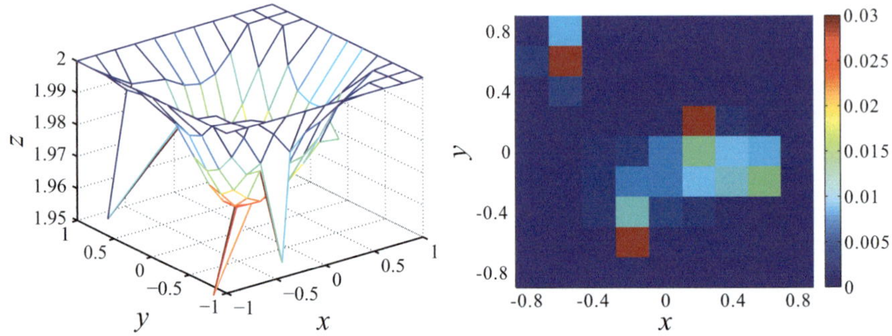

Fig. 10.4 Simulated result of reconstruction (including measuring error). *Left*: reconstruction result, *right*: distribution of errors

method can reconstruct the shape almost correctly. Thus, our proposed optimization method provides a robust reconstruction of the deformation shape in many cases.

10.2.3 Active Sensing

Furthermore, to achieve more precise reconstruction and to minimize procedural errors introduced by large deformations (left panel of Fig. 10.5), we employ active-image patterning whereby an optimum image pattern for which the reflected image has an equal distribution of reflection points. Hence, we propose a method of generating sequential optimum patterns. The method uses the estimated reconstructed shape of the previous frame and generates an optimum pattern with evenly distributed reflection points (right panel of Fig. 10.5). From the reflected image of the generated optimum pattern, the reconstruction proceeds successively. We call this pattern an active pattern. Results of two examples are shown in Figs. 10.6 and 10.7.

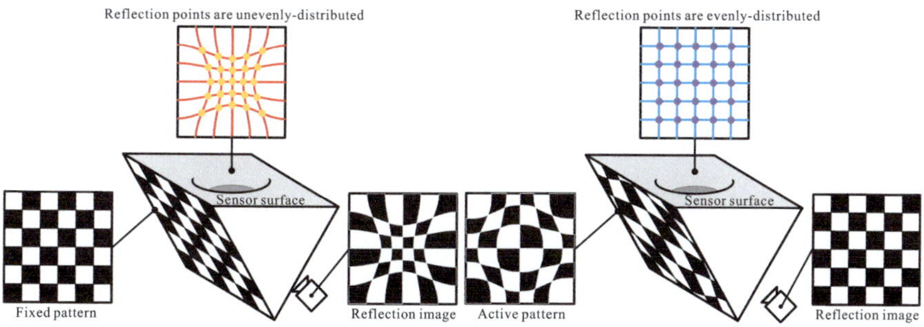

Fig. 10.5 Active sensing: (*left*) tracking feature points from fixed image pattern, (*right*) tracking feature points from dynamic image pattern

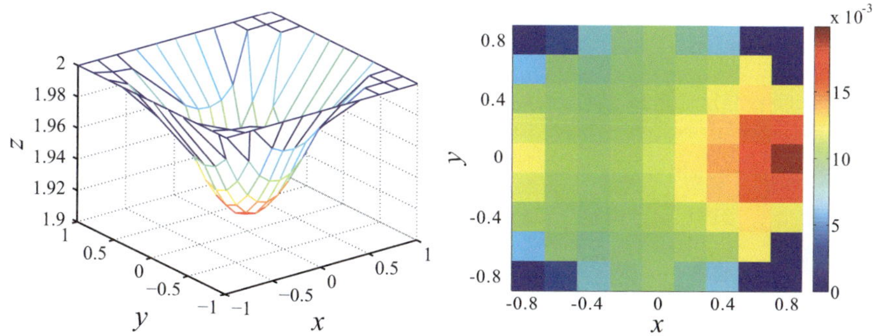

Fig. 10.6 Simulated reconstruction result of large deformation from fixed image pattern: (*left*) reconstruction result, (*right*) distribution of errors

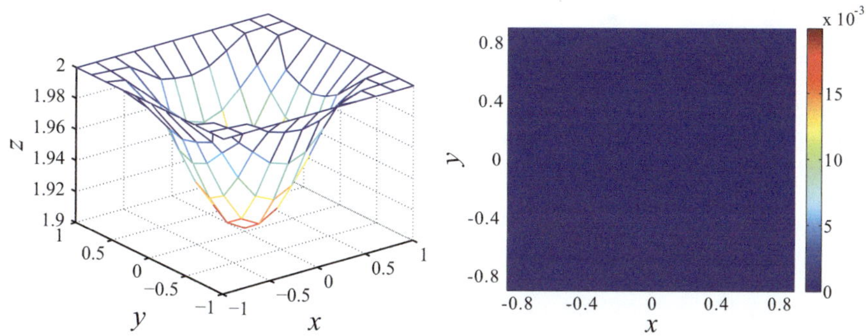

Fig. 10.7 Simulated reconstruction result of large deformation from dynamic image pattern: (*left*) reconstruction result, (*right*) distribution of errors

Fig. 10.8 System overview: The system consist of a LCD, a camera and a sensor body made of silicone rubber. The display shows deformable image pattern, and the pattern is refreshed by the estimated sensor shape from previous frame

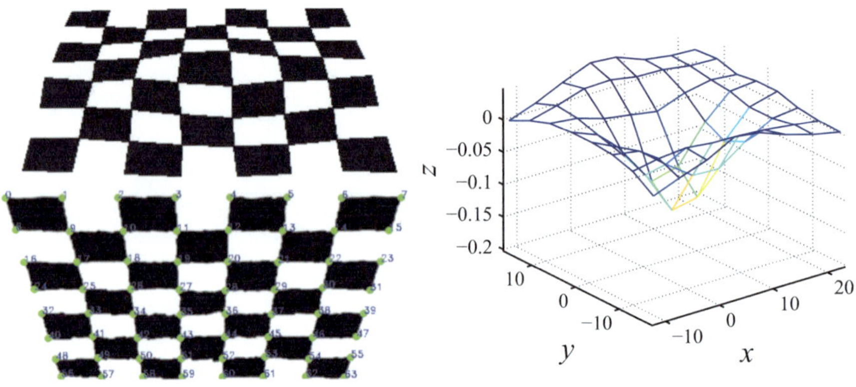

Fig. 10.9 Reconstruction result of large deformation from dynamic image pattern: (*left*) reconstruction result, (*right*) distribution of errors. (*top left*) dynamic image pattern, (*bottom left*) reflection image, (*right*) reconstruction result

10.2.4 Implementation

To evaluate the performance of the proposed sensor, we implemented both the reconstruction method and the active image patterning to the system. Figure 10.8 presents a photo of the sensor setup which is composed of a silicone rubber block for sensory input, a camera (HD Pro Webcam, Logicool Inc.), and a dynamic pattern display (LCD-8000U, Century Inc.). In addition, we employed in the analysis of reflected images the method of Lukas and Kanade for tracking feature points. A reconstructed example is shown in Fig. 10.9.

10.3 Several Applications of Proposed Method

10.3.1 Free Form Sensor

The proposed sensor can be used in various forms and shapes and because of its simple construction. For example, by attaching feathers to the surface of the sensor, we can construct a feather-based tactile sensor (Fig. 10.10). The sensor can measure very light touches and rubbing of the feathers. By imbedding the sensor within toys such as teddy bears, we can produce intuitive and interactive haptic toys. Furthermore, the rubber shape of sensor body itself can be freely designed to some extent. If images from reflection can be measured, cylindrically shaped sensors also can be produced. Such sensors would be useful as robot fingers that offer greater dexterity in manipulating objects.

10.3.2 Several Measurement Styles

As our sensor employs reflection, changes in reflection carry tactile information. By employing light-emitting diodes and photodetectors, instead of camera and image pattern, we can construct thin touchscreen-like tactile sensors (Fig. 10.11). The measurement employs not image reflections but rather the distribution in illuminance. The method with illuminance is based on high-speed sensor, such as diodes photodetectors. According to the sensors, the method can evaluate fast transform of illumination. Thus the deformation also can be detected in a high frequency. Although spatial resolution may deteriorate, high-frequency scanning and flexibility is achievable. Furthermore, we can combine the two measurement methods, reflection imaging and illuminance.

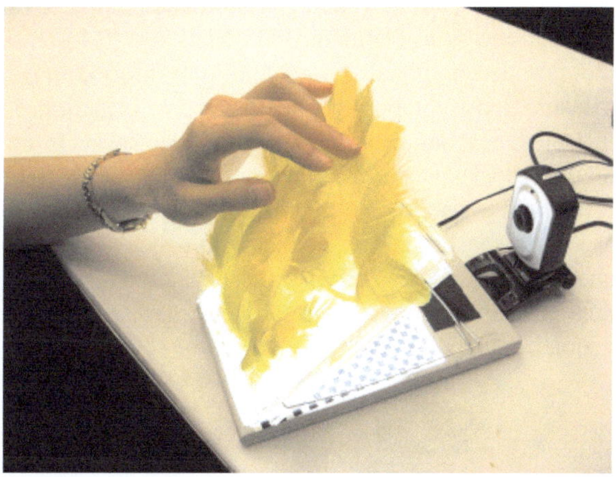

Fig. 10.10 Feather-based tactile sensor

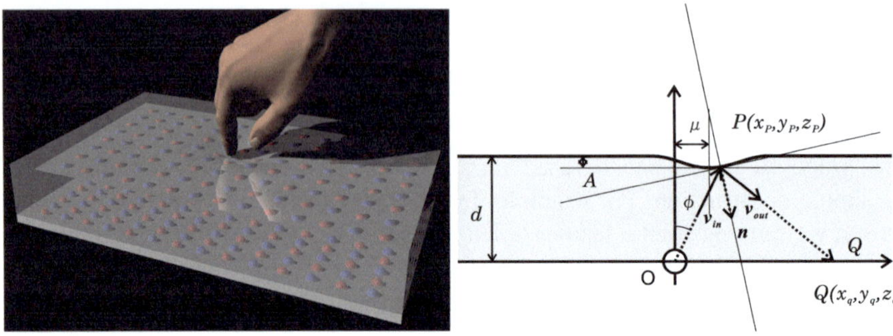

Fig. 10.11 Thin touchscreen-like tactile sensor: (*left*) overview of sensor, (*right*) optical path inside the sensor

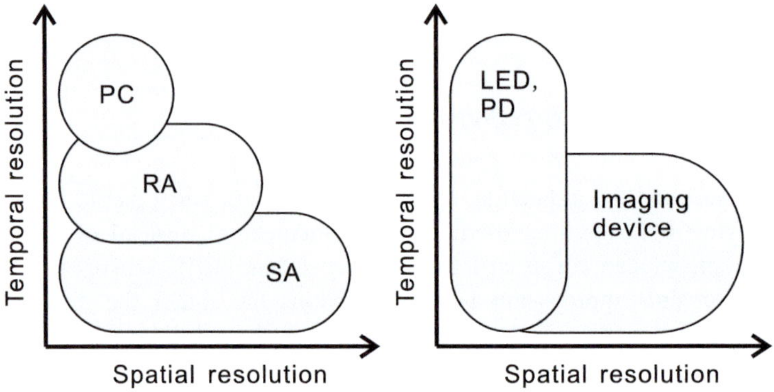

Fig. 10.12 Comparison of spatial and temporal characteristics between (*left*) human and (*right*) proposed reflection-based tactile sensor

The left panel of Fig. 10.12 illustrates the characteristics associated with human cutaneous sensation. Touch sensations are conveyed by three types of corpuscles found in the skin: Pacinian (PC), Meissner (RA) and Merkel (SA1). Merkel cells measure dense spatial, sparse temporal information, whereas Pacinian corpuscles measure dense temporal, sparse spatial information. Meissner corpuscles measure intermediate characteristics of both corpuscles. In comparison, the measurement methods also have corresponding characteristics with respect to spatial and temporal resolution. As shown on the right of Fig. 10.12, the imaging method covers spatial aspects of sensory input, and the illuminance method covers temporal aspects. These characteristics correlate with those in human skin. Thus by employing the two methods simultaneously, we can construct a tactile sensor that behaves in a similar manner.

10.4 Conclusion

In this chapter we described a versatile tactile sensor that supports intuitive input and dexterous manipulation, and proposed a reflection-based tactile sensor. Our sensor is simple but sensitive. Furthermore, various kinds of sensing can be realized (e.g., feather touch sensing, fast sensing, and sensitive tactile sensing with human-like characteristics). Based on the above result, a more dexterous robotic hand or more intuitive tactile input devices can be developed.

References

1. Begej, S.: An optical tactile array sensor. SPIE Intell. Robot. Comput. Vis. **521**, 271–280 (1984)
2. Yamada, K., Goto, K., Nakajima, Y., Koshida, N., Shinoda, H.: A sensor skin using wire-free tactile sensing elements based on optical connection. In: Proceedings of the 41st SICE Annual Conference (SICE 2002), vol. 1, pp. 131–4. The Society of Instrument and Control Engineers (SICE), Tokyo (2002)
3. Lee, M.H., Nicholls, H.R.: Tactile sensing for mechatronics – a state of the art survey. Mechatronics **9**(1), 1–31 (1999)
4. Nicholls, H.R., Lee, M.H.: A survey of robot tactile sensing technology. Int. J. Robot. Res. **8**(3), 3–30 (1989)
5. Ferrier, N.J., Brockett, R.W.: Reconstructing the shape of a deformable membrane from image data. Int. J. Robot. Res. **19**(9), 795–816 (2000)
6. Kamiyama, K., Vlack, K., Kajimoto, H., Kawakami, N., Tachi, S.: Vision-based sensor for real-time measuring of surface traction fields. IEEE Comput. Graphics Appl. Mag. **25**(1), 68–75 (2005)
7. Johnson, M.K., Adelson, E.H.: Retrographic sensing for the measurement of surface texture and shape. In: Computer Vision and Pattern Recognition (CVPR), Miami, pp. 1070–1077 (2009)
8. Yamaguchi, I., Yamamoto, A.: Surface topography by wavelength scanning interferometry. Opt. Eng. **39**(1), 40–46 (2000)
9. Hidetoshi, M., Kazutoshi, K., Takehito, Y., Tatsuji, K., Kitou, M., Naoshisa, T.: Measuring surface shape from specular reflection image sequence – quantitative evaluation of surface defects of plastic moldings. Jpn. J. Appl. Phys. Part 2 **34**(12A), L1625–L1628 (1995)
10. Massig, J.H.: Deformation measurement on specular surfaces by simple means. Opt. Eng. **40**(10), 2315–2318 (2001)
11. Saga, S., Taira, R., Deguchi, K.: Precise shape reconstruction by active pattern in total-internal-reflection-based tactile sensor. IEEE Trans. Haptics **7**(1), 67–77 (2014)

Chapter 11
Thermal Displays and Sensors

Katsunari Sato

Abstract A thermal display, which presents the user a sensation of changes in skin temperature, allows the user to touch an object in a virtual or remote environment with more realistic feeling and permits the design of stimulation in the investigation of the characteristics of human thermal perception in psychophysical experiments. As part of the thermal display, the thermal sensor, which measures the surface temperature of the human skin and environment, is important in controlling the output of the thermal display and presenting an arbitral thermal sense. This chapter briefly introduces such thermal devices mainly used as human interfaces. Furthermore, as an example of research on state-of-the-art thermal devices, this chapter presents thermal devices developed by the author: a vision-based thermal sensor using thermo-sensitive paint and a camera, a thermal display using spatially divided thermal stimuli presenting a rapid temperature change, and a head-phone-type interface that changes impressions of music and individual feelings using a thermal stimulus.

Keywords Thermal display • Thermal sensor • Human interface • Virtual reality

11.1 Introduction

Presentation of the thermal sense by which we perceive an object's temperature and material and recognize a dangerous situation has been a great challenge in the development of haptic displays. The thermal sense has been revealed to be an important property in human texture perception [1–3], and many haptic displays that present both mechanical skin deformation and temperature change in the skin have thus been developed [4–6]. Additionally, in the field of virtual reality, it is desirable for the thermal display to represent thermal changes generated by the environment such as changes due to the position of the sun [7].

Other applications of the thermal display are psychophysical experiments and interactive systems. Researchers in the field of psychophysics have used thermal

---

K. Sato (✉)
Nara Women's University, Nara, Japan
e-mail: katsu-sato@cc.nara-wu.ac.jp

H. Kajimoto et al. (eds.), *Pervasive Haptics*, DOI 10.1007/978-4-431-55772-2_11

displays to produce arbitral thermal stimulations. The use of a thermal display not only makes it easy to provide the settings of an experiment but also helps the researcher find new perceptual phenomena. The thermal display also has the potential to be used as information feedback [8] such as in the case of the vibration of mobile phones. Because a thermal display is static (i.e., no displacement or sound is produced), it can effectively produce information in vibrating or noisy environments. Furthermore, the affective characteristics of the thermal sensation allow us to develop a novel interactive system that promotes emotional activities.

This chapter introduces the basis of thermal devices and examples of thermal devices that the author has developed. The next section describes the thermal sensor, which is important to the control of a thermal display. Thereafter, the thermal representation method and applications to interactive systems are presented.

11.2 Thermal Sensor

The role of the thermal sensor used as part of a human interface is to measure the temperature of the skin surface of a human or the temperature of the environment to control the output of a thermal display and thus present an arbitral thermal sensation to a user. The thermal sensor is also important in acquiring information about changes in skin temperature when a user touches an object in a remote environment in the field of telexistence (or telepresence).

11.2.1 Method of Thermal Sensing

Two main methods are adopted to measure temperature information: electrical and optical methods.

The electrical method translates the temperature information to the resistance or voltages of circuit elements. The resistance of a thermistor changes according to its temperature; i.e., the resistance of the thermistor decreases or increases as the temperature increases. A thermocouple is a thermal sensing element that employs the Seebeck effect to produce a voltage when a temperature variation is applied between the two ends of an element that consists of different metals. Because such electrical elements are relatively small and have high resolution, they are applied by a number of systems.

The optical method can measure the surface temperature of an object without physical contact. Thermography detects infrared radiation radiated from the surface of an object and estimates the temperature of the surface. Thermo-sensitive paint, which changes color in response to a thermal change, is also applicable for the optical measurement of temperature. Such paint can visualize thermal information and is thus applied in the temperature management of products such as food, the temperature measurement of industrial machines, and the analysis of fluids [9]. The advantage of the optical method is that we can obtain a two-dimensional temperature distribution using a camera system.

11.2.2 Finger-Shaped Thermal Sensor

Both electrical and optical methods are practical for not only human interfaces but also industrial purposes. The thermal sensors effectively measure changes in temperature to control the devices and machines. However, the thermal sensors still face the greatest challenge in the field of human interfaces: the development of a finger-shaped sensor for telexistence systems.

There are three main difficulties in realizing such a sensor. First, the sensor has to measure both deformation and changes in temperature on its surface. Because a human perceives both tactile and thermal sensation on his/her skin surface, both forms of information have to be measured on the sensor surface and presented to the skin surface. Second, the performance of the sensor has to be better than that of a human's fingertip. The performance parameters are the measuring range, measuring accuracy (resolution), spatial resolution, and time response. If these performance parameters are not satisfactory, a human may have the feeling that he/she is touching an object through a glove. Finally, the physical properties of the sensor, such as shape, compliance, and temperature, have to be similar to those of a human fingertip to simulate the interaction generated by the contact of an object with human skin (Fig. 11.1). If the physical properties of the sensor are not similar to those of human skin, the sensor cannot measure deformation correctly. Furthermore, the contact area and contact time affect the thermal change on the surface of the sensor. The thermal change is also affected by the thermal properties of the fingertip; i.e., temperature and thermal conductivity.

Previous studies [10–12] and the design of commercially available sensors (such as a BioTac tactile sensor, SynTouch LLC) have attempted to measure thermal information along with tactile information. These sensors integrate an electrical thermal sensor with a tactile sensing unit. They can mimic the shape and compliance of human skin through the placement of a sensing element in an elastic body. In this case, however, the thermal change on the sensor surface in the contact area cannot be measured, and the spatial and time responses of the sensor decrease.

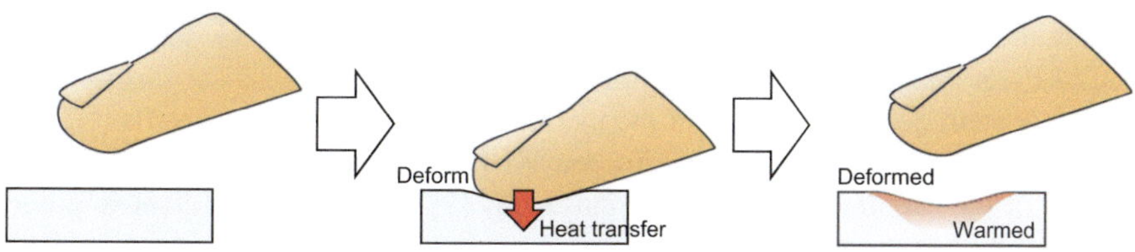

Fig. 11.1 Interaction generated by the contact of an object with human skin

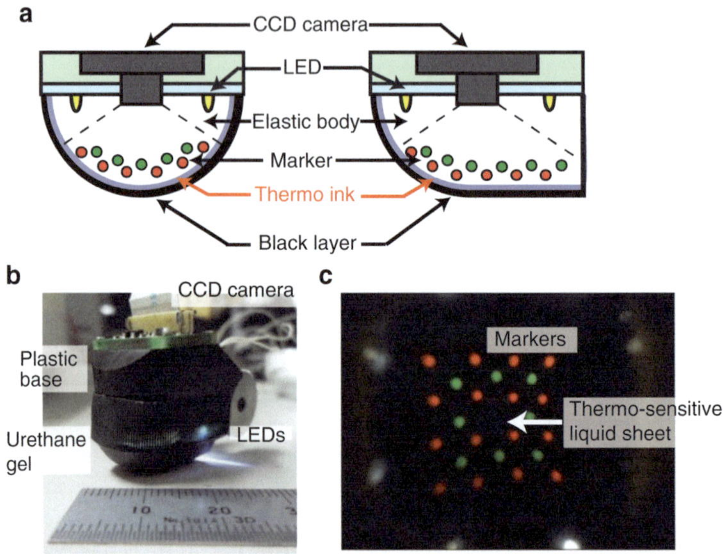

Fig. 11.2 (**a**) Configuration of the vision-based haptic sensor. (**b**) Constructed prototype. (**c**) Image of the markers and thermo-sensitive liquid sheet captured by the CCD camera [13]

11.2.3 Vision-Based Thermal Sensor

We have proposed a vision-based thermal sensor [13] using thermos-sensitive paint and a camera to measure the thermal change on the surface of a haptic sensor for telexistence.

11.2.3.1 Configuration And Measurement Method

Figure 11.2a shows the configuration of the proposed vision-based thermal sensor. The sensor consists of a transparent elastic body, two-layers of markers, a charge-coupled device (CCD) camera, an elastic sheet with thermo-sensitive paint, and light-emitting diodes (LEDs) as the light source. We can select an elastic body for which the compliance is the same as that of the human fingertip. Therefore, the sensor can mimic the deformation of the skin caused by contact. The light source is also used to maintain the temperature of the sensor the same as that of a human fingertip.

The thermo-sensitive paint is printed on the inner side of the sensor surface so that its color changes in response to the thermal change on the sensor surface. The camera system detects the color of the thermo-sensitive paint and converts it to the temperature of the sensor surface. To convert from color to temperature, we use the hue of the captured image. A previous study [9] indicated that a change in hue corresponds to a thermal change and we thus consider that the temperature of the sensor surface can be calculated using an equation developed for converting hue to temperature.

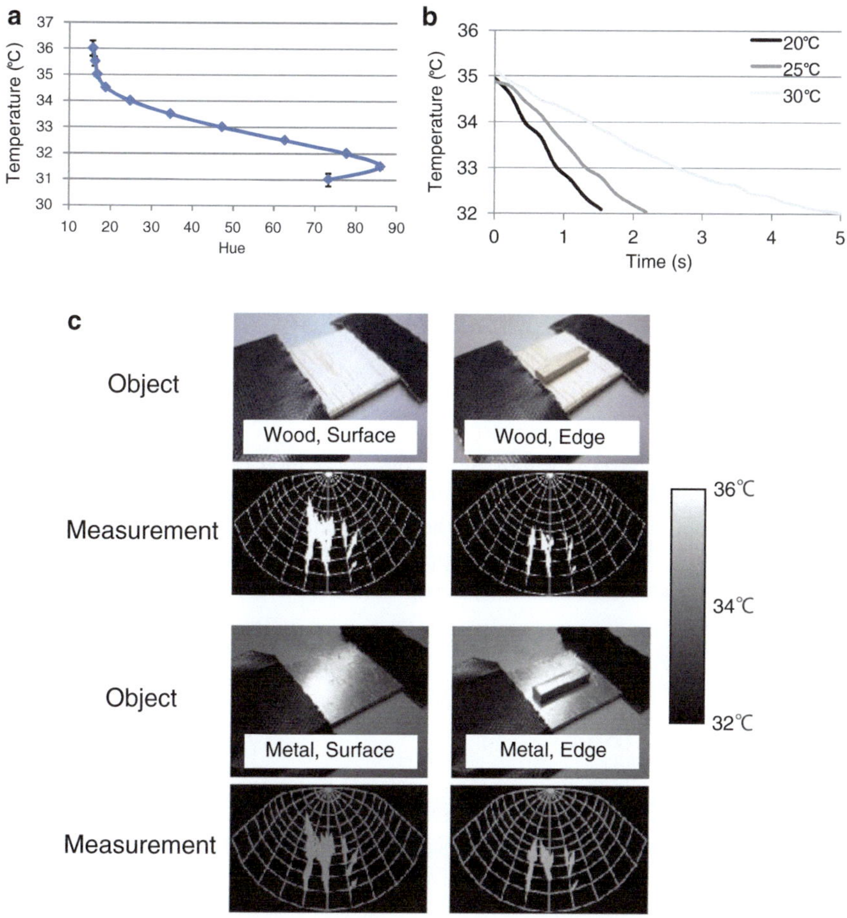

Fig. 11.3 (**a**) Relationship between the hue of the captured thermo-sensitive paint sheet and the temperature of the sheet [13]. (**b**) The thermal change recorded by the sensor for each temperature of an aluminum board [13]. (**c**) Measurements of the force distribution and temperature when the prototype sensor is in contact with objects that differ in shape and material (RGB colors that represent temperature in the original image [5] were converted to gray-scale colors)

Deformation is measured employing technologies of a vision-based force sensor [14]. When a force is applied to the sensor, the markers are displaced. A CCD camera captures the displacement of the markers and the distribution of the applied force is then calculated from these displacements.

Figure 11.2b, c shows the constructed prototype and an image captured by the CCD camera, respectively.

11.2.3.2 Measurement of Thermal and Force Information

Figure 11.3a shows an example of the relationship between the captured hue of a captured sheet of thermo-sensitive paint and the temperature of the sheet. Because the temperature range of standard thermo-sensitive paint is approximately 5–10 °C, to cover the measurement range of 15–45 °C, we must use multiple thermo-sensitive paints that have different temperature ranges. Figure 11.3b shows the thermal change recorded by the sensor when it is in contact with aluminum boards having temperatures of 20, 25, and 30 °C. When the temperature of an aluminum board is low, the rate of thermal change becomes large. From these basic evaluations [13], we confirm that the proposed sensor can be applied to thermal measurement for telexistence by further improving the measurement range.

Figure 11.3c shows the force magnitude and temperature measured by the prototype sensor when it is in contact with a wood surface, wood edge, metal surface, and metal edge [5]. In the figure, the applied forces and temperatures are represented by arrows and colors, respectively. The prototype sensor effectively measures the differences in material and shape in terms of a temperature change and distribution of force vectors, respectively.

The conventional prototype still suffers problems such as its measurement range and the time of its response to temperature changes because of the reaction range of the thermo-sensitive paint and the low heat conductivity of the elastic material. In future work, the authors will select an appropriate thermo-sensitive paint and elastic material to accurately simulate the thermal interaction between an object and the human fingertip.

11.3 Thermal Display

To represent thermal sensation using human interfaces, it is important to select a thermal display (representation method) appropriate to the application.

11.3.1 Presentation of Thermal Sensation

Four main methods are used to present the sensation of temperature change: electrical, fluidic, radiation, and chemical methods.

The simplest method used to present thermal sensation is the use of Joule heat generated by an electrical wire. Although this method can only be applied to heating, it is widely used because of its low implementation cost and stability. Another method is to use a Peltier device, where heat flux is created between the two sides of the device by applying a voltage so that the surface temperature changes. It works as both a cooler and a heater depending on the polarity of the voltage, and the rate of temperature change is controlled by adjusting the voltage. The Peltier device can be easily implemented in human interfaces.

The fluidic method uses water or air, the temperature of which is controlled before the fluid reaches the user. The water flow can change the temperature of the skin surface rapidly and widely compared with the electrical method. If the initial temperature of the water is high or low, a rapid temperature change can be presented [15]. Air flow can present thermal sensation without a sense of contact on the skin surface of the user.

Far infrared radiation can warm the skin in a non-contact manner and is thus mainly used in the field of psychophysical experiments. When we incorporate this method with the air flow method, we can also present a sense of coldness [16].

Previous studies have revealed that there is a thermal sensation when skin is in contact with a chemical material; e.g., capsaicin and menthol produce heat and cold sensations, respectively [17]. These chemical effects have the potential to be applied to thermal displays.

11.3.2 Spatially Divided Thermal Stimuli

A thermal display using a Peltier device was developed as a glove-type or grip-type device to be used in both research and practical applications. In such displays, a Peltier device is continuously in contact with the skin and may be unable to represent the sensation of a rapid temperature change or the sensation of a sudden change of the skin surface, especially if a person touches hot and cold objects continuously in the virtual world, because the boundary temperature changes gradually (Fig. 11.4).

To present a rapid temperature change via a glove or grip display using a Peltier device, the authors focus on the characteristics of human thermal perception [18]. If the thermal display is designed effectively on the basis of these characteristics, its performance could be improved using conventional Peltier devices.

11.3.2.1 Characteristics of Thermal Perception

Among the characteristics of human thermal perception, we focus on spatial characteristics (spatial summation or thermal referral) and thresholds depending on the adapting temperature.

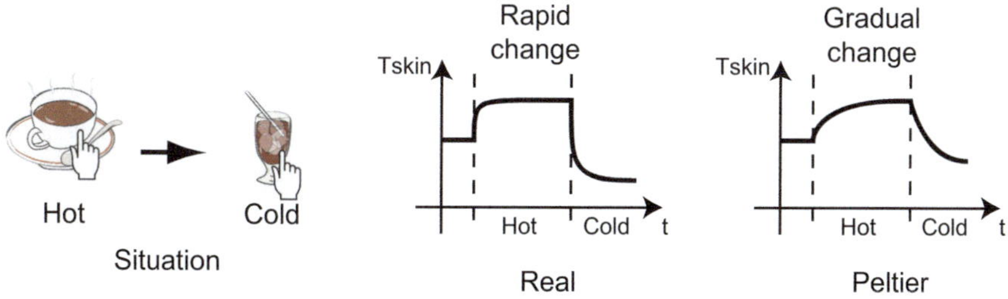

Fig. 11.4 Slow change in skin temperature for a conventional thermal display using a Peltier device [18]

If different thermal stimuli are applied to the skin, the skin summates them over space. It is thus difficult to detect the position of thermal stimuli [19]. Moreover, the perceived intensity of a thermal stimulus increases with an increase in stimulation area [20]. The threshold is also affected by other thermal stimuli; e.g., the warm threshold decreases when the area of warm stimulation increases, whereas the cold threshold decreases with an increase in area of cold stimulation [21]. Both hot and cold thresholds reduce when hot and cold stimuli are presented simultaneously. A human perceives the intermediate strength of thermal stimuli when the stimuli are applied simultaneously; i.e., the strengths of thermal stimuli are summated and redistributed to all areas in contact with the thermal stimuli [22].

The thermal threshold depends on the adapting temperature. When the skin temperature is around 28 °C, the thresholds for detecting hot and cold stimuli are around 1.0 and 0.1 °C, respectively. As the skin temperature increases, the hot and cold thresholds decrease and increase, becoming 0.2 and 1.1 °C, respectively, when the skin temperature reaches approximately 40 °C [23].

11.3.2.2 Representation of Thermal Sensation

On the basis of the characteristics of spatial summation and adapting temperature, we propose a thermal display that consists of spatially divided hot and cold stimuli. Figure 11.5 presents the configuration of the proposed thermal display. Four Peltier devices, each approximately 10×10 mm, are located on a heat sink. Two of the devices act as hot stimuli and two as cold stimuli.

Figure 11.6 illustrates an example of the presentation of thermal information using the proposed display. The basic idea here is to control hot and cold stimuli independently and thus present the sensation of a rapid temperature change virtually. To represent the hot sense, the temperature of the skin in contact with Peltier devices for hot stimuli $T_{skin,\,hot}$ is increased while that for cold stimuli $T_{skin,\,cold}$ is maintained. To represent the cold sense, the temperature of Peltier devices for cold stimuli $T_{skin,\,cold}$ is decreased and that for hot stimuli $T_{skin,\,hot}$ is maintained. In both cases, the temperature perceived by the user $T_{perception}$ could be intermediate of the surface temperatures of Peltier devices for hot and cold stimuli because of

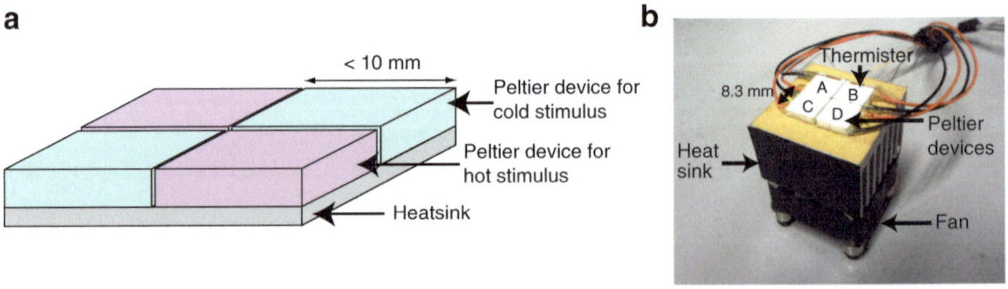

Fig. 11.5 (**a**) Configuration and (**b**) prototype of the proposed thermal display [18]

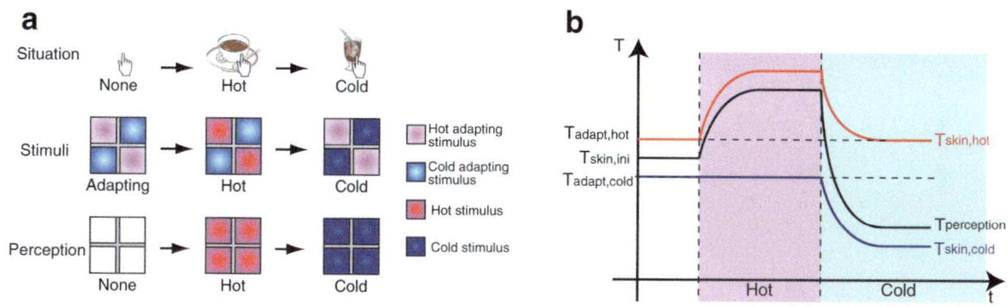

Fig. 11.6 Example of thermal presentation using spatially divided hot and cold stimuli. (**a**) Spatial patterns of thermal presentation. (**b**) Temperature of each stimulus [18]

spatial summation. We also expect users to feel a rapid temperature change when activated Peltier devices are switched from hot to cold and vice versa because stimuli perceived by the user may also be switched.

Note that the area of each hot and cold stimulus is halved compared with that for the conventional method that controls temperatures on all Peltier devices similarly, thus increasing the thresholds of both hot and cold stimuli [21] and making it difficult to perceive a temperature change. To compensate for this, we decrease the threshold for thermal stimuli by controlling the adapting temperature. Each Peltier device that provides a hot or cold stimulus respectively increases or decreases its surface temperature from the initial skin temperature $T_{skin,\ ini}$ so that the skin is sensitive to the thermal stimulus.

When there are no thermal stimuli (i.e., the user does not touch an object in the virtual world), the surface temperatures of Peltier devices are set at $T_{adapt,\ hot}$ and $T_{adapt,\ cold}$, respectively. These adapting temperatures are determined so that the user perceives the same temperature as the initial temperature because of the effect of spatial summation.

If the presentation of a hot stimulus begins, such as when a user touches a hot object in the virtual world, Peltier devices acting as hot stimuli are activated and the user feels a hot sensation. Because the temperature of skin in contact with the hot Peltier device increases from the initial skin temperature, the user would quickly perceive hot stimuli even though the area of stimulation is halved.

When a user touches a cold object in the virtual world soon after the hot stimulus, Peltier devices acting as cold stimuli are activated and the temperature of the Peltier devices acting as hot stimuli returns to the adapting temperature. Because the temperature of the skin in contact with the cold Peltier device maintains the adapting temperature even though hot stimuli are present, the user can quickly perceive subsequent cold stimuli. People are sensitive to changing stimuli, and user perception can be switched from a hot stimulus to a cold stimulus and the user can feel as if the skin temperature has rapidly changed.

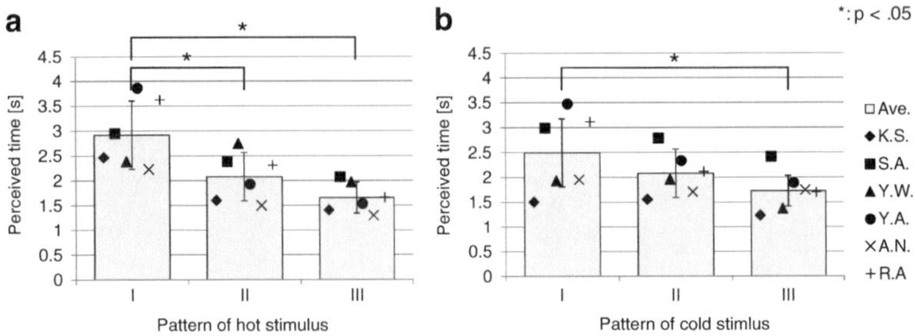

Fig. 11.7 Perceived timing of (**a**) hot and (**b**) cold stimuli when a user continuously touches hot and cold objects in the virtual world. (I) stimuli from all Peltier devices (uniform stimuli), (II) stimuli from two Peltier devices without adapting stimuli, and (III) stimuli from two Peltier devices with adapting stimuli (proposed method) [18]

11.3.2.3 Perceived Thermal Sensation

The results of a basic evaluation of the proposed method show that participants perceived separate thermal stimuli as a single stimulus when the thermally stimulated area was small. Spatially distributed warm and cold stimuli allowed participants to perceive thermal sensation rapidly even when the cold stimulus was presented rapidly after the warm stimulus and vice versa (Fig. 11.7). Participants also perceived a difference in temperature of 2 °C for each hot or cold stimulus. From these results [18], we conclude that the proposed method satisfies the requirements for the thermal display in that it presents both an arbitrary temperature sense and rapid temperature change, and can be implemented as a compact device using Peltier devices.

Our recent studies have suggested that spatially divided thermal stimuli reduce the energy required to represent thermal sensation. The experimental results showed that participants perceived temperature approximately intermediate of hot and cold stimuli for the static (constant temperature) state [24]. For the dynamic (temperature changing) state, however, participants perceived the same or larger temperature changes for spatially divided stimuli relative to the case of uniform stimuli, even though the consumed energy of spatially divided stimuli was less than that in the uniform case [25].

The above is an example of a thermal presentation method based on the characteristics of human thermal perception. We expect to develop a thermal display more efficiently if we appropriately use these characteristics.

11.4 Thermal Devices for Interactive Systems

11.4.1 Affective Interactive Systems

The thermal sense is employed in interactive systems that promote emotional activities. Kushiyama [26] proposed a thermal display called "Thermoesthesia" that allows a user to touch and interact with a visually displayed object that has a changing temperature. ThermoPainter is a temperature-change based pallet that allows a user to draw objects by placing their hands or blowing their warm breath on the interface [27]. As in the case of these systems, many thermal devices tend to be linked to artistic expression or emotional events. We believe that this is because the thermal sense inherently affects human emotions. For example, in an experiment focused on the emotional effect of thermal stimuli, Barph determined that physical warmth (or coldness) affects the feeling of interpersonal warmth (or coldness) [28]. Employing these emotional characteristics of thermal sense, we can develop a novel affective interactive system.

11.4.2 ThermOn: Enhancement of Music Impressions

As an example of an affective interactive system having a thermal device, this section introduces the ThermOn system [29, 30] which is based on the hypothesis of new emotional compatibility between the auditory and thermal senses. It is likely that thermal stimuli function synergistically with sound stimuli because these stimuli are emotionally compatible.

11.4.2.1 Overview of the ThermOn System

ThermOn is a system that dynamically applies warm or cold stimuli, synchronized to music, to users (see Fig. 11.8a). The system is expected to modify the listening experience and produce stronger emotions in the user. For example, by applying cold and warm stimuli during the calming and exciting passages of music, respectively, the listener's impression can become richer than it is with the original music alone.

Fig. 11.8 (**a**) Concept of ThermOn (**b**) Headphone-type device using a Peltier device [29]

We assume two principal target users of the system: listeners and musicians. Listeners will be able to enjoy a new type of emotional experience. Musicians, as the creators of music, will be able to expand their range of expression by employing thermal stimuli as a new instrument. For example, a musician might add a warm stimulus to the chorus of a music passage to produce greater excitement, whereas another musician could randomly apply warm and cold stimuli to achieve the impression of chaos.

We use headphones as the wearable and portable music interface. An overview of the device is presented in Fig. 11.8b. For the current prototype, a Peltier device that can both warm and cool is implemented. The large size of the headphones allows the Peltier device to be easily embedded. The user perceives the thermal stimuli clearly, because the threshold for thermal stimuli on the face is the lowest in the human body [31]. Testing a number of sites near the ear, we chose the area between the sideburn and the tragus.

11.4.2.2 Effects of the ThermOn System

We evaluated the effectiveness of the thermal stimuli synchronized to music by first conducting a physiological experiment based on the measurement of the skin potential response (SPR) and then conducting a psychometric experiment using a Likert-scale questionnaire [29]. The experiment used the rock song, *I Don't Want to Miss a Thing* (Aerosmith, Columbia). This music was selected because it possesses a clear shift in music context and was known to all the participants.

We confirmed that adding warm stimuli synchronized with the chorus of the rock song passage has a biological effect (Fig. 11.9a) and can enhance excitement impressions and change other impressions of the music, including likability and

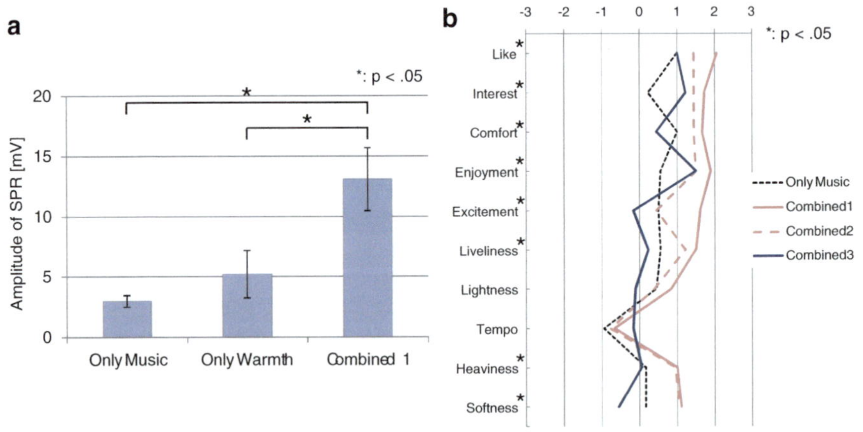

Fig. 11.9 (**a**) Averaged peak SPR values. (**b**) Average scores of all results across words and stimuli type. Combined 1, warming was applied at the change in music context. Combined 2, warming was applied periodically without synchrony. Combined 3, cooling was applied at the change in music context, as for Combined 1

comfortable listening (Fig. 11.9b). Therefore, a strong reaction in the physiological experiment can be interpreted as a positive effect possibly involving impressions of excitement and entertainment. From these results, we conclude that adding warm stimuli corresponding with the chorus of a rock song passage altered the impressions of the music, especially impressions of excitement and entertainment.

In a more practical applied experiment [30], we evaluated the ability of the ThermOn system to produce a common effect for listeners, and determined the kind of effect, when thermal stimuli were applied to music in a subjective manner to enhance the shift in the context of the music. We conducted psychological evaluations to investigate the changes in the impression of the music and individual impressions with and without thermal stimuli.

We selected four music passages that differed in type: M1 and M2 were Japanese pop songs (rap-rock and ballad, respectively), M3 was a song from a musical; and M4 was the opening music (no vocals) from a film. We subjectively determined four different types of thermal stimuli (with varying temperature type and timing) for each music passage to enhance the shift in context. A psychometric experiment using the semantic differential method was conducted.

Figure 11.10a, b presents the factor spaces of the impressions of music and individual feelings, respectively. The blurry factors of each music passage decreased by adding thermal stimuli. The individual's feeling of relief when listening to the music M1 was significantly enhanced by adding thermal stimuli. The direction of this change seems to correspond with the original impressions of the music and the type of temperature. The impressions of music M1 were relatively rough (or wild) and cold stimuli were not used for this music. This condition corresponds to that of the basic experiment.

The results show that subjectively adding thermal stimuli corresponding to changes in the music context has the potential to make the impression of the music sharper than is the case without thermal stimuli. Furthermore, presenting thermal stimuli with music could produce greater feelings of relief that seem to occur when the impression of the music is rough and only warm stimuli are applied.

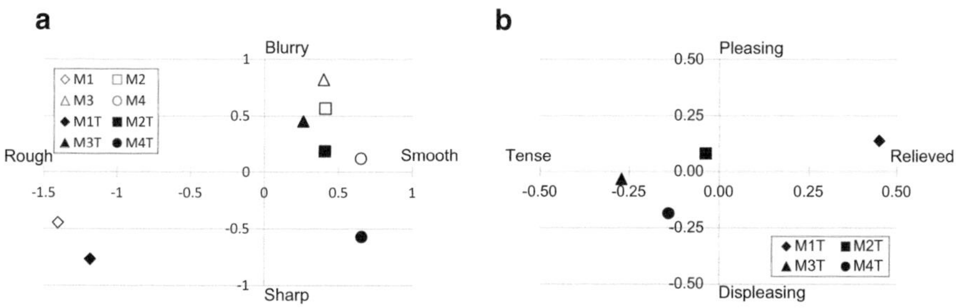

Fig. 11.10 Factor spaces for (**a**) original impressions of the music and changes in impressions of the music and (**b**) changes in individual feelings. The shape of a symbol represents the type of music passage; the color represents the condition: *white* and *black dots* are scores for only music and music with thermal stimuli, respectively [30]

Although the effectiveness of ThermOn depends on the type of thermal stimuli and music, the author used only five music passages in the study. Therefore, in future work, the effect of thermal stimulation on the music experience will be evaluated in more depth. The results of the present study imply that the timing and type of thermal stimuli are influencing factors. Furthermore, effects were found not only for the original impression of the music but also for individual feelings. These effects and their conditions must be clarified to improve the effectiveness of the ThermOn system. Other genres of music and visual content such as movies will also be tested.

11.5 Conclusion

This chapter provided a brief introduction of thermal devices used as a human interface and examples of the research that the author has conducted. There remain a number of challenges that face the realization of the finger-shaped thermal sensor mainly for telexistence applications. A thermal display based on human thermal perception could improve the time response and energy consumption of the display. Furthermore, the emotional characteristics of thermal devices show the potential of developing a novel interactive system. Because thermal sensation is essential in not only haptic devices but also our daily life, further research on thermal devices is expected.

References

1. Yoshida, M.: Dimensions of tactual impressions (1). Jpn. Psychol. Res. **10**(3), 123–137 (1968)
2. Shirado, H., Maeno, T.: Modeling of human texture perception for tactile displays and sensors. The First Joint Eurohaptics Conference and Symposium on Haptic Interface for Virtual Environment and Teleoperator Systems. pp. 629–630 (2005)
3. Okamoto, S., Nagano, H., Yamada, Y.: Psychophysical dimensions of tactile perception of textures. IEEE Trans. Haptics **6**(1), 81–93 (2013)
4. Kammermeier, P., Kron, A., Hoogen, J., Schmidt, G.: Display of holistic haptic sensations by combined tactile and kinesthetic feedback. Presence. MIT Press J. **13**(1), 1–15 (2004)
5. Sato, K., Shinoda, H., Tachi, S.: Design and implementation of transmission system of initial haptic impression. SICE Annu. Conf. **2011**, 616–621 (2011)
6. Kurogi, T., Nakayama, M., Sato, K., et al.: Haptic transmission system to recognize differences in surface textures of objects for telexistence. Proc. IEEE Virtual Real. **2013**, PO–042 (2013)
7. Lécuyer, A., Mobuchon, P., Mégard, C., et al.: HOMERE: a multimodal system for visually impaired people to explore virtual environments. Proceedings of IEEE VR. pp. 251–258 (2003)
8. Wettach, R., Behrens, C., Danielsson, A., Ness, T.: A thermal information display for mobile application. Proceedings in MobileHCI '07. pp. 182–185 (2007)
9. Ozawa, M., Miller, U., Kimura, L., Takamori, T.: Flow and temperature measurement of natural convection in a hele-shaw cell using a thermo-sensitive liquid crystal. Exp. Fluids **12**, 213–222 (1992)
10. Siegel, D., Garabieta, I., Hollerbach, J.M.: An integrated tactile and thermal sensor. Proc. of Int'l Conf. on Robotics and Automation (ICRA'86), pp. 1286–1291 (1986)

11. Monkman, G.J., Taylor, P.M.: Thermal tactile sensing. IEEE Trans. Robot. Autom. **9**(3), 313–318 (1993)
12. Taddeucciand, D., Laschi, C., Magni, R. et al.: An approach to integrate tactile perception. Proceedings of IEEE International Conference on Robotics and Automation (ICRA'96), pp. 3215–3221 (1996)
13. Sato, K., Shinoda, H., Tachi, S.: Finger-shaped thermal sensor using thermo-sensitive paint and camera for telexistence. IEEE International Conference on Robotics and Automation (ICRA 2011), pp. 1120–1125. (2011)
14. Sato, K., Kamiyama, K., Kawakami, N., Tachi, S.: Finger-shaped gelforce: sensor for measuring surface traction fields for robotic hand. IEEE Trans. Haptics **3**(1), 37–47 (2010)
15. Sakaguchi, M., Imai, K., Hayakawa, K.: Development of high-speed thermal display using water flow. Human interface and the management of information. Information and knowledge design and evaluation. Lect. Notes Comput. Sci **8521**, 233–240 (2014)
16. Marks, L.E., Stevens, J.C.: Perceived cold and skin temperature as functions of stimulation level and duration. Am. J. Psychol. **85**(3), 407–419 (1972)
17. Patapoutian, A., Peier, A.M., Story, G.M., Viswanath, V.: ThermoTRPs and beyond: mechanisms of temperature sensation. Nat. Rev. Neurosci. **4**, 529–539 (2003)
18. Sato, K., Maeno, T.: Presentation of rapid temperature change using spatially divided hot and cold stimuli. J. Robot. Mechatron. **25**(3), 497–505 (2013)
19. Yang, G., Kwon, D., Jones, L.A.: Spatial acuity and summation on the hand: the role of thermal cues in material discrimination. Psychon. Soc. Inc. Atten. Percept. Psychophys. **71**(1), 156–163 (2009)
20. Marks, L.E., Stevens, J.C., Tepper, S.J.: Interaction of spatial and temporal summation in the warmth sense. Sens. Processes **1**, 87–98 (1976)
21. Greenspan, J.D., Kenshalo, D.R.: The primate as a model for the human temperature-sensing system: 2. Area of skin receiving thermal stimulation. Somatosens. Res. **2**, 315–324 (1985)
22. H-Ni, H., Wanatabe, J., Ando, H., Kashino, M.: Mechanisms underlying referral of thermal sensations to sites of tactile stimulation. J. Neurosci. **31**(1), 208–213 (2011)
23. Kenshalo, D.R.: Correlations of temperature sensitivity in man and monkey. A first approximation. In: Zotterman, Y. (ed.) Sensory Functions of the Skin with Special Reference to Man, pp. 305–330. Pergamon Press, Oxford (1976)
24. Sato, K., Maeno, T.: Perceived temperature from spatially divided warm and cold stimuli (I). VRSJAC2012, in Japanese (2012)
25. Sato, K., Maeno, T.: Perceived temperature from spatially divided warm and cold stimuli (II). SI2012. In Japanese (2012)
26. Kushiyama, K., Inose, M., Yokomatsu, R. et al.: Thermoesthesia: about collaboration of an artist and a scientist. ACM SIGGRAPH Sketches (2006)
27. Iwai, D., Sato, D.: Heat sensation in image creation with thermal vision. ACM SIGCHI International Conference on Advances in Computer Entertainment Technology, pp. 213–216 (2005)
28. Williams, L.E., Bargh, J.A.: Experiencing physical warmth promotes interpersonal warmth. Science **322**, 606–607 (2008)
29. Akiyama, S., Sato, K., Makino, Y., Maeno, T.: ThermOn – thermo-musical interface for an enhanced emotional experience. Proceedings of ISWC 2013, pp. 45–52
30. Sato, K., Akiyama, S., Takagi, H., et al.: ThermOn—thermal interface to enhance the impression of music and individual feeling. Int. J. Adv. Comput. Sci. **4**(6), pp. 270–280 (2014)
31. Stevens, J.C., Choo, K.C.: Temperature sensitivity of the body surface over the life span. Somatosens. Mot. Res. **15**, 13–28 (1998)

Part III
Solid Application

Chapter 12
TECHTILE Workshop for Creating Haptic Content

Masashi Nakatani, Yasuaki Kakehi, Kouta Minamizawa, Soichiro Mihara, and Susumu Tachi

Abstract In recent times, haptic technology has advanced to the point where it can be used to manipulate tactile perception. Although haptics research is very popular in academia, the relevant technologies have not yet been implemented to devices that are commonly used, except for small electronic devices. In this chapter, we propose a basic haptic device called the "TECHTILE toolkit" that can design haptic content. We conducted a usability study of our toolkit by hosting more than 50 workshops worldwide that involved professional researchers as well as members of the general public, from children in elementary schools to college students. We classify haptic experiences suggested by participants during the workshops, and also discuss potential haptic content.

Keywords Haptic content • Haptic experience • Vibratory stimuli • TECHTILE

12.1 Introduction

The field of haptics has attracted attention from a wide variety of audiences. Research on haptic technology is conducted vigorously in the 1990s in order to provide force feedback in Virtual Reality (VR) environments [1]. Such force feedback systems were subsequently extended to develop teleoperation systems capable

M. Nakatani (✉)
Institute of Gerontology, The University of Tokyo, Bunkyo, Japan

Graduate School of Media Design, Keio University, Minato-ku, Japan
e-mail: nakatani@tachilab.org

Y. Kakehi
Faculty of Environment and Information Studies, Keio University, Minato-ku, Japan

K. Minamizawa
Graduate School of Media Design, Keio University, Minato-ku, Japan

S. Mihara
Bunkyo, Japan

S. Tachi
Institute of Gerontology, The University of Tokyo, Tokyo, Japan

H. Kajimoto et al. (eds.), *Pervasive Haptics*, DOI 10.1007/978-4-431-55772-2_12

of conducting dexterous manipulations, such as robotic surgery. Haptics technology has also garnered interest from companies that manufacture everyday supplies, such as textile for clothes, paper wipers, and cosmetic products. Consumer electronics are among the biggest targets for the application of haptic technology to our daily lives. The outcomes of research in haptics are now emerging in multiple disciplines.

Several haptic technologies have already been implemented in mass-produced consumer products, especially in small electronic devices. A representative example is the use of vibrators in smartphones. These vibrators can notify users when they receive a phone call or a text message. The game controller is also a popular application of haptic feedback technology. The controller of the Nintendo Wii contains a vibrator that provides haptic feedback according to gaming content. However, the haptic feedback employed in these examples tends to have fixed, standardized patterns such that users cannot generate synthetic/artificial haptic feedback. Therefore, there is no method to utilize these vibrators to develop haptic content with various kinds of vibratory patterns.

Creating attractive haptic content can help accelerate the use of haptic technology in daily life. As seen from the instance of the Apple Watch, haptic technology has begun to be commonly used in consumer electronics. However, creating haptic content typically requires specialized skills because haptic feedback systems tend to be complex. It is becoming increasingly more important to provide content production environments in haptics rather than haptic technology itself.

With regard to audio-visual media, a variety of tools have been developed to generate attractive content. This helps expand and distribute digital content to mass audiences. Moreover, a variety of hardware systems that can play audio-visual content is also commonly available. As a result, a large number of developers are currently working in the audio-visual content market.

There are three possible directions in order to successfully popularize haptic content development: (1) the development of an easy-to-use content production environment for haptic content, (2) the establishment of a studying community to help users learn to use tools to develop haptic content, and (3) the provision of an opportunity for users to practice what they learn by creating haptic content. This direction is analogous to the approach taken by the game developer Unity Technologies, which provides a game development environment as well as a knowledge-base forum (documentation and the Unify Script Wiki) by creating study communities called "Unity Community." An analogous approach should be applicable to the development of haptic content.

Several proposals have been made in haptic research to satisfy at least one of the three directions mentioned above. PHANToM is the most striking example that contributes to providing general users with accesses to haptic technology [2]. Sample programs that control PHANToM can present haptic feedback of material properties, such as the hardness and the roughness of an object in VR environments. The Penn Haptic Texture Toolkit recently employed PHANToM to play back different kinds of haptic texture information [3]. This provides an opportunity for users to experience the possibilities achieved by haptic devices. A disadvantage of PHANToM is that it requires professional knowledge in order to attain real-time control of actuating motors in case users wish to create their own haptic content. PHANToM is also expensive, such that non-specialists may not be able or willing to pay for it.

Hapkit is another toolkit intended to transmit the outcomes of research in haptics to general users [4]. It was initially called the "Haptic paddle," and provided force feedback with one degree of freedom [5]. Hapkit is open source, and allows general users to develop a haptic device from scratch. This toolkit is also associated with the online lecture "Introduction to Haptics," which explains how to develop and use the Hapkit. This set of information contributes to making haptic research more visible and accessible. Hapkit satisfies directions (1) and (2) for haptic content development mentioned above. However, the range of haptic content that it offers is limited because of the size and the degree of freedom of the Hapkit.

Tactile Labs Inc. provides hardware that can present haptic feedback with a simple configuration. A vibrator called a "Haptuator" is now actively used in various research labs. They currently provide instructions on how to assemble a haptic toolkit based on their products upon request, and general users can benefit from having open-source information available to them.

All of the above technologies provide opportunities to integrate haptic devices into various aspects of our daily lives. However, none of them is intended to provide concrete information regarding how users can employ the relevant devices to create haptic content.

In this chapter, we propose an approach to promote the creation of haptic content by general users. In particular, we discuss a framework for thinking about haptic content in terms of information design, and consider the hardware specifications of an introductory haptic toolkit created along these lines. In Sect. 12.4, we show how we employed our toolkit in more than 50 workshops that we hosted worldwide in order to provide users with the opportunity to attempt to create haptic content. We present the results of the workshop attendees, and we also classify those results in order to discuss methods for creating haptic content.

12.2 Three Major Components of Haptic Content: Object, Body, and Mental Imagery

We experience haptic events, i.e. "haptic experience," rather than pure tactile sensations in our daily lives. Pressure, flutter-vibration, slip, temperature, and pain are representative subclasses of tactile sensations. These tactile sensations are integrated as needed to generate tactile perception, which leads to the subjective quality of touch modality. Haptic experience is similar to tactile perception but includes a sense of agency as well as the sense of ownership (i.e., my body experiences this tactile perception) [6]. We assume that haptic experience is elicited by a haptic content, which provides haptic feedback in a concrete context.

We devised a framework where haptic experience content is decomposed into three major components: object, body, and mental imagery (Fig. 12.1). We created this framework based on past research on information design [7].

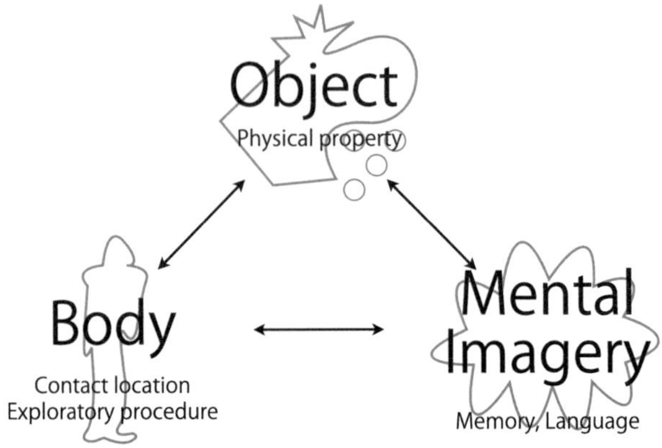

Fig. 12.1 The three major components of haptic content

Objects directly provide haptic feedback to the body, and they have physical characteristics that affect haptic experiences. Several studies have examined the relationship between the physical properties of an object and the consequent tactile perceptions [8–10].

The body is the site of physical interactions between the skin and the object. Mechanoreceptors embedded in the skin detect physical contact with external stimuli. The body also represents procedures regarding how observers touch objects, called "exploratory procedures [11]." Therefore, the body acts as an active touch sensor, instantiated by stroking or tapping, to obtain haptic experiences.

Mental imagery provides haptic experience based on prior knowledge or memory, without the aid of contact with objects. Mental imagery is affected by other modalities, such as vision and hearing. It is a cognitive process that was originally proposed in the context of vision research [12], and was subsequently applied to haptic research on blind people [13]. We extend this concept by adding prior associations between haptic experience and multimodal integration to mental imagery. For example, visual capture is a well-known phenomenon whereby visual information dominates perceptual processing in interpreting multimodal information [14]. Hence, a delay in visual feedback leads to the perception of elasticity or viscosity [15]. Similarly, auditory information can also bias haptic experiences in a certain manner. The parchment-skin illusion is an example of auditory-haptic integration [16]. Perceivers feel varying roughness on the surface of the skin upon hearing sounds with and without the high-frequency components filtered, even though the skin condition has not changed. These multimodal integrations happen because we have a learned association between visual feedback and haptic experience.

The above three factors (object, body, and mental imagery) closely interact with one another. Tanaka et al. discussed the relationship between objects and the body. They pointed out that the interaction between the contacting object and the skin produces tactile perception that affects observers' exploratory movement [17]. Classical psychological studies in the field of haptic research have investigated the manner in which exploratory procedures may affect our perception, suggesting that objects and the body influence each other to create haptic experiences [11, 18].

Object and mental imagery also complement each other. The appearance of objects has learned associations with haptic experiences, such as the relationship between the size appearance and the weight of an object. This haptic memory of the association of objects' haptic experience with other modalities is being studied. The similar relationship in auditory system holds for the association of the perception of the water temperature with sound [19].

Body perception is associated with mental imagery. The body schema is a representative of the mental image of the body. It is updated during body movement, and affects haptic experience. For example, a patient with astereognosis cannot recognize the shape of an object because the body schema and the haptic perception of objects have deteriorated [20].

This three-factor framework provides a starting point for the analysis of haptic content into its components, and helps provide the desirable haptic experiences. Alternatively, we can combine all three factors to achieve realistic haptic experiences rather than adhere to a single factor. Based on this, we developed an introductory haptic device that provides a development environment for haptic content.

12.3 The TECHTILE Toolkit

The TECHTILE (TECHnology based tacTILE design) toolkit aims to be an easy-to-use content production environment to create haptic content, which is discussed in Sect. 12.1. The toolkit should also be able to provide haptic experience in terms of the three major components (object, body, and mental imagery) discussed in Sect. 12.2.

In order to satisfy these criteria, we broke down the hardware specifications of the toolkit as follows: (i) The toolkit can record and replay information regarding haptic experience through vibratory signals in an intuitive manner without the use of a computer. (ii) The toolkit can provide haptic feedback from any part of the body without restricting body movement. (iii) The toolkit can combine sensory information with other modalities, especially vision and hearing. The first aspect is important for the design of a haptic device for beginners' use. The second specification is intended to include the contributions of the body to haptic experience in the toolkit, whereas the third specification helps provide an association with mental imagery in interpreting haptic feedback.

The first prototype of the TECHTILE toolkit used a microphone (Audio-Technica AT9904) as a sensor for vibratory signals, and a vibrator (ALPS Force Reactor AF, L-type) as a haptic feedback actuator. These devices were small and sufficiently light to attach to arbitrary objects or parts of the body such that users could move freely during their haptic exploration. Electric signals transduced by the microphone were connected to an audio amplifier (Rasteme Systems, RSDA202) and directly connected to a vibrator. We chose an audio amplifier that could increase the intensity of signals in flat frequency characteristics to up to about 1000 Hz. The audio amplifier had manual volume control for changing the intensity of haptic feedback. This setup is similar to a conventional audio amplifier-speaker set, such that users do not require any explanation before they begin using the TECHTILE toolkit.

The advantage of using audio signals to store haptic information is that most users know how to handle audio information. The TECHTILE toolkit can also digitize audio signals directly from the microphone by using an analogue-digital converter (PLANEX PL-US35AP). Following digitization, users can use a commercially available audio authoring software to store the audio data in digital format. This also provides a method for users to edit haptic information using digital audio filters, such as by cutting off a certain range of frequencies or adding a delay/echo, etc.

We created sample variations of content using our toolkit. A pair of paper cups was prepared. A microphone was attached to one cup, the sender cup, and a vibrator was attached to the other – the receiver cup. Once an experimenter placed objects (water, glass beads, coins, etc.) into the sender cup, a user could perceive its content in the receiver cup. A popular example involved transferring the tactile sensation of rotating glass balls in a cup (Fig. 12.2, right). In this demonstration, a user held the receiver cup and, by observing the sender, who was holding the sender cup with the microphone attached to it, was able to correctly determine its content. In this setup, the TECHTILE toolkit provided audio-visual feedback as well as vibratory playback through the actuator, which made receivers feel as if they were holding the sender cup. Moreover, several modalities of tactile perceptions (hardness, softness, granularity, crispiness) could be determined.

The above example involved a room to study multimodal interactions in perceiving tactile perceptions of material properties. The cup demo can also present the haptic experience of a glass ball rotating in a cup, even though the actuator we used

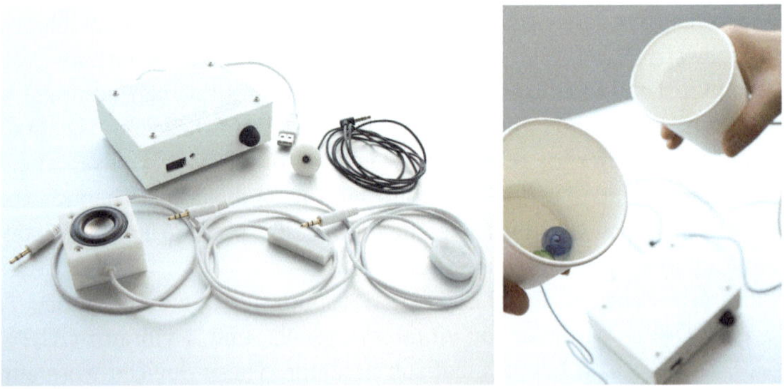

Fig. 12.2 The TECHTILE toolkit that can record and play back haptic experiences (*left*). An example of the transmission of the haptic content related to feelings a glass ball in a cup (*right*)

had only one degree of freedom in the vibratory direction which is normal to the bottom of a paper cup. This phenomenon suggests that visual feedback or mental imagery associated with past experiences may also bias perceivers' subjective reports. This also implies that our toolkit is successful in eliciting mental imagery of tactile perceptions by combining the three major components of haptic content. Detailed information containing instructions on how to develop the TECHTILE toolkit is available on our website [21].

12.4 Workshop Implementation

We introduced our toolkit in introductory haptic workshops. This step was crucial toward rendering haptics research and the implementation of haptic technology more visible to the general public. Moreover, we could test whether our toolkit was functional in actual workshops.

Most of our target audience was not aware of any haptic technology from daily life. Even the term "haptics" was not well understood in a general context. Considering this fact, we thought that a haptics workshop ought to be sufficiently engaging to attract attention of the audience by providing information to help understand this little-known area.

We designed a workshop using the TECHTILE toolkit to familiarize people with haptic experiences and provide the basics of haptics research. To this end, we hosted over 50 workshops in universities, academic conferences, research institutes operated by private companies, and science museums in Japan, United States, Finland, Austria. We now describe the basic structure of our workshop.

12.4.1 Workshop Flow

The timetable of each of our workshops was as follows:

1. Awareness of the sense of touch (ice-break session and short lecture)
2. Introduce the TECTILE toolkit
3. Session 1: Record and play back haptic experiences
4. Session 2: Create haptic content (modulation, augmentation, and subtraction)
5. Presentation and short demonstration
6. Reflection

Each workshop began with an introduction of all participants, followed by a short lecture and a demonstration on the sense of touch. For example, a blindfolded participant was asked to explore an object in a box using his/her hand. Another exercise involved participants communicating their tactile experiences using only words, without being presented with the relevant stimulus. It was also interesting to ask them to consider how we can make other people ticklish without actually tickling them. We routinely conducted a small set of haptics-related games in the beginning in order to make participants aware of the modality of touch.

After the first session, participants were introduced to the TECHTILE toolkit. We explained the purposes and operation of the toolkit, followed by the time for experiencing haptic content described in Sect. 12.3. This demonstration was also effective in providing an opportunity for the introduction of the three components of tactile experience, by changing objects or the manner in which users were allowed to move their bodies. The mental imagery of haptic experiences was provided through mainly visual and auditory information. This was our best opportunity to communicate the concrete form of tactile experiences to a general audience (Fig. 12.3, top left).

We used two studies. The first study involved recording and playing back haptic experiences, where the participants used the TECHTILE toolkit to explore the best practices of generating realistic haptic experiences. The second study involved applying the recorded vibratory information to different materials and haptic experiences. The objective of this task was to generate situations where haptic content increased affective interaction with others. In both cases, participants were divided into small groups consisting of four or five members in order to guarantee sufficient conversation among participants.

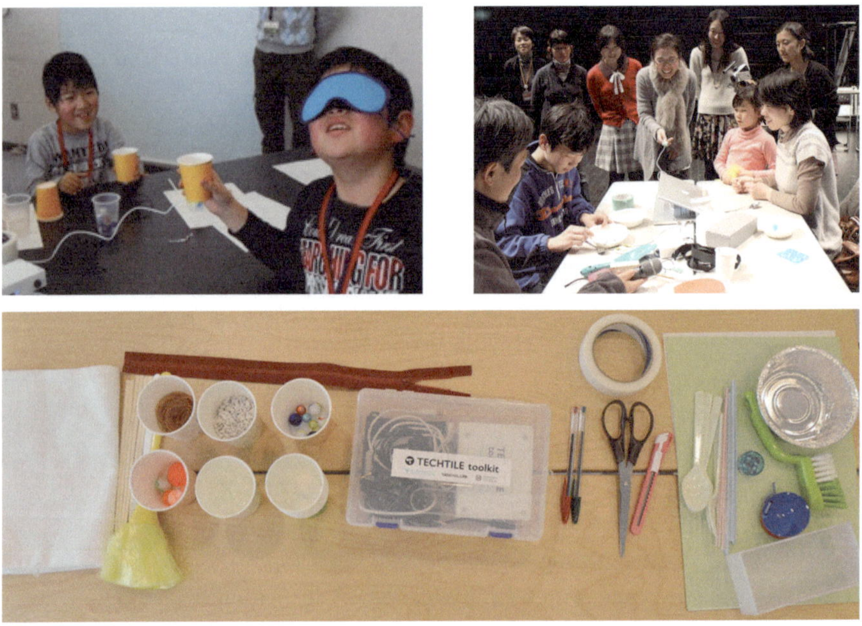

Fig. 12.3 Snapshots from the TECHITLE workshops. *Top left*: haptic experience of the cup demo using the TECHTILE toolkit. *Top right*: demonstration of the user-generated haptic content during the workshop. *Bottom*: prepared items for the workshop

The first session was intended to enable the participants to generate haptic content while exploring the toolkit (Fig. 12.3, top right). Most participants had no prior experience in analyzing and creating haptic content. Therefore, we ask them with exploring the best scenario for plausible haptic experiences. We prepared items of everyday use, such as scissors, razor blades, pens, kitchen cutlery (spoons, forks, knives), as well as various kinds of papers and clothes to provide diverse tactile sensations (Fig. 12.3, bottom). By attaching a microphone to each of these objects, we enabled the participants to collect contact-evoked (frictional and collision) sounds. Since the TECHTILE toolkit can provide haptic feedback through a vibrator in real time, participants could easily examine what the collected contact-evoked sounds felt like. The vibrator was sufficiently small to attach to an arbitrary object so that participants could examine the vibration through the object. A workshop facilitator kept reminding participants that body movements were also necessary to have better haptic experiences.

The second session was intended as an advanced course on creating haptic content. By attaching the vibrator to an object, the haptic experiences of using the object were modulated. For example, a pen attached with the vibrator may provide multiple kinds of haptic experience in writing characters. We also prepared custom-made software that provided an easy-to-use editing environment for recorded audio optimized for vibratory signals (this is also available from [21]).

In the presentation session, participants provided a short demonstration of the haptic content that they had developed during the workshop. We provided an instruction sheet consisting of questions that contained several tips for the presentation, such as

(a) What object did you use?
(b) To which part of body did you attach the vibrator(s)?
(c) How did you move your body to experience the haptic content?
(d) How do you combine haptic feedback with other modalities?

Presentation was followed by a short demo session, in which the presenters showed conceived haptic experiences to the audience. And each group offered comments for the presenters' benefit. This step was important because no one single group could have examined all possible haptic experiences. By experiencing the haptic content generated by other groups, the participants could realize different perspectives in creating haptic content.

In the reflection session, the facilitator reminded participants of what they had accomplished in the workshop. We routinely took pictures during the workshops, and pointed out the process of trial-and-error during the workshop. We stressed the importance of the three components of haptic content. Each workshop lasted between 1 and 2 h.

12.4.2 Case Study

Figure 12.3 shows several pictures from our workshops. We prepared various objects for the workshop, and participants were allowed to examine these objects. In the following, we classify representative outcomes of the workshop and discuss some implications of them.

12.4.2.1 Friction-Evoked Sound Provides Realistic Haptic Experience

The simplest haptic content through the TECHTILE toolkit involved recording frictional sounds through the microphone. In this type of trial, participants scratched the surface of objects to produce vibratory feedback by the actuator. This type of example was helpful for the participants to understand how several variations of haptic experience in object property can be presented through the TECHTILE toolkit. Some participants simply held the vibrator in their hands to examine how it felt. Others divided a single output into two using an audio splitter cable in order to share their haptic experiences with multiple observers. This idea was interesting because a haptic experience is usually assumed to be subjective rather than shared, objective experience.

12.4.2.2 Object-Mediated Sound Provides Context for Haptic Experience

Most proposed examples of haptic content in the literature involve object-mediated haptic feedback. The microphone in our workshops was attached to objects of the participants' choice, and was used to strike the surface of another object. This was effective in obtaining frictional vibrations associated with material properties of an object, such as surface roughness. Participants preferred pairs of hard objects because they produced audio signals of larger magnitudes.

A representative configuration was the one where the participants attached both the microphone and the vibrator to a stylus-like shape, such as a pair of chopsticks. They struck relatively rough textures (coarse cloth, or a perforated steel sheet) with one chopstick, and transferred the signal to another chopstick using the vibrator. A group of participants pointed out that the synchronized haptic movement, as played by the recorder, needed to provide realistic haptic experience in the receiver. This observation implied that participants appreciated the importance of the body in creating haptic experiences.

The most striking example of this configuration was the haptic content of playing badminton (Fig. 12.4, top right). A microphone was attached to one racket and a vibrator to the other of a pair of badminton rackets. With this setup, participants could transfer the haptic experience of hitting a shuttle to the receiver. Interestingly,

Fig. 12.4 Representative haptic content proposed from workshop participants. *Top left*: friction-evoked sound produces material perception of denim cloth. *Top right*: object-mediated sound transmits the experience of playing badminton. *Bottom left*: body-generated sound shares what one feels in swallowing water. *Bottom left*: vibratory stimuli augment the haptic experience of opening a box

good and bad shots were detectable on the receiver side, suggesting that the haptic experience of hitting a shuttle can help judge a player's skills at badminton.

Another interesting example involved the use of kitchen containers (e.g., a soup bowl or cup). Besides the chopstick example, participants recorded sounds using a microphone attached to the sender cup, and transferred the signal to a bigger or smaller object than originally recorded object. This trial tested whether scale change can produce varying haptic experiences even if the original vibratory signal is the same.

Finally, some participants were quite interested in feeling an object rotating in a container, such as the glass ball mentioned in Sect. 12.3. They found that the shape and the size of an object attached to the vibrator affected their haptic experiences. This was associated with the contribution of the visual input in modulating haptic experience, suggesting that sensory inputs other than the sense of touch were considered in our workshop.

12.4.2.3 Movement-Generated Sound Provides a Clue to Sharing What One Feels

Several groups were attracted by the use of body movement in haptic content. They pasted the vibrator to different parts of the body to provide haptic feedback.

An interesting example in this regard involved transferring one participant's haptic experience of swallowing to another. The microphone was attached to a participant's throat and played back to another (Fig. 12.4, bottom left). The receiver had the feeling of drinking the same kind of liquid as the sender drunk (water with and without gas, jelly-like liquid, etc.) by recording and replaying the information from the vibrator. This example was reminiscent of the study involving a straw-mediated haptic feedback system [22].

Another group attempted to record body sensations during dance movements. This trial was interesting to watch, but the amplitude of the vibrator's output was barely perceivable during body movement, suggesting that a more powerful actuator was required to directly provide vibratory feedback onto the skin. We did not prepare any special devices to capture vibrations in elastic media, and we need to research this possibility in the future workshop.

12.4.2.4 Vibratory Information Augments Haptic Experience

Adding vibratory information regarding an object provides an additional haptic experience in using that tool. For example, a group attached the vibrator to scissors and provided tactile feedback according to the cutting motion. When an appropriate vibratory stimulus was chosen, it provided various kinds of haptic experiences, even though the object with which the participants interacted was the same. A group attached the vibrator to a box and provided haptic feedback during the opening motion of the lid of the box. If vibratory feedback was provided at the appropriate time, participants felt as if the box were more or less precious than they did without haptic feedback (Fig. 12.4, bottom right).

12.4.3 Discussion

The TECHTILE toolkit was sufficiently intuitive for the workshops' participants to create haptic content that provided haptic experiences. Most participants understood the "cup" demo described in Sect. 12.3, and actively explored haptic experiences elicited through vibratory stimuli. Some participants attached the vibrator to their faces, throats, and arms to test parts of the body suitable for receiving vibratory feedback. Interestingly, haptic experience was enhanced by providing additional vibratory stimuli to an original object with daily behavior (Sect. 12.4.2.4). This implies that our toolkit provided haptic content that could expand our experiences in daily life. These aspects suggest that the TECHTILE toolkit satisfies the hardware specifications listed in Sect. 12.3.

The TECHTLE toolkit provides a method to prototype haptic experiences. Haptic experience is not widely assumed that it is manipulable. People usually design a physically tangible consumer product and evaluate how it feels like. However, if we have a method to manipulate haptic experience in the process of product design, we can cut off multiple costs during prototyping a product. The development of an introductory toolkit provides an alternative option to create haptic experience for consumer products.

The haptic content proposed in the workshop provided some insights into scientific research in haptics. Even though we used a vibrator with one degree of freedom, the observers successfully experienced, say, a glass ball rotating in a cup. This phenomenon happened because the visual feedback of the sender's movement in this content provided context to interpret ambiguous haptic vibratory signals. The use of the TECHTILE toolkit provided an opportunity to explore unknown haptic experiences in a time-effective manner. Using the TECHTILE toolkit, we provided both a methodology to consider haptics and familiarized general audiences with haptic technology, and finally to create haptic content.

12.5 Future Work

The use of the TECHTILE toolkit promotes the sharing of subjective haptic experiences. Some workshop participants arbitrary noticed that haptic information can be shared with multiple perceivers. A possible haptic content is the broadcast of a tennis match. A small microphone is attached to the players' rackets to record sounds generated by hitting the ball. Touch-evoked sound is superimposed on conventional audio-video content and simultaneously broadcasted to multiple receivers. An observer receives the data and enjoys the TV program with holding a racket with a haptic vibrator attached to it. In this configuration, people can enjoy the same haptic experiences as the athletes during the tennis match, which makes watching a tennis match a far more visceral experience.

We developed a prototype system that attained haptic broadcast, including audio-visual information (Fig. 12.5). The captured videos were presented on an iPad, and the haptic signals were stored in an audio channel and played back through the vibrator. Most users could intuitively understand how they perceived haptic broadcasts without instruction, suggesting that this kind of content is easy to use for general audiences. Besides sports broadcasting, scientific TV programs, such as the Discovery Channel, may benefit from including haptic content in their programs. The advantage of our developed toolkit is that it can enable viewers to enjoy such "haptic TV programs" without requiring any further hardware development. We have begun providing haptic content through the website YouTube to inspire further research on this issue.

Fig. 12.5 Two examples of haptic content that provides haptic experience of an object in a cup (*left*) and playing badminton (*right*)

We are working on creating a website that shares haptic content. In this service, we ask users to upload their haptic content, recorded through the TECHTILE toolkit. Once sufficient numbers of content has been collected, we can also create a featured issue for best practices in haptic content. We may need to develop software that can edit haptic content. We also plan to provide information regarding how to develop haptic content, as in the Make culture [23].

The use of our toolkit also provides several benefits for teaching. Conventional classrooms accommodate the sense of touch in education by, say, providing students opportunities to interact with small animals on school trips, or tangible objects in science and art classes. If the TECHTILE toolkit is provided in schools, it provides experience-based education in the classroom. In contrast to reading a textbook, experience-based education can provide rich information to learn body-associated knowledge. If a school also provides physical exercise, such as a gym class, the toolkit can be used here as well. A picture textbook, whose content is associated with haptic experiences, adds embodied understanding of what they are meant to learn. We are planning to design such a teaching material.

We have begun offering the TECHTILE toolkit to educators interested in employing haptic experiences to the classroom. We plan to collect the best practices in teaching through the haptic experiences of teachers in order to research the concept of haptic-oriented teaching. This may lead to a paradigm shift in teaching also increase a variety of teaching methods in classrooms.

12.6 Conclusion

In this chapter, we discussed state-of-art haptics technology and the need to develop an easy-to-use haptic tool for creating haptic content. We developed an introductory toolkit for popularizing haptic content creations among the general public as well as researchers. Our developed toolkit was intended to record and play back sounds elicited by physical interactions through a vibrator. We hosted workshops to allow participants familiarize themselves with experience-based design for haptics. In future work, we plan to prototype haptic content in various scenarios in order to promote the use of haptic technology. By employing both the development of introductory toolkit and its use in the workshop, we plan to facilitate translational research of haptics in the field of product design, art, entertainment, and education.

Acknowledgement This work was in part supported by YCAM InterLab camp vol. 2, the JST-ACCEL "Embodied Media" Project, and the JST-CREST "Harmonized Inter-Personal Display Based on Position and Direction Control" Project.

References

1. Iwata, H.: History of haptic interface. In: Grunwald, M. (ed.) Human Haptic Perception: Basics and Applications, pp. 355–361. Birkhäuser Basel, Berlin (2008)
2. Massie, T.H., Salisbury, J.K.S.: The PHANToM haptic interface: a device for probing virtual objects. In: ASME Haptic Interfaces for Virtual Environment and Teleoperator Systems in Dynamic Systems and Control, pp. 295–301 (1994)
3. Culbertson, H., Lopez Delgado, J.J., Kuchenbecker, K.J.: [D11] The Penn Haptic Texture Toolkit. In: Haptics Symposium (HAPTICS), 2014 IEEE, pp. 1–1 (2014)
4. Morimoto, T.K., Blikstein, P., Okamura, A.M.: [D81] Hapkit: an open-hardware haptic device for online education. In: Haptics Symposium (HAPTICS), 2014 IEEE, pp. 1–1 (2014)
5. Richard, C., Okamura, A.M., Cutkosky, M.R.: Getting a feel for dynamics: using haptic interface kits for teaching dynamics and controls. In: ASME IMECE 6th Annual Symposium on Haptic Interfaces, pp. 15–21 (1997)
6. Gallagher, S.: Philosophical conceptions of the self: implications for cognitive science. Trends Cogn. Sci. **4**, 14–21 (2000)
7. Jacobson, R.: Information Design. The MIT Press, Cambridge, MA (2000)
8. Bergmann Tiest, W.M., Kappers, A.M.L.: Analysis of haptic perception of materials by multidimensional scaling and physical measurements of roughness and compressibility. Acta Psychol. **121**, 1–20 (2006)
9. Shirado, H., Maeno, T.: Modeling of human texture perception for tactile displays and sensors. In: Eurohaptics Conference, 2005 and Symposium on Haptic Interfaces for Virtual Environment and Teleoperator Systems, 2005. World Haptics 2005. First Joint, pp. 629–630 (2005)
10. Okamoto, S., Nagano, H., Yamada, Y.: Psychophysical Dimensions of Tactile Perception of Textures. Haptics, IEEE Transactions on 6, pp. 81–93 (2013)
11. Lederman, S.J., Klatzky, R.L.: Hand movements: a window into haptic object recognition. Cogn. Psychol. **19**, 342–368 (1987)
12. Finke, R.A.: Levels of equivalence in imagery and perception. Psychol. Rev. **87**, 113–132 (1980)
13. Hollins, M.: Haptic mental rotation: more consistent in blind subjects? J. Vis. Impair. Blindness **80**, 950–952 (1986)
14. Pavani, F., Spence, C., Driver, J.: Visual capture of touch: out-of-the-body experiences with rubber gloves. Psychol. Sci. **11**, 353–359 (2000)
15. Lecuyer, A., Coquillart, S., Kheddar, A., Richard, P., Coiffet, P.: Pseudo-haptic feedback: can isometric input devices simulate force feedback? In: Virtual Reality, 2000. Proceedings. IEEE, pp. 83–90 (2000)
16. Jousmaki, V., Hari, R.: Parchment-skin illusion: sound-biased touch. Curr. Biol. **8**, R190 (1998)
17. Tanaka, Y., Nagai, T., Sakaguchi, M., Fujiwara, M., Sano, A.: Tactile sensing system including bidirectionality and enhancement of haptic perception by tactile feedback to distant part. In: World Haptics Conference (WHC), 2013, pp. 145–150 (2013)

18. Lederman, S.J., Klatzky, R.L.: Haptic classification of common objects: knowledge-driven exploration. Cogn. Psychol. **22**, 421–459 (1990)
19. Velasco, C., Jones, R., King, S., Spence, C.: The sound of temperature: what information do pouring sounds convey concerning the temperature of a beverage. J. Sens. Stud. **28**, 335–345 (2013)
20. Grunwald, M.: Haptic Perception in Anorexia Nervosa. Human Haptic Perception: Basics and Applications, pp. 326–351. Springer, Berlin (2008)
21. http://www.techtile.org/
22. Hashimoto, Y., Nagaya, N., Kojima, M., Miyajima, S., Ohtaki, J., Yamamoto, A., Mitani, T., Inami, M.: Straw-Like User Interface: Virtual Experience of the Sensation of Drinking Using a Straw. Proceedings of the 2006 ACM SIGCHI international conference on Advances in computer entertainment technology, p. 50. ACM, Hollywood (2006)
23. http://makezine.com/

Chapter 13
Computational Aesthetics: From Tactile Score to Sensory Language

Yasuhiro Suzuki and Rieko Suzuki

Abstract This chapter presents an algorithmic consideration of aesthetics of the tactile sense. For this purpose we have developed a notation method for massage, which we have termed the Tactile Score. Massage, which we have used for many years, is a form of haptic engineering, with the intention of creating comfort in the tactile sense. The Tactile Score is used to analyze massages and to extract the "grammar" of massages. The use of the grammar enables us to compose grammatically correct and incorrect massages and compare their effectiveness. In addition, we also apply the Tactile Score to texts to extract the tactile sense.

Keywords Tactile sense • Tactile score • Sensory language • Massage

13.1 Prolusion

There is something beyond words. The same is equally true of artistic and sensibility expressions, and in some cases, we can consider that there is something which cannot be expressed with linguistic rationality. At the same time, the expression of high-level artistic sensibility is performed by means of linguistic rationality in linguistic arts such as poems and literature.

There is a claim that we cannot lump painting, music, and dancing together because they are not linguistic arts. However, if we accept that there is a world in which we cannot understand non-linguistic arts, such as painting, music, and dancing, with linguistic rationality, it is similar to considering that life consists of vitalistic forces as the answer to the question "What is life?"

Music has a language, and is composed of musical scores. Even though there are differences in performing techniques, a listener is able to recognize that the performed songs are all the same, with the same musical score. To put it the

Y. Suzuki (✉)
Graduate School of Information Science, Nagoya University, Furo-cho Chikusa, Nagoya 464 0814, Japan
e-mail: ysuzuki@nagoya-u.jp

R. Suzuki
Facetherapie Co. Ltd., Tokyo, Japan

H. Kajimoto et al. (eds.), *Pervasive Haptics*, DOI 10.1007/978-4-431-55772-2_13

other way around, musical scores have been developing in such a way. The system of musical scores has been established for a long time, and composers have been developing the language referred to as a musical score and exploring the possibility of new expression until the present day. In the meantime, a musical score cannot define the way in which the music is performed. Various representations are expressed with the same musical score, depending on the technique and interpretation of the performer. Performers have freedom in this regard, but modern composers need to verbalize the part corresponding to the expression that has been thought to be the performer's freedom.

When we create a song by fusing electronic music with performance by an acoustic instrument, it is possible to derive the freedom of the expression of the performer by notating a traditional score for the performance with the acoustic instrument; however, the traditional notation is insufficient for electronic music. Therefore, composers have to describe "the playing expression of a musical score" which has been in the hands of performers.

This problem does not occur only in music. We have developed the Tactile Score [1] as a notation for aesthetic massage[1] [1, 2]; for example, the notation representing hand movements such as: "Massage the cheeks in a circular motion twice, and the corners of the eyes in a small circular motion twice" is not a sufficient description for effective aesthetic massage. Based on our experience, we know that this kind of description cannot be used to describe our massage.

The use of hand movements as a notation for massage corresponds to a musical score in music, and the actual massage with this score corresponds to the performance. Both a musical score and its performance are important; moreover, the way the performance is conducted is crucial to achieve a certain effect with massage. Before we systematized the Tactile Score, the learners of massage simply observed the teacher performing a massage and attempted to copy it.

However, contrary to music, massage does not have audible sounds; therefore, it is difficult to visually confirm whether the learner is able to correctly perform the massage as instructed. Therefore, the teacher could only evaluate the learner's skill by being massaged by the learner. On the other hand, teaching the massage by using the Tactile Score, including the description of the massage, corresponds to musical expression as well as hand movements. Application of the Tactile Score has enabled us to evaluate learners' skills without actually being massaged, by determining the extent to which the massage complies with the Tactile Score.

[1] FaceTherapy massage

13.2 Computational Aesthetics: Computational Understanding of Aesthetics

There is something beyond linguistic rationality. This may be true; however, if we regard artistic vitalistic force as an explanation of the principle of sensibility expressions, it means that we despair of understanding beauty. Since the criticism by Kant [3], people have given up thinking about beauty rationally. However, for example, the Tactile Score can be used to rationally express tactile comfort and discomfort, considering that many people have been trying to construct beauty in various fields such as composition, dancing, and graphics.

We attempt to consider the activities of performers, in which they challenge such constructions, as an aesthetic algorithm. The sense of touch is irreducible for living systems; we are unable to see and hear it directly but it strongly affects our deepest emotions. Sometimes we experience an unpleasant feeling with someone but we are unable to explain what it is and why the feeling occurs; yet, we experience an unpleasant tactual atmosphere from such a person. Alexander [4], an architect, proposed a concept for the atmosphere of architecture, which he termed "quality without name," and he presented a method, named pattern language, to create such an atmosphere. He explored architectures with a good quality, but without names; he decomposed such architecture into basic components and we can reproduce such an architecture having quality without name by composing the basic components and he named this method as the pattern language.

We share the same tactile sense experienced by a person who is either able or unable to exist in harmony with others or that is conveyed by architecture that is either agreeable or not agreeable. From the viewpoint of the sense of touch, pattern language is not only the design language for architecture, but also the language for the sense of touch. Quality without name is also a matter of the sense of touch; similarly, his pattern language can also be regarded as the language for the sense of touch in the broad sense.

Similar to the atmosphere produced by architecture, we cannot see and hear the sense of touch, although we are able to strongly feel its existence. Thus, acquiring the language of the sense of touch would enable us to not only visualize the sense of touch, but also experience quality without name.

How can we acquire the language of the sense of touch? One of the forms of haptic engineering, which we have developed for more than 2,000 years, is massage; a massage is composed of various actions associated with tactile sensing and is mainly generated by hand movements, in which the sequence of tactile sensing is essential. As we are able to compose the tactile sense by designing the sequence of hand movements, we have been able to develop a method for composing tactile sense by analyzing the hand movements of massages.

13.3 Tactile Score

Tactile sensing has mainly been investigated in medical and cognitive science and has also been of interest to engineering, especially haptic engineering. In haptic engineering, various haptic devices and materials have been proposed and one of the challenges is to develop a method for haptic composing.

We have developed approaches for composing methods in both the visual and auditory sense; for example, in the visual sense, the Munsell color chart is a method for composing colors, whereas in the auditory sense, the various music theories constitute methods for composing sounds. However, in the haptic sense we do not have such methods.

Therefore, a notation was developed for describing massage, which we named the Tactile Score® [5]. Because massage is an art consisting of sequences of tactile sense, we use the notation for music, i.e., the art of the sequence of sounds. By investigating massage, we have found that a massage is composed of three main basic elements: pressure, contact area, and rhythm; hence, we are able to describe these elements.

The pressure experienced as a result of massage varies from person to person; thus, we describe the degree of pressure in a relative sense. This was done by specifying the basic pressure, which was obtained by the "calibration" of pressure, and the calibration is performed by imagining he pressure when something very precious, such as a new baby, is held, i.e., the baby is neither held with very strong nor with weak pressure.

Furthermore, the contact areas of hands are important for expressing tactile sense; for example, small contact areas, such as fingertips, exert strong pressure, similar to touching the face with a pointer, or wrapping the face in the palms of hands.

We use a music score for the Tactile Score in which the pitch of tones illustrates the degree of pressure; the third line illustrates the basic pressure, and the line is sandwiched between two upside-down triangles or the Alto clef. We describe the pressure below the third line (low-pitched sound) when it is stronger than the basic pressure, or above the third line (high-pitched) when it is weaker than the basic pressure. Here the beat is quadruple time, but triple and double time are also acceptable.

Next, we number the areas of a hand to describe the size of the dimension (Fig. 13.1, Upper). In addition, we encode the spatial position and the movement of the stroke as curves, lines, dots, each with its respective size, i.e., small, medium, and large as tactile steps. This is done similarly to the sol-fa of a musical score on a face (Fig. 13.1, Lower).

When a human hand creates tactile sense [5], the contact area corresponds to the touching dimension of the hand. We describe the rhythm in the same manner as a musical score.

In the Tactile Score shown in Fig. 13.2, at the first count in the A_5 part, circles are drawn on both sides of the cheeks using the center of the palm by applying pressure that is weaker than the basic pressure. At the second count, the hands are moved to

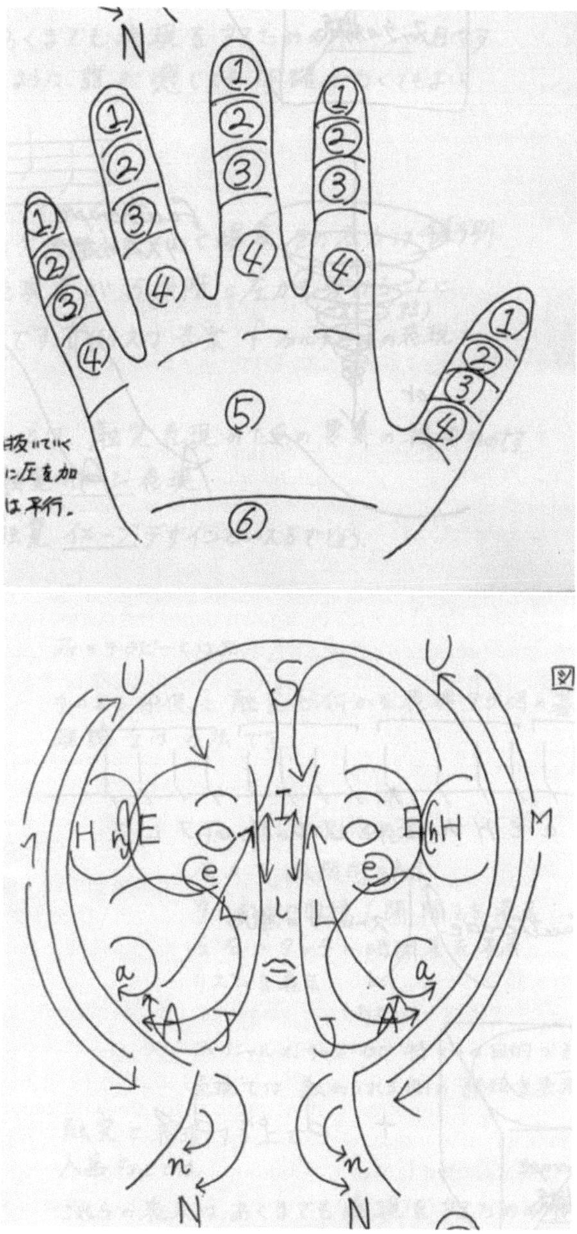

Fig. 13.1 (*Upper*) Usage of parts of the hand. (*Lower*) Strokes of massage on a face; these strokes are obtained from massage experiences in aesthetic salons; strokes that pass uncomfortable areas are excluded

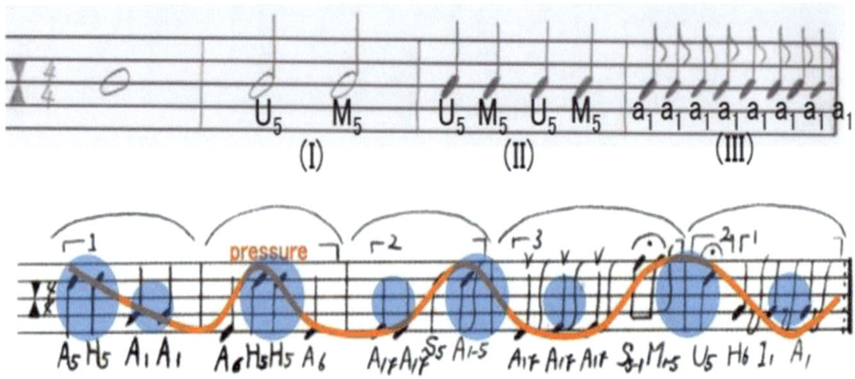

Fig. 13.2 (*Upper*) Example of the Tactile Score, with a special marking on the staff notation. (*Lower*) *1* denotes both hands moving at the same time, *2* indicates a small gap between hands, and *3* indicates a large gap between hands, the Sulla-like marks illustrate a unit component of massage, the integral-like marks illustrate releasing pressure, and the breath-like marks correspond to short pauses in massage, much like breaths when playing music

the tails of the eyes and small circles are drawn using the center of the palm while maintaining the same pressure that was applied at the first count, at the third and fourth counts, upon which the hands are moved to both sides of the cheeks and circles are drawn using the fingertips with a stronger pressure than the basic pressure.

13.4 Tactile Sense and Technologies

Almost all parts of our bodies are covered with skin. If something touches the skin, we can feel it. The skin connects our outside and inside; in other words, it connects our perception and mind through tactile sense, even though some parts, such as the eyes, are not covered with skin.

For example, an eye is an organ for visualization and an ear is an organ that enables us to hear. Therefore, these parts of the body connect our outside with our inside. Speaking in broad terms, all the sensing organs, such as the skin, eyes, and ears, constitute the membrane connecting our inside with our outside. When something touches the membrane, we feel its touch. For example, it is a baby's soft cheek, mellow music, a gentle word, delicious food, and tender solicitude. It has various meanings such as touch, sound, image, taste, and concern. We are able to sense being touched on the membrane, in other words, tactile sense.

The tactile sense of skin (skin sensibility) consists mostly of a membrane that wraps the entire body. Tactile sense (skin sensibility) is a unique organ which is unable to "close" among the senses consisting of our membranes. We can cover our ears, close our eyes, hold our nose, and close our mouth, but not close our tactile sense (skin sensibility). The importance of this tactile sense has become recognized and research pertaining to tactile sense has been conducted in various areas.

Tactile sense has been of interest to basic science, such as psychology, psychophysics, cognitive science, and recently it has also been of interest to engineering and design. In engineering, technologies of tactile sense have been developed in virtual reality, e.g., robotics and ergonomics. One of the main issues they have addressed is "how to regenerate the tactile sense mechanically" (for example, [5]) and the necessary and desirable applications of such technologies have been explored. Reviewing the large amount of research on tactile engineering should preferably be included elsewhere, in another book. In the past, tactile engineering has mainly been applied to entertainment, such as the integration of a tactile sensing device into a video game controller, and to communication technology, such as in mobile phones.

For example, Chang and O'Sullivan (2008) proposed the ComTouch [6], which is a device that augments remote voice communication with touch by converting hand pressure into vibrational intensity among users in real-time. They used 24 people to observe possible uses of the tactile channel when it was used in conjunction with audio. By recording and examining both audio and tactile data, they found strong relationships between the two communication channels. Their studies show that the users developed an encoding system similar to that of Morse code, as well as three original uses: emphasis, mimicry, and turn-taking [6].

There have been attempts to connect tactile sense and linguistic expressions; however, there is a problem in that the intuition may be lost, because we tend to arbitrarily connect tactile meaning with general language. We therefore tried to create tactile impressions from phonological aspects that were text media.

13.5 Language of Tactile Sense

Tactile perception conveys messages that differ from speech language [4].

A person, who is patted on the shoulder once, might think of accidental collision, yet when patted twice, it has meaning and is interpreted as though someone is trying to initiate communication. Moreover, a mother gently taps her baby at a steady rhythm in caressing, and the steady rhythm evokes a sense of security in the baby.

In other words, counts and rhythm are important in tactile perception; a single circular stroke could not be distinguished from a mere rubbing, whereas double strokes or more would be recognized as massage. Hence, this "(more than) double strokes" is an "alphabet" of a language of tactile sense and a set composed of these alphabets is a "word"; a massage is composed of these words as though they form a sentence. A poem is composed of sentences and these sentences generate "rhythm" similar to a poem, sentences composed of tactile sensing words also generate rhythm.

As with a mother's gentle tap, a steady rhythm adds meaning and a sense of security to massage; therefore, a steady rhythm such as this would be considered the same as a measure in music.

By imposing a rhythm on a tactile sense, we are able to create "impressions"; a rhythm of touching provides a "theme" for the impression provoked by tactile sense. We suppose that our image caused by tactile sense emerges from the temporal relationship; we always compare the tactile sense from the past and the present. If we touch something hard, and then touch something harder, we regard the former as soft. Thus, the image will be determined by comparing what/how we touched in the past and touch in the present; hence, we can generate the tactile sense created by a mother's hands by pairing senses, such as hardness and softness, to generate a rhythm of tactile sense by designing a pair of tactile senses.

13.6 "Grammar" of Composing Tactile Sense

We have analyzed the grammar of massage [1, 2], and have divided massage into 42 basic techniques. We then examined the tactile impressions of these techniques by using the semantic differential method (SD method), which has been used in basic psychology, through principle component analysis, and we could classify these basic techniques into six categories. By using this result, we extracted the grammar of massage as the transformation rule within the categories.

Speaking correctly brings "comfort": By using the acquired grammar, we can either "speak" massage grammatically correct or not. We examined and compared our feelings caused by grammatically correct or incorrect massages. The stress marker was examined by using a device to analyze salivary alpha amylase. It has been reported that salivary alpha amylase increases when a person suffers from stress. We then composed Tactile Scores that were grammatically correct and incorrect (Fig. 13.3).

We compared the values of the stress marker before and after the massage and established that, when the massage was grammatically correct, the value of the

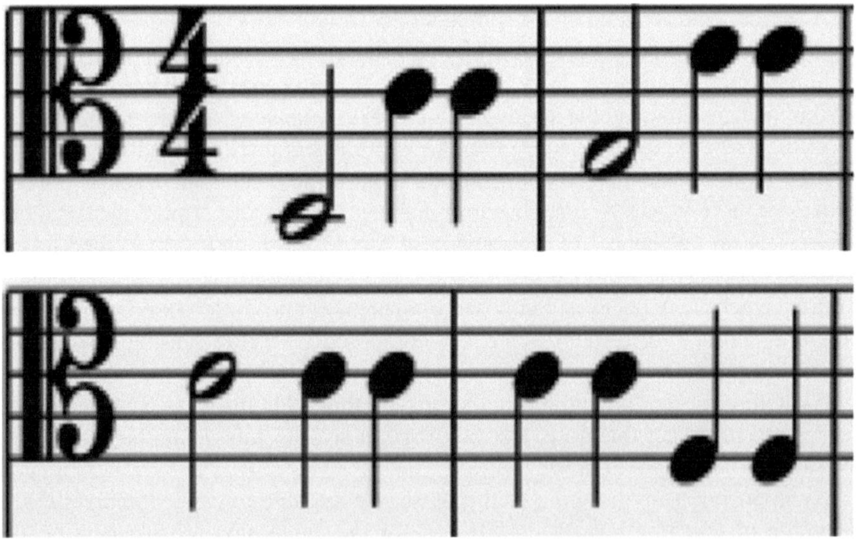

Fig. 13.3 (*Upper*) Grammatically correct Tactile Score. (*Lower*) Grammatically incorrect Tactile Score

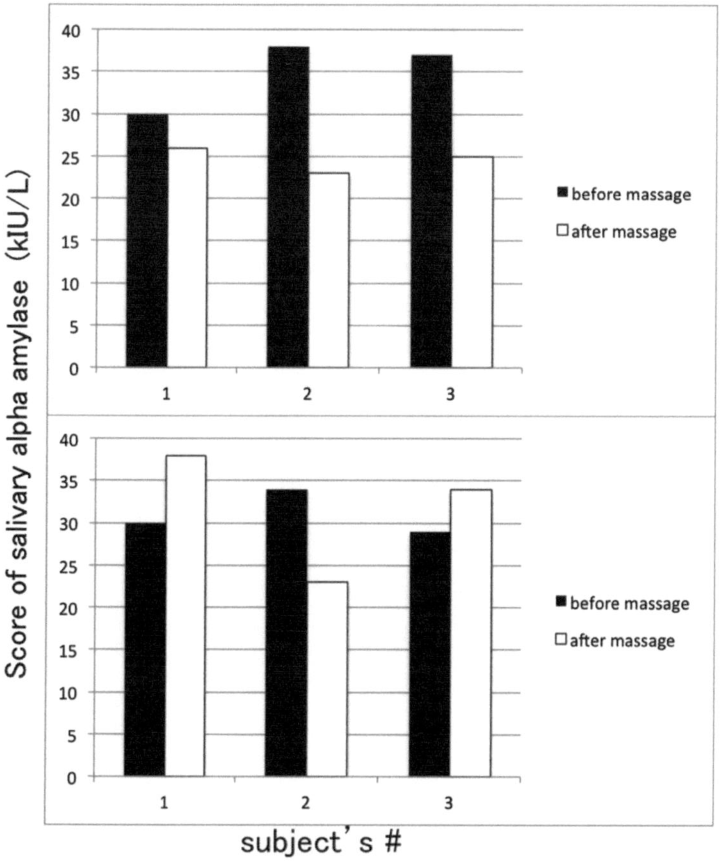

Fig. 13.4 The result of analyzing salivary alpha amylase when subjects received massages that were grammatically correct (*Upper*) and incorrect (*Lower*); The number of subjects were three; *black* shows the measurement before the massage and white, after. The *vertical axis* illustrates the score of salivary alpha amylase (kIU/L): scores from 0 to 30 indicate a small amount of stress, 40–45: fair, 46–60: large, and more than 60: huge and the horizontal axis illustrates the number of subjects

stress marker decreased; on the other hand, when it was grammatically incorrect, the value increased. This indicates that we feel comfortable when massage is grammatically correct and that we are able to recognize grammatically incorrect massage (Fig. 13.4). This is our preliminary result. The experiment was subsequently performed using a larger number of subjects to obtain strong confirmation of this result.

We additionally confirmed the following effect resulting from massage produced by a grammatically correct Tactile Score: we captured an image of the person before and after the grammatically correct massage and compared the shape and color of their face by using image processing. This preliminary result shows that the shape of the face tends to shrink, especially around the cheeks, and the brightness of the facial image tends to increase (Fig. 13.5).

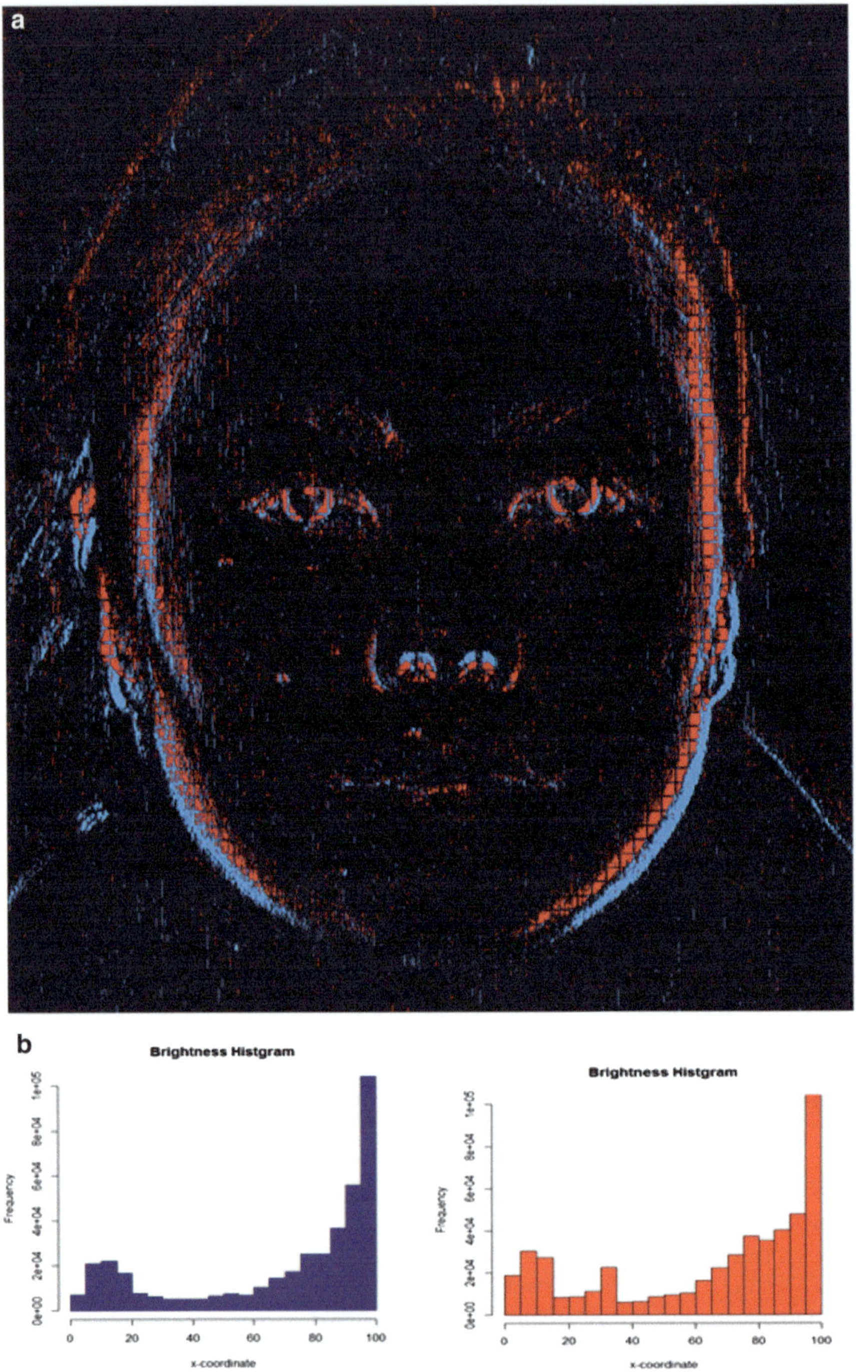

Fig. 13.5 (**a**) *outside line* illustrates the shape of a face before massage, and *inside line*, after massage. (**b**) Histograms illustrate the brightness of the facial image before (The average of three subjects was 67.06) and after (73.21) massage, indicating that the total value of brightness increased

13.7 Tactile Score as Sensory Language

We mainly developed the Tactile Score for describing tactile sensing produced by massage. We can feel tactile sense not only by touching via skin but also by seeing and hearing. Hence, we are able to feel tactile sense by reading a text. Whenever we read a text, we visualize and read it aloud, even if we do not use voice, we visualize and read it aloud by using our inner voice. Therefore, when we read a text, we perceive visual and aural stimulation from the text and feel tactile sense. This is referred to as "literary style."

13.8 Extraction of Tactile Information from Texts

Two modalities, namely text (vision) and voice (audio) are fused in a language. For example, "Though I know how busy you must be, please ∼" is a more polite text than "Please ∼." Although we can use many other methods and feature amounts, in this preliminary study, we focus on the length of the text because it is easy to measure [7].

We use Japanese texts by separating sentences with a comma, which is a feature of Japanese. The location of a comma is often arbitrary and depends on the author's emotion. The Japanese language does not consist of single words; therefore, it has to be separated into words and these words comprise a semantic unit, which is a minimum unit of semantics; for example, the semantic unit of "I go to school" is "I / go to/ school", where "to" is not a sematic unit. However, providing the perfect definition of a semantic unit is challenging and may be ambiguous in general.

For example, the following Japanese sentence,

私は, 学校に行きます.
Watashi wa, gakkou ni ikimasu.,

which means "I go to school", has been divided into single words. However, as the Japanese language does not consist of single words, we preprocess a sentence by dividing it into words by using Morphological Analysis to obtain:

Watashi / wa /, / gakkou / ni / iki / masu /.

and its semantic units are

Watashiwa (I) /, / gakkouni (to school) / ikimasu (go);

thus, this sentence is composed of six words and three semantic units. We may define various types of "semantic units" and in this study a sematic part is a unit of which the meaning is independently understandable.

13.9 Method for Extracting Tactile Sense from a Text

The algorithm of the method has two parts: preparation and transformation. The preparation algorithm comprises the following:

Divide a text into sentences.
Divide one sentence from beginning to end between a comma and period and term each section of divided text "a part".
Divide each part into semantic units.

For example, the text

"今日は日曜日. だけど, 私は, 学校に行く."
"Kyowa nichiyoubi. Dakedo, watashiwa, gakkouni iku."

"I know today is Sunday. But I go to school." is transformed into the following two sentences by step 1:

Kyowa nichiyoubi.
Dakedo, watashiwa, gakkouni iku.

Each sentence is divided into parts by step 2;

Kyo / wa / nichiyoubi.
Dakedo / watashi / wa / gakkou / ni / iku.

Each part is recomposed into semantic units:

Kyowa / nichiyoubi.
Dakedo / watashiwa / gakkouni / iku.

The semantic units obtained in this way are transformed into "tactile notes" by the following transformation algorithm, for which we define the operation and as follows:

$$a\&d = a + d$$

when a and d are positive or zero and

$$-a\&d = -a - d$$

when a and d are negative.

1. Count the number of words in each semantic unit.
2. Count the number of phonemes (vowel and consonant sounds) in each semantic unit.
3. Calculate the average number of phonemes in each semantic unit, ave_p and set this equal to the basic pressure; when ave_p is not an integer, it is rounded off to an integer.
4. Calculate the value
   (phonemes of a semantic unit–ave$p$) $\&1 = d$;

d is regarded as the tone interval from ave_*p*, which we refer to as a "tactile tone".
5. When a modification relation exists between semantic units A and B, and the phonemes of A and B are different, such a relation is expressed by denoting B as a chord, which is composed of tactile tones of A and B.

Through this algorithm we obtain tactile tones, which are transformed from a text.

For example, the tactile tones of the semantic units of

Kyowa / nichiyoubi and Dakedo / watashiwa / gakkouni / iku;

are found by determining the number of words of each semantic unit (step 1):

今日は/ 日曜日is 3 / 3

だけど　私は　学校に　行く is 3 / 2 / 3 / 2;

Kyowa / nichiyoubi is 3 / 3

Dakedo watashiwa gakkouni iku is 3 / 2 / 3 / 2;

and the number of other phonemes is 3 / 4 / 4 / 2 (step 2); thus,

the ave_$p = (2 + 5 + 3 + 4 + 4 + 2) / 6 = 3.33\ldots$

and it is rounded off to 3 (step 3); for example, the tactile tones of "kyowa" are (2–3) and $1 = -2$; thus, the tone interval from ave_p is -2; hence, the tactile tone of each unit is 3, 1, 3, 1, −2. Next, we write these tactile tones into the Tactile Score.

13.10 Method: Tactile Scoring

In this algorithm, the beat is quadruple time, but this can be changed to another time and beat. For example,

Kyowa / nichiyoubi. and Dakedo / watashiwa / gakkouni / iku;

are "half note / half note" and "quarter note / quarter note / quarter note / quarter note/". We obtain the Tactile Score of "kyowa nichiyoubi dakedo watashiwa gakkouni iku" as shown in Fig. 13.6.

Fig. 13.6 Tactile score of "kyowa nichiyoubi dakedo watashiwa gakkouni iku"

Gionsyozyano kanenokoe
Syogyomuzyono hibikiari
Sarasouzyuno hananoiro
Jyosyahissuino kotowariwoarawasu
Ogoreruhitomo hisasikarazu
Tadaharunoyono yumenogotoshi
Takekimonomo tsuiniha horobinu
Hitoeni kazeno maeno chirini onazi

祇園精舎の鐘の声、
諸行無常の響きあり。
娑羅双樹の華の色、
盛者必衰の理をあらはす。
おごれる人も久しからず。
ただ春の夜の夢のごとし。
たけき者も遂にはほろびぬ、
ひとへに風の前の塵に同じ。

Fig. 13.7 The beginning part of *The Tale of the Heike* in Japanese

When a semantic unit is long, such a unit is divided by nouns or verbs; for example, "when I find myself in times of trouble" is divided into "When I / find / myself / in times of trouble", which means that each part is assigned to a quarter note, where the part of "in times of trouble" is long and it is divided as "in times of / trouble" and each part is assigned to an eighth note.

13.11 Material

We apply this method to the beginning of a historical novel entitled *The Tale of the Heike*, which has been handed down in Japan over a thousand years by troubadours who passed it down orally when letters had not yet been developed. A study of the story shows that the troubadours not only handed down the text, but also the rhythm.

We decided to use the story as our material because the text is based on the rhythm, and many people have read it for a thousand years to the present day, and the beginning is well known (Fig. 13.7).

English translation of Fig. 13.7: "*The sound of the Gion Syozya bells echoes the impermanence of all things*; *the color of the sala flowers reveals the truth that the prosperous must decline. The proud do not endure*, *they are like a dream on a spring night*; *the mighty fall at last*, *they are as dust before the wind*. – Chapter 1.1, Helen Craig McCullough's translation [8]."

13.12 Tactile Score of The Tale of the Heike

By using the transformation algorithm, this text is transformed into:

Gionshoja no / kane no / koe.
Shogyomujo no / hibiki / ari.
Sarasoju no / hana no / iro.
Joshahissui no / kotowari wo / arawasu.
Ogoreru / mono mo / *hisashi kara zu*.

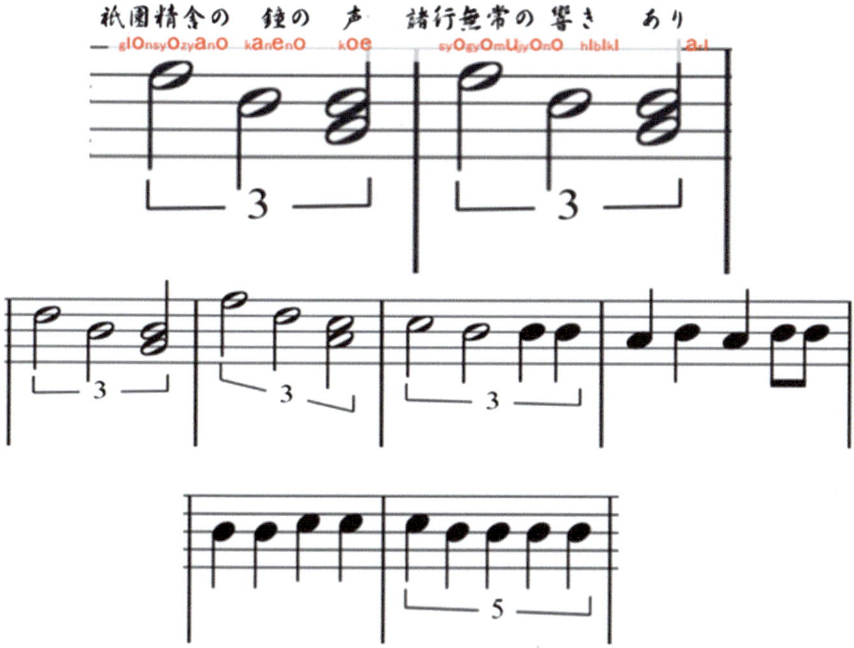

Fig. 13.8 Tactile Score of *The Tale of the Heike*. (*Top*) The first and second parts of the text; (*Middle*) From the third to the sixth parts; (*Bottom*) The seventh and eighth parts

tada / haru no / yo no / *yume no goto shi*.
Takekimono mo / tsuwini wa / horobi nu.
hitoe ni / kaze no / mae no / chiri ni / ona ji.

In these units, "hisashi kara zu" and "yume no goto shi" are long, therefore they are divided as "hisashi / karazu" and "yume no / gotoshi". The average phoneme of the units is 3.09, which is rounded to $ave\_p = 3$, in which case we obtain the Tactile Score transformed from this text as shown in Fig. 13.8.

13.13 Tactile Score of Hamlet

We attempted to apply our method to English text; however, counting the phonemes of English words is a challenge; thus, in this study, we counted vowel sounds. Moreover, defining a semantic unit is a challenge; thus, in this paper, we divided text into small chunks of which we could understand the meaning. We used the English language version of *The Tale of the Heike* for this purpose. By using our algorithm, we obtained the Tactile Score of the English version as shown in Fig. 13.9.

The sound of /the Gion Syozya bells/echoes /the impermanence of /all things;
the color of /the sala flowers/reveals/the truth /that the prosperous must decline.

Fig. 13.9 Tactile Score of the English version of The Tale of the Heike

Fig. 13.10 Tactile Score of *Hamlet*

The proud/do not endure,
they are/like a dream/on a spring night;
the mighty/fall /at last,
they are/as dust/ before the wind.,

where the average of the number of vowels in every semantic unit is 3.71, which is rounded to ave_$p = 4$. Because of the length of the clause in the 2nd paragraph, *that the prosperous must decline*, it was divided as "that the prosperous / must / decline".

We also applied this method to English literature, namely *Hamlet* by W. Shakespeare. By using the algorithm, we obtain:

to be, or
not to be
that is/the question;
whether 'tis/nobler/in the mind/to suffer
the slings/and/arrows of/outrageous/fortune
or/to take/arms/against a sea of troubles
and/by opposing/end/them
To die,
to sleep
no more
and/by a sleep/to say/we/end;

where part of the 5th unit "against a sea of troubles", is comparatively longer than the others; thus, we divided it into "against / a sea of trouble". The average of the number of vowels is 2.53, which is rounded to ave_$p = 3$, which produced a Tactile Score of *Hamlet* as shown in Fig. 13.10.

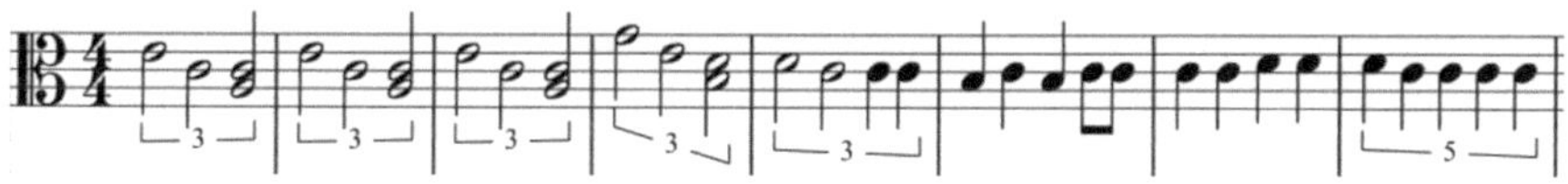

Fig. 13.11 Tactile Score of *The Tale of the Heike* (original)

Fig. 13.12 Tactile Score of the English version of *The Tale of the Heike*

This Tactile Score illustrates that the speed of rhythm is composed of "slow (1st measure) – (gradually faster, from 2nd to 4th measure) → fast (5th and 6th measures) → (gradually slower, 7th measure) → slow (8th–10th measure) → fast (11th measure)" and there are similar patterns (same rhythms in the same note intervals) in the 2nd and 3rd, 8th to 10th measures.

The speed of rhythm in the Tactile Score of *The Tale of the Heike* is as shown in Fig. 13.11.

It is almost the same, and the changes of pressures are large (from the 1st to the 4th measures) to small (from the 5th to the 8th measures); and there are similar patterns from the 1st to the 4th measures (same rhythm in the same note intervals). The tactile note of the English version of *The Tale of the Heike* does not have a clear structure, and the patterns shown in Fig. 13.12 are absent.

This result indicates that "tactile impression" of *The Tale of the Heike* and its English version may be different and we suppose that if the English version were to have a similar Tactile Score, its impression would also have been similar to that of the original.

13.14 Sensory Language

In the previous section we showed that, in the integrated visual and audial sense, a text is translatable into a tactile sense via the Tactile Score. We also showed that the Tactile Score is translatable into the visual or audio sense.

13.15 From Tactile Score to Visual and Auditory Sense

In terms of the performing arts we selected *MatchAtria* by Yui Kawaguchi (dancer) and Yoshimasa Ishibashi (movie director), and translated the text, which was written by Kawaguchi, by obtaining its Tactile Score. Kawaguchi then re-translated the

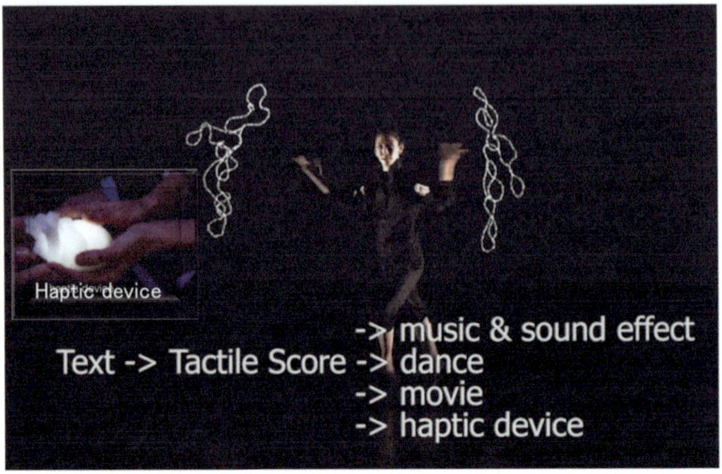

Fig. 13.13 A picture of *MatchAtria* [9], of which the text was translated into the Tactile Score before it was re-translated into a dance notation, computer graphics, music, and sound vibration; the sound vibration produced in this way caused the haptic device to vibrate

Tactile Score into a notation for dancing and Ishibashi translated the Tactile Score into computer graphics, after which we translated it into music and sound vibration.

Therefore, we translated the text into a notation for dancing, computer graphics, music, and sound vibration, after which they were synthesized as performing arts. In this art performance, Kawaguchi integrated the dance performance with the three-dimensional computer graphics produced by Ishibashi, whereas a haptic device shaped as a human heart was produced by Ando (Osaka University) and Watanabe (NTT) to enable audiences to feel the sound vibration (Fig. 13.13).

13.16 From Designed Tactile Score to Visual, Audio and Haptic Sense

In the *MatchAtria* [9], the Tactile Score was obtained through translation from a text. Next, we designed a Tactile Score according to the grammar of the Tactile Score.

When we say something, we feel a tactile sense ourselves as a result of the sound vibration of the voice; hence, using the voice can be regarded as a "self-massage."

A massage is composed of the area of touching, pressure, and rhythm, which can be composed by using the grammar that exists for this purpose. This means we are able to compose a text for producing vocal sounds according to the grammar of the tactile sense. Composing this text was done by regarding the extent to which the mouth opens as the area of toughing, and the strength of the voice as the pressure and rhythm for transforming the voice into rhythm.

Fig. 13.14 From a tactile story telling, the picture book is composed of Tactile Score, which is composed based on the grammar; the Tactile Score in the book is a simpler version, which does not require skill to read, and is easy to understand and massage

We have created massage picture books for children and parents. The picture books are composed of a Tactile Score and pictures and texts that are translated from the Tactile Score. Parents can read the book to children and can also massage children, vice versa (Fig. 13.14).

We have proposed a Tactile Score for creating a tactile sense produced by massage. The composition of massage was investigated and it was established that a "grammar" existed. Moreover, we have confirmed the effectiveness of the score for beauty treatment by using a "grammatically correct" massage.

Based on these results, we have attempted to utilize it as a sensory language, not only for composing a massage. We extracted the tactile sense from texts and translated it into multimodal media (dance notation, computer graphics, sound vibrations, and so on), and composed a text for "self-massage" by using voice in reading, and creating picture books of "tactile story-telling".

The Tactile Score can be used as a sensory language, which is able to extract tactile sense from multimodal media and describe it as "intermediate forms" by using the Tactile Score. We are able to analyze, edit, and compose the intermediate forms and re-translate them into multimodal media again. Through these media translations, we will be able to name the things that have been named as quality without name.

Acknowledgments This research was supported by JSPS KAKENHI (Grant in aid for scientific research (C)) Grant Number 21520113 and 15K00268.

References

1. Suzuki Y., Suzuki R.: Tactile Score. Springer, Tokyo (2013a)
2. Suzuki Y, Suzuki R.: Introduction to Facetherapie, ("Facetherapie Nyumon," in Japanese), Syun-jyu-sya (2013b)
3. Kant, I.: Kritik der reinen Vernunft (1787)
4. Alexander, C.: A Pattern Language: New York Towns, Buildings, Construction. Oxford University Press (1977)
5. Margolis, E., Samuels, R., Stich, S.P. (eds.): The Oxford Handbook of Philosophy of Cognitive Science. Oxford University Press, Oxford (2010)
6. Chang, A., O'Sullivan, C.: An Audio-Haptic Aesthetic Framework Influenced by Visual Theory, pp. 70–80. Springer, Berlin (2008)
7. Suzuki Y., Suzuki R., Watanabe, J.: Translation from Text to Touch – Touching a "Japanese Old Tale". ACM CHI 2014, U.S. (2014)
8. McCullough, H.: The Tale of the Heike. Stanford University Press, Stanford, California, U.S. (1990)
9. Mendora (12 Jan 2014) http://mendora.com/works/matcha/matcha_JP.htm

Chapter 14
Haptic Aids for the Visually Impaired

Han SungMin

Abstract Haptic aids for persons with visual impairment are introduced. First, the devices are categorized by their purposes; reading a braille, viewing a 2D information, and enabling input and output. Then they are classified from the viewpoint of the necessity of users' active motion, followed by the discussion of benefits and drawbacks of active and passive devices. The chapter concludes with current and future development of haptic aids for the visually impaired.

Keywords Haptic aid • Tactile display • Assistive technology • Visually impaired

14.1 Introduction

It is believed that tactile skills for persons with visual impairment (PVI) can be improved by tactile-display technology, and also by pedagogies and display systems that utilize cross-modal perception. During the 1980s, many research groups developed paperless braille-display systems that are used by many PVI. The piezoelectric bimorph design is widely considered the most effective design for the pin-motion actuators used to form braille characters.

This paper categorizes diverse field of tactile devices to support PVI. Furthermore, it discusses haptic aids for PVI, and, in particular, braille (tactile) displays on which braille characters are formed by piezoelectric bimorphs.

14.2 Haptic Aids

Because PVI cannot use visual information, they rely on haptic perception systems in their daily lives [1]. A haptic perception system is an integral perception system in which tactile sense works together with kinesthetic sense (constituting the somatosensory system).

H. SungMin (✉)
Department of Special Education, Fukuoka University of Education, Fukuoka, Japan
e-mail: hanshin7153@nifty.com

H. Kajimoto et al. (eds.), *Pervasive Haptics*, DOI 10.1007/978-4-431-55772-2_14

When PVI use a so-called "white cane" for mobility, tactile paving stimulates the soles in their feet, indicating when to stop and helping to determine orientation. The white cane is used to determine the state of the ground and to check for obstructions, and PVI discern the presence or the absence of objects by sensing many modal cues such as wind direction, differences in temperature and air pressure. At metro stations, they use their fingers to read fare displays, check the price, and purchase tickets [2]. Traditionally, most implements used by PVI have been low-tech, and they tend to serve merely as visual substitutes, including white canes, braille, and perforated panels. In recent years, information and communications technology and virtual reality technology have produced a number of hi-tech sensory substitutes to help PVI in terms of mobility and information processing.

Braille was originally developed in the eighteenth century by a school for the blind in France. Researchers have come up with ways to render braille in six-bit character code for digital haptic characters, and many PVI use this technology for both reading and writing. In particular, technology for forming braille characters (viz., braille-information terminals and braille displays) focuses on information processing, and such technology has enjoyed considerable popularity. These include passive systems, active systems, and systems that make use of a multi-modal interface.

In the field of sensory substitution, devices that substitute the sense of hearing with the sense of touch are called tactile aids. Whereas tactile aids are primarily used to receive information, devices that substitute the sense of sight with the sense of touch are used to collect information about the outside world through active hand movements. On this basis, the latter devices are called haptic aid.

14.2.1 Braille Displays

The purpose of developing visual-substitution technology is to aid with the mobility and communication of PVI. Traditional implements, such as white canes (for mobility) and braille (for communication), have evolved into electromechanical technology, examples of which include the OPtical-to-Tactile-CONverter (Optacon) [3] and the Tactile Vision Sensory Substitution (TVSS) [4], and recent years have seen the development of information and communications technology such as braille-based personal digital assistants (PDAs) and the Forehead Sensory Recognition System (FSRS) [5]. Unlike white canes, TVSS and FSRS allow objects to be perceived tactilely from some tens of meters away. These technologies convert photographic information captured by a camera into electrical or physical signals, which can then be displayed as tactile information. As such, they are helping to realize an environment that is functionally equivalent to a visual environment, and they are expanding the mobility sphere of PVI from as far as the hand can reach to as far as the eye can see.

Braille displays and Optacon enable the transmission of information based on the locations of dots or vibrations. Optacon can be used to read the shape of black-ink letters on a white background. Braille systems, on the other hand, use a six-pin array that conveys the written information through elevations and retractions of the pins. At one time, braille was exclusively available on paper media, but there are now many information devices that use refreshable braille (such as braille-based electronic organizers and PDAs), The dissemination of electronic braille data created for such devices has prompted the emergence of reading technologies, whereby the braille is not simply read by hand; it is also read aloud by a screen reader.

There are two main types of braille-information devices: "braille displays" convert writing that appears on PC screens into braille and display it as such; and "braille memo devices" function as stand-alone braille note-taking devices that provide editing and search functions. Braille displays and braille electronic organizers are not usually distinguished from each other in Japan. In the United States, however, braille PDAs and braille electronic organizers (when they function as stand-alone braille memo devices) are called note-takers, and these are strictly distinguished from braille displays, which have no memory or braille note-taking functions, and can only provide a display.

Japan is currently leading the world in the development of braille-display modules (cells). It is not only braille-display modules for linguistic information that are being developed; there are also modules for tactile displays of PC screen information (e.g., SC5 and SC10: KGS Co. Ltd.).

Braille-display actuators use piezoelectric bimorphs, and they have been incorporated in Optacon, braille displays, tactile television, and tactile computer mice. Unlike other actuator designs, such as solenoid actuators, the piezoelectric bimorph design enables easy densification at low electrical power, and, as such, it has been applied in tactile displays in many portable sensory substitution devices [6].

14.2.2 Optacon

Optacon is a well-known example of haptic aid that utilizes piezoelectric elements. Optacon was designed to allow the reader to perceive by touching the shape of letters. The reader uses one hand to scan the print with a small camera module, and the index finger of the other hand for tactile reading. The significance of Optacon lies in the fact that, unlike braille, it allows PVI to read the same letters (printed letters) that are read by people without visual impairment. John Linvill, who was inspired to design the system for his visually-impaired daughter, invented Optacon [3]. Its name, the OPtical-to-Tactile CONverter is derived from the fact that it substitutes opticus with tactus (i.e., it substitutes optical perception with tactile perception).

The tactile unit consists of a 24×6 (total = 144) array of tactile stimulators (20×5 [=100] in OPTACON II) that vibrate at 230 Hz (fixed). The total weight of the device is 450 g, and it can be run on batteries for portability. With some practice, the material can be read at an average of 40 words per minute (wpm), and highly skilled users can reach reading speeds of 100 wpm [7].

The average reading speed for braille is 80 wpm [8], and highly skilled braille readers can reach reading speeds of 250 wpm [9]. Reading with Optacon therefore proceeds at approximately half the speed of braille. There are also some users whose tactile reading performance does not improve even with practice. In addition, unlike the Latin alphabet, Japanese script has a mixture of kanji and kana characters, and this make tactile reading difficult. There are around 600 Optacon units in circulation in Japan, but manufacturing has now been suspended.

14.2.3 Braille-Information Devices

Among the most popular types of haptic aid are braille-information devices (i.e., braille displays and braille PDAs). There are around 10,000 such devices in circulation in Japan. In principle, PVI can purchase the devices at only 10 % of the cost, because these devices are classed as "equipment covered by public aid" under the Long Term Care Insurance Act (there is some variance between local authorities) [10]. Because braille involves perforations in thick paper, it is awkward to carry around a quantity of braille text equivalent to that in a book. However, braille PDAs weigh approximately 1 kg, and with the inclusion of information-processing capabilities in braille displays, the user can write scripts for (ink) printing, send and receive email, and connect to the Internet. Braille PDAs also include audio-recording and -playback functions, meaning that they can function similar to a laptop. For people with both visual and aural impairment, braille PDAs represent an especially important tool for gaining information about the outside world, and they are also used as translation and conversation systems.

14.3 Tactile (Braille) Displays Based on Active and Passive Touch

14.3.1 Definition of Active and Passive Touch

James J. Gibson, the American perceptual psychologist who expounded the theory of affordances, used the terms "active touch" and "haptics" to refer to situations where people purposely move their hands to touch an object. He used the term "passive touch" to refer to the sensation of an object coming into contact with a person (who does not physically move), and he distinguished passive touch from active touch. Active touch is a purposive act in which the person gets an impression of the object as well as spatial information [11, 12].

In psychology and biology, passive touch has traditionally meant the receptive sense of being touched, and active touch has traditionally referred to the perception of touching an object, as opposed to the skin that is in contact with the object.

Phenomenological psychologist David Katz discussed the bipolarity of tactile phenomena (i.e., objectification and subjectification). Based on the fact that touching requires part of the body to be in direct contact with the object, whereas the eye can obtain information about objects at relatively "astronomical" distances, Katz classed vision as a "distant sense" and touch as a "near sense." He then argued that color in visual perception (a distant sense) involves objectification, whereas the perception of tactual phenomena (a near sense) involves subjectification [13]. This model of bipolarity corresponds to Gibson's model of active and passive touch.

According to Katz, research has been conducted individually into passive touch, stimulated touch, and pressure and temperature receptors. This research understands active touching in daily life to involve the perception of the outside world, as opposed to body sensations, and movements of the hand are essential for determining roughness and texture.

Research into active information processing, such as that carried out by Katz and Gibson, has made use of tachistoscopes, devices that display visual stimuli, such as images or letters, for an extremely brief period of time and, as such, can be used as an experimenting device in vision studies. Tachistoscopes have been used in parallel processing and serial processing for the purpose of linguistic comprehension, alongside eye-focus simulations and real-time control of comprehension. Such experimental research is known as active vision studies.

Today, research on the tactile mode focuses on 2D patterns and 3D objects; there are very few tactile studies that have used linguistic stimuli. The reason for this scarcity is that, unlike vision studies, there are no experimenting devices for measuring tactile linguistic processing. Moreover, it is difficult to secure (PVI) participants who are capable of tactile linguistic processing for experimentation.

Examples of tactile-mode research on 2D graphics and 3D objects include Loomis [14] and Lederman [15]. Tactile-mode research has yielded many findings, specifically regarding studies into the priority of tactile and active touch.

14.3.2 *Tactile Display That Uses Active Touch*

In the case of braille displays, users move their hands to read the text in the same way as braille print, and as such, braille displays can be considered tactile displays for the purpose of linguistic processing based on active touch.

On the other hand, in the case of Optacon, it comprises a camera module moved by one hand, and a tactile array in which tactile letters are presented and read with the index finger of the other hand. Is this reading process passive, or is it active? Optacon and TVSS involve active touch in the sense that the information can only be obtained during muscle movements and in the sense that users purposively obtain the information. However, users do not explore an object directly with their hand.

To analyze string information with a temporal hierarchical structure (and language is one example of such a structure), data-driven and conceptually driven processing must take place interactively [16]. In other words, both a perceptual and a cognitive system are required, and as for the process of reading by hand movements, there must be an elucidation of both the sampling method for the braille pattern being read and the linguistic comprehension process. People with no visual impairment identify visual letters through the visual system, but PVI perceive braille through the mediation of the somatosensory system. Accordingly, "braille reading involves converting simple tactile information into meaningful patterns that have lexical and semantic properties [18]".

14.3.3 Tactile Display That Uses Passive Touch

Traditional braille displays for PVI involve braille reading through hand movements similar to braille print. In other words, most braille displays involve active touch. However, with the recent developments in assistive technology, there has been an emergence of braille displays that allow users to read without moving their hand (e.g., disc-type systems), thus revealing the possibilities for tactile linguistic information processing through passive touch. The disc-type braille display produced in 2003 by ASKK Co. Ltd. enables braille reading by passive touch. The compact disc-type display includes a disc that rotates continuously to enable the user to read text. Users place their index fingers on the disc, and the rotary disc in the center displays braille by elevating and retracting the stainless steel pin protrusions.

The device holds 4 MB of memory, and it can be set at seven different reading speeds, ranging from 100 to 400 cells/min (or 50–500 cells/min with the improved version). The reading area comprises 15 cells. The device weighs around 430 g, approximately one third the weight of a traditional braille display.

Traditional braille displays use hundreds of bimorphs, and they have actuators positioned at each dot location. By contrast, the disc-type device is controllable with three driving actuators, thus achieving an efficient miniaturization. Accordingly, this rotating-wheel design has been gaining attention around the world. In 2000, rotating braille displays were researched and developed by the United States' National Institute of Standards and Technology, and Belgium's Catholic University of Leuven. However, the devices were never made commercially available. During the 1980s, TSI Co. Ltd. launched Braille Mate, a compact and lightweight system allowing braille input and voice checking. Braille Mate uses a single cell to display braille, enabling continuous braille reading. It was the first display based on passive touch. Nevertheless, the production of this device has been suspended.

14.3.4 Difference Between Active and Passive Tactile Displays

The systems for obtaining written (linguistic) information by touch that are currently available include braille displays such as Optacon, body-braille, and finger-braille. However, it is essential for further research to explore linguistic-information processing in tactile mode (i.e., with passive and active touch) to enable the production of sensory substitution devices with a user-centered design. One way to facilitate linguistic comprehension by tactile perception is to provide a means for active real-time scanning that can match the comprehension rate (in the case of visual perception, real-time scanning involves eye movements, such as saccades).

In the case of Optacon, it is possible to perform touching actions in tandem with the comprehension rate, making it much the same as reading in the visual modality. In this sense, it can be classified as active touch or as active information processing. On the other hand, linguistic processing based on body-braille, finger-braille, or disc-type braille displays cannot be adjusted to the comprehension rate of the reader, making it similar to information processing in the auditory modality. In this sense, it can be classified as passive touch or as passive linguistic processing.

When PVI read braille, they scan the line of print from left to right, keeping pace with their rate of comprehension. Naturally, the speed will vary depending on the difficulty level of the text and the reader's acuity. Research into visual perception has demonstrated that eye movement is affected by the difficulty level of the text [19]. Frequent up-down movements of the index finger (i.e., the finger touching the text) are observed among novice braille readers, whereas expert braille readers tend to slide their fingers smoothly from left to right. The tactile reading of braille entails complex cognitive processes, including precise control of the fingers, an acute perception of protrusions, patter recognition, and a comprehension of words and their meanings [20].

Reading requires an interplay between perceptual and cognitive systems, but with regard to reading by active touch, there is a need to clarify how braille patterns are sampled and how they are involved in the tactile perception mode.

On the premise that braille reading by active touch is similar to reading by sight, and that braille reading by passive touch is similar to listening, Sung-min Han defined the former as "braille for reading" and the latter as "braille for listening" [10]. According to Han, the technology for the former (traditional, widely-used braille displays) is frequently used among students and researchers, while the technology for the latter (e.g., disc-type braille displays) is used more for light reading, such as reading novels and commentaries. This tendency is comparable to the tendency whereby visually impaired students and researchers invariably require braille prints, whereas general PVI typically resort to audio devices when reading magazines and novels.

Currently, the Japanese firm KGS produces over 70 % of the world's share of braille cells used in piezoelectric bimorph actuators capable of displaying braille for active-touch reading. As for disc-type braille displays, the Japanese firm ASKK has nearly a 100 % share.

14.4 Tactile Displays for Linguistic and Graphic Information

Tactile displays for PVI can be broadly divided into two categories, based on their intended use. There are displays such as braille displays and braille-information terminal devices (braille PDAs) that have been developed for the purpose of receiving linguistic information. Moreover, displays have been developed for obtaining non-linguistic information, such as palpable graphics, 3D graphics, and virtual reality. Braille displays and braille PDAs remain among the most commercially oriented types of tactile displays, but there is currently ample research into tactile displays for communicating graphic information and 3D information.

Examples of the recent tactile displays for communicating non-linguistic information include tactile television and the tactile mouse. These displays have been developed for the tactile reading of printed figures, printed words, and other such information. Sakai et al. [17] have developed an interactive display system based on a touch-operated 2D tactile display, and they are piloting a tactile spatial information display for television-data transmission and an Electronic Program Guide (EPG).

A tactile mouse is a general PC mouse fitted with a miniature tactile display that feeds back tactile information according to the movement of the mouse, thereby communicating graphic information. Research is exploring ways to increase the capabilities of tactile mice, including the ability to represent on-screen 3D displays by providing a sense of convex/concave shapes and undulations, vibration actions, and simulated weight sensations. As part of this research, Watanabe et al. [21] have developed a tactile mouse capable of recognizing letters and diagrams in printed material.

KGS's tactile graphic-display system, which enables redrawing, was adapted by Kobayashi et al. [22] to develop an electronic "raised-writer" system in which the user can write and erase as many times as desired by touching graphics in real time. They are researching its applications not only for education, but also for entertainment.

In recent years, virtual reality research has prompted the development of haptic displays capable of artificially recreating the sense of touch. For example, a force-feedback device PHANToM [23] allows users to feel virtual force in their fingertips, and to touch the object, apply force to it, and to feel its texture. Such haptic displays that communicate 3D information have great potential as tactile displays for PVI.

14.5 Conclusion

Until now, the braille displays used by PVI have proved effective among people with visual and aural impairment, and digitally recorded books have proved effective for people with learning disabilities. As is evident by the fact that the United States' Recordings for the Blind (RFB) changed its name to Recordings for the Blind and Dyslexic (RFB&D), it is now known that audio books are effective tools for persons with learning disabilities as well as PVI.

As Japan braces itself for a super-aging society, universal-design research is thriving. For example, Japanese telecommunications company NTT launched a series of easy-to-use phones designed by a PVI group. These phones have been praised for their easy operation, and have been well received in the general market.

As interdisciplinary research continues to permeate across all areas of academia, even the field of neuroscience is adopting terminology such as "the social brain" and progressing from a basic science to a science concerned with social interaction.

The piezoelectric bimorph actuator-based braille-display technology discussed in this paper is being applied in such fields as tactile physiology, neuroscience, and virtual reality.

References

1. Jansson, G.: Haptics as a substitute for vision. In: Assistive Technology for Visually Impaired and Blind People, p. 135. Springer, London (2008)
2. Shingledecker, C.A., Foulke, E.: A human factors approach to the assessment of the mobility of blind pedestrians. Hum. Factors **20**(3), 273–286 (1978)
3. Linvill, J.G., Bliss, J.C.: A direct translation reading aid for the blind. Proc. IEEE **54**(1), 40–51 (1966)
4. Collins, C.C.: Tactile television – mechanical and electrical image projection. IEEE Trans. Man-Machine Syst. **MMS-11**(1), 65–71 (1970)
5. Kajimoto, H., Kanno, Y., Tachi, S.: Forehead Electro-tactile Display for Vision Substitution, EuroHaptics (2006)
6. Bliss, J.C.: A relatively high-resolution reading aid for the blind. IEEE Trans. Man-Machine Syst. **10**, 1–9 (1969)
7. Craig, J.C.: Vibrotactile pattern perception: extraordinary observers. Science **22**(196(4288)), 450–452 (1977)
8. Nolan, C. Y., Kederis, C. J.: Perceptual Factors in Braille Word Recognition. American Foundation for the Blind Research Series, 1969, No. 20. American Foundation for the Blind, New York (1969)
9. Grunwald, A.P.: A braille-reading machine. Science **7**(154(3745)), 144–146 (1966)
10. Han, Sung-Min: Present state and challenges concerning welfare systems of the assistive technology device. Braill Times, No.632, pp. 28–37 (2010)
11. Gibson, J.J.: Observations on active touch. Psychol. Rev. **69**, 477–490 (1962)
12. Gibson, J.J.: The Senses Considered as Perceptual Systems. Houghton Mifflin, Boston (1966)
13. Katz. D.: The World of Touch. Lester E. Krueger (Trans.), L. Erlbaum (1989)
14. Loomis, J.M., Klatzky, R.L., Lederman, S.J.: Similarity of tactual and visual picture recognition with limited field of view. Perception **20**(2), 167–177 (1991)
15. Lederman, S.J., Klatzky, R.L.: Hand movements; a window into haptic object recognition. Cogn. Psychol. **19**, 342–368 (1987)
16. Lindsay, P.H., Nouman, D.A.: Human Information Processing, 2nd edn, p. 278. Academic, New York (1977)
17. Sakai, T., Handa, T., Matsumura, K., Kanatsugu, Y., Hiruma, N., Ito, T.: Information Barrier-free Presentation System for Visually Impaired Users. CSUN Technology & Persons with Disabilities Conference (2007)
18. Sadato, N.: How the blind "see" braille: lessons from functional magnetic resonance imaging. Neuroscientist **11**(6), 577–582 (2005)
19. Rayner, K., Pollatsek, A.: The Psychology of Reading. Prentice-Hall, Englewood Cliffs (1989)
20. Sadato, N., Pascual-Leone, A., Grafman, J., Deiber, M.P., Ibanez, V., Hallett, M.: Neural networks for Braille reading by the blind. Brain **121**, 1213–1229 (1998)
21. Watanabe, T., Kobayashi, M., Ono, S., Yokoyama, K.: Practical use of interactive tactile graphic display system at a school for the blind. In: Current Developments in Technology-Assisted Education (2006)
22. Kobayashi, M., Watanabe, T.: Communication system for the blind using tactile displays and ultrasonic pens – MIMIZU. In: Computers Helping People with Special Needs, pp. 731–738. Springer, New York (2004)
23. Massie, T. H., Salisbury, J. K.: The PHANToM haptic interface: a device for probing virtual objects. Proceedings of the ASME Dynamic Systems and Control Division, (1994)

Chapter 15
Simple Tactile Technologies Utilizing Human Tactile and Haptic Characteristics

Yoshihiro Tanaka and Akihito Sano

Abstract Human tactile and haptic characteristics promote the development of tactile devices that are simple and effective. In addition, issues and ideas for tactile devices can be found from present needs and conditions of manufacturing and medical fields. This chapter presents tactile technologies that are based on human tactile and haptic characteristics. First, wearable and hand-held tactile sensors, tactile enhancing devices without an electronic element, and tactile sensing with a human finger are introduced. They were developed for surface irregularity inspections and texture evaluations. Then, a palpation system with sensory feedback for laparoscopic surgery is introduced. Finally, tactile design based on tactile illusion is introduced. It can combine a hard material and soft feel. These tactile technologies utilize the human haptic ability and structural enhancing mechanism, and human tactile perception.

Keywords Bidirectionality • Mechanical principles • Self-reference • Tactile enhancement • Tactile illusions

15.1 Introduction

The tactile sense is important in manufacturing and medical operations. The touch of a well-trained worker is still needed for complex operations although it has not been quantified. In medical fields, minimally invasive surgery has been developed for its merits of easy recovery and less scarring, but such surgery limits the tactile sense of the surgeon. The tactile sense is one available modality for the diagnosis of an affected area such as the area of a tumor, which is generally stiffer than normal tissue, or the discrimination of tissues such as blood vessels. For example, if surgeons can palpate tissue in endoscopic surgery, they can determine a more precise cutting line of tissue to remove the tumor and can operate on an early-stage tumor that cannot be detected only visually in surgery. The quantification and design of tactile feelings is also important in various fields. The tactile feelings of products

Y. Tanaka (✉) • A. Sano
Nagoya Institute of Technology, Gokiso-cho, Showa-ku, Nagoya 466-8555, Japan
e-mail: tanaka.yoshihiro@nitech.ac.jp; sano@nitech.ac.jp

H. Kajimoto et al. (eds.), *Pervasive Haptics*, DOI 10.1007/978-4-431-55772-2_15

have been designed by trial and error since the mechanism of tactile perception has not been revealed. The design of tactile feelings might develop and expand the quality and flexibility of product design. Diagnosis by palpation and evaluation in rehabilitation has been employed subjectively by surgeons and patients. The quantification of tactile sensations might result in operations that provide better outcomes.

Such applications of tactile technologies require many issues to be addressed according to the situation. Here, human haptic and tactile characteristics often provide good solutions. These characteristics promote the development of tactile devices that are simple and effective. This chapter presents several tactile sensors, devices, design utilizing the human haptic ability and structural enhancing mechanism, and human tactile perception. First, characteristics of haptic perception focused in this chapter are explained. Next, several tactile sensors and devices for surface irregularity inspections and texture evaluations in manufacturing are introduced. Then, a palpation system with sensory feedback for laparoscopic surgery is introduced. Finally, tactile design based on tactile illusion is introduced.

15.2 Characteristics of Haptic Perception

15.2.1 Self-Reference

Tactile sensations are derived from skin deformations and thermal changes generated by mechanical interactions between the skin and an object. Whereas neural signal processing is important for tactile and haptic perception, exploratory properties such as the contact force and scanning velocity and mechanical properties such as the stiffness and shape of the skin and nails are also important factors affecting tactile sensations. For example, as shown in Fig. 15.1, the texture information of the contact object is transferred on the skin. Even for the same object and exploratory movements, the mechanical stimulation of the skin might differ among individuals. Thus, tactile sensation has the characteristic of the self-reference.

15.2.2 Bidirectionality

In haptic perception, the relationship between tactile sensations and exploratory movements is bidirectional as shown in Fig. 15.1. Tactile sensations are generated by exploratory movements, but tactile sensations also influence exploratory movements. Lederman and Klatzky demonstrated that humans can select different exploratory movements to obtain objective information such as the roughness, shape, and temperature of the target [1]. This shows a qualitative aspect of the

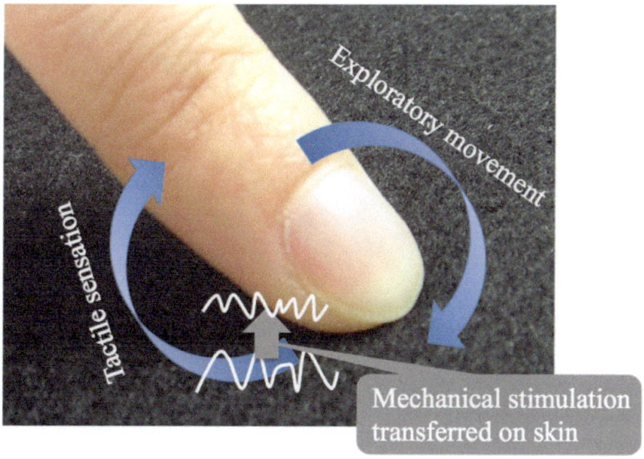

Fig. 15.1 Self-reference and bidirectionality on haptic perception. We uses skin deformations and thermal changes for tactile sensations, and mechanical stimulation by the contact object is transferred on the skin (self-reference). Tactile sensations are generated by exploratory movements, but tactile sensations also influence exploratory movements (bidirectionality)

bidirectionality. Related works showed a quantitative aspect of the bidirectionality. Humans can adjust exploratory movements to improve accuracy and effectiveness. Smith et al. demonstrated that humans use a larger force for a convex detection than for a concave detection [2]. Drewing et al. reported that repetition of the same movements enhances the sensitivity in roughness discrimination [3]. Kaim and Drewing reported that humans adjust their exerted indentation force depending on the task difficulty and object stiffness [4]. Tanaka et al. demonstrated that humans use a larger variation of the exerted force for smooth roughness discrimination than for rough roughness discrimination [5]. These behaviors are reasonable for the characteristics of mechanoreceptors and signal processing. Haptic bidirectionality contributes to make the best use of mechanoreceptors. This indicates that the sensor's capability might be enhanced or expanded by applying human haptic bidirectionality.

15.3 Tactile Inspection in Manufacturing

Surfaces are often inspected to detect irregularity and to evaluate texture in manufacturing; e.g., bump detection for automobile body and roughness measurement for product surface. Depending on the shape, size, and conditions of the target surface, the touch of a well-trained worker is still required to detect small irregularities and surface undulations and to evaluate texture and roughness. Conventional surface inspection devices such as laser displacement sensors and dial gauges and image processing are generally difficult to apply in practice since the setup and/or measurements are often time consuming or the measurements are sensitive to surface conditions such as a gloss condition. In addition, target surfaces are often curved or small. Inspections using a robot arm are also difficult since devices tend to be large.

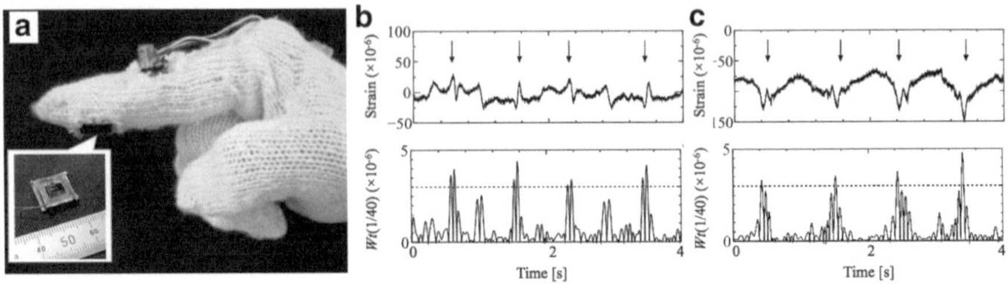

Fig. 15.2 Finger-mounted tactile sensor for surface irregularity detection. (**a**) Overview of the sensor. Two strain gauges are attached to the center of a flexible rubber sheet surrounded by a rubber wall. The user, wearing the sensor-embedded work glove, touches the rubber wall with his/her fingertip and scans an object surface. (**b**) and (**c**) Examples of the sensor output and its wavelet transformation for a flat surface (*left panels*) and a curved surface of $R = 35$ mm (*right panels*) with a small bump (about 60 μm in height and 20 mm in diameter). *Arrows* show passes over the small bump

15.3.1 Wearable Tactile Sensor

Hand-held or finger-mounted tactile sensors are relatively easy to apply in practice and are intuitive to use [6, 7]. Devices can be small and use a human's ability to make clever manipulations since the user moves the sensors. Real-time sensing and display can provide an intuitive understanding of the location of the target object. Hand-held tactile sensors are suitable for applications in large areas and finger-mounted tactile sensors are suitable for applications involving complex structures.

Figure 15.2a shows a finger-mounted tactile sensor developed by Tanaka et al. [7]. The sensor consists of a flexible rubber sheet on which two strain gauges are attached and a rubber frame that is set on the rubber sheet, and the sensor is embedded in a work glove. The sensor can detect surface irregularity generated on the rubber sheet by recording the difference between the outputs from the two strain gauges as the sensor scans the target surface. The user touches the rubber frame with a finger, and the force exerted by the user is applied to the rubber sheet through the surrounding frame. A flexible sheet enclosing silicone oil (0.3-mm thick) mediates the contact between the sensor and object surface. The sensor thus makes stable contact owing to relatively uniform contact pressure and is applicable to curved surfaces. The thickness of the sensor is equivalent to that of the work glove. The sensor employs a wavelet transform to detect surface irregularity from the transient response of the sensor, as shown in Fig. 15.2b. Thus, the finger-mounted tactile sensor is available for the surface irregularity detection. However, this sensor demanded for users to keep the stable contact with the finger during the rubbing the objective surface. The sensor did not sufficiently include the user's haptic bidirectionality since the user perceived tactile sensation diminished by the sensor. If sensors can allow users to touch an object directly, usability and detection capability might be enhanced.

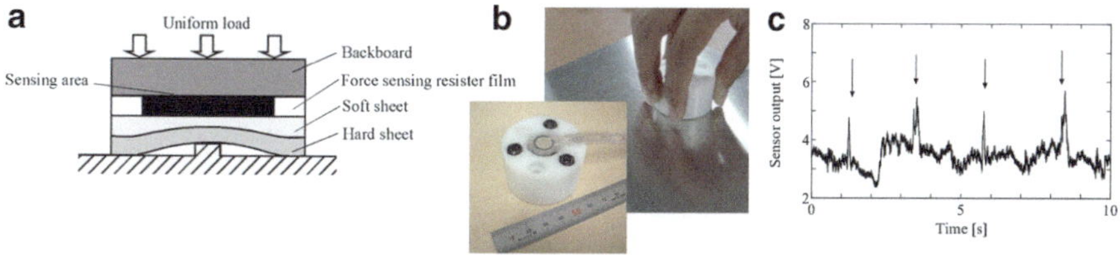

Fig. 15.3 Hand-held tactile sensor for small convex detection. (**a**) Principle of geometric enhancement of a small convex feature using a hard elastic medium. (**b**) Overview of the sensor. A polypropylene sheet and silicone rubber sheet were used as a hard elastic medium and soft medium, and a force-sensing resistor film was used to detect the pressure change by scanning the sensor over the object surface. (**c**) Example of the sensor output for a flat surface with a small convex feature (about 12 μm in height and 0.3 mm by 0.3 mm). *Arrows* show passes over the small convex feature

15.3.2 Mechanical Enhancing of Elastic Medium

Skin is an important factor for tactile sensations from the characteristic of the self-reference. Considering the relation between the skin and mechanoreceptors, the skin mediates the transmission of mechanical stimulation to mechanoreceptors. The skin can be described as a mechanical filter. Enhancing or coding functions of the fingerprint [8, 9], dermal papilla [8, 10], and collagen fibers [11] have been revealed previously. These skin structural mechanisms can be applied to tactile sensors by, for example, the use of an elastic medium, and they can provide simple and robust sensors based on mechanical principles.

A hard elastic medium is effective in detecting small convex features [12]. As shown in Fig. 15.3a, a hard elastic medium can geometrically enhance the profile of a small convex feature. In addition to the use of the hard elastic medium, a soft elastic medium can reduce much smaller irregularities that are not the focus of detection, such as surface roughness in the detection of small convex features. The use of adequate elastic mediums selectively enhances the sensitivity of tactile sensors with a bandpass filter. Figure 15.3b shows an assembled tactile sensor. A polypropylene sheet is used as a hard medium (0.2-mm thick), a silicone rubber sheet as a soft medium (0.2-mm thick), and a force-sensing resistor film to detect pressure change. Experimental results showed that this sensor responds strongly to a small concavity having a height of 12 μm and square dimensions of 250 μm by 250 μm, as shown in Fig. 15.3c [13].

15.3.3 Tactile Contact Lens

The tactile contact lens presented in Fig. 15.4a is a unique tactile enhancing device [14]. The device does not have electronic elements such as sensors and actuators. Users scan surface undulations with the device, as shown in Fig. 15.4a,

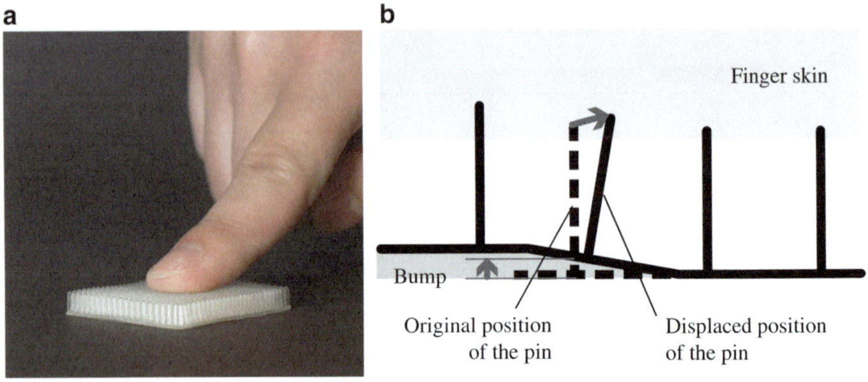

Fig. 15.4 Tactile contact lens. (**a**) Overview of the tactile contact lens composed of a flexible thin sheet and pins. The user presses against the pins of the tactile contact lens and scans the object surface. A small surface undulation is perceived to be larger. (**b**) Enhancement principle of the tactile contact lens. Normal deformation by surface undulation is enhanced and transformed to shear deformation by the lever mechanism and large shear stimulation is provided to the skin

and they perceive small surface undulations to be larger than they actually are. The enhancement principle of the device is the lever mechanism. The device has a pin array and a flexible sheet on the bottom. As shown in Fig. 15.4b, normal deformation by a small surface undulation is transformed and enhanced as shear deformation of the contact skin. It has been reported that humans perceive small bumps through shear forces [15]. Therefore, mechanical enhancement by the device derives a strong perception of surface undulation. We see a similar lever mechanism in the dermal papilla of our skin [10]. In addition to the mechanical enhancing mechanism, we can see that haptic bidirectionality between tactile sensations and exploratory movements is employed by the device. Users perceive tactile sensation and adjust their exploratory parameters such as their contact force and scanning velocity. General tactile enhancing devices often provide extra mechanical stimulation using a tactile display. However, the tactile contact lens functions as both a sensor and display. The tactile contact lens is a device that well utilizes human haptic perception and a simple mechanical enhancing principle.

15.3.4 Tactile Nail Chip

Inspired by the tactile contact lens, a tactile device called the tactile nail chip has been developed [16]. The tactile nail chip slightly bends the nail to change tactile sensitivity. From the characteristic of the self-reference, there is a possibility that the tactile sensitivity can be changed by changing mechanical properties of skin. Jeong et al. proposed tactile enhancement via stiffening of the fingertip by pressing the proximal phalange of the finger [17]. Tanaka et al. noted that tactile perception is

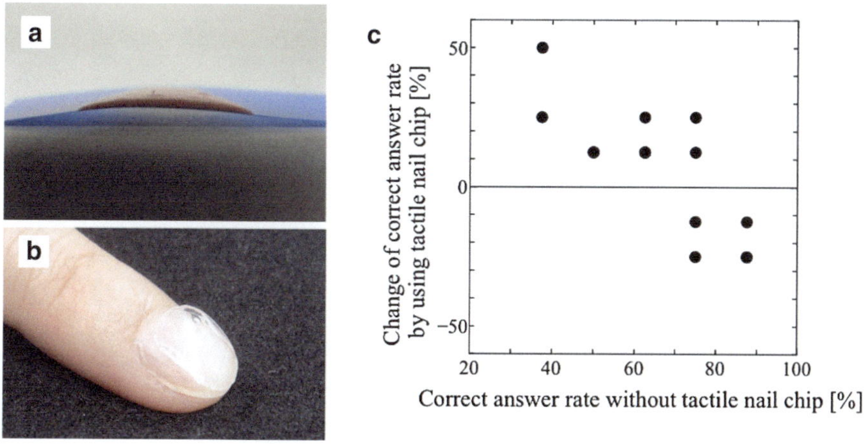

Fig. 15.5 Tactile nail chip. (**a**) Flexible sheet bent inward with the application of nail polish. (**b**) Overview of the tactile nail chip. The tactile sensation is affected by bending the nail inward. (**c**) Result on discrimination test on passively applied loads. Four loads, 15, 20, 30, and 60 [gf], were applied on the forefinger. Each *plot* indicates participant

sometimes affected when nail polish is applied to a nail [16]. Experiments revealed that a polypropylene sheet bent inward when nail polish in general use was applied to the sheet in a raised state. This indicates that the tactile perception on the fingertip can be affected by nail deformation due to the application of nail polish. The skin properties including the nail might affect tactile perception from the consideration of the self-reference. Thus, Tanaka et al. proposed the tactile nail chip that changes the tactile sensitivity of the fingertip by bending the nail [16]. This device has a very simple principle and allows the user to touch objects with their bare fingertip. The bidirectionality is also established. In the case of the tactile nail chip presented in Fig. 15.5a, a commercial nail tip was attached at an angle of 90° relative to its orientation in general use with double-sided tape to bend the nail inward. By bending the nail inside, almost all participants in Tanaka et al.'s study felt a change in tactile perception. Some participants reported a feeling that stimuli felt on the finger pad transferred to the nail and a feeling that the tactile sensitivity of the finger pad was diminished. Although there were individual differences, experimental results showed changes in weight perception as shown in Fig. 15.5c [16] and enhancement of texture identification [18].

Influences of the skin properties on tactile sensations and sensitivities have not been sufficiently revealed yet. The stress distribution in the fingertip including the nail was investigated in a photoelasticity experiment. Artifactual human fingers were fabricated from silicone rubber using a mold of a human finger, as shown in Fig. 15.6. A finger without a nail (Fig. 15.6a), a finger with a nail attached to the finger (Fig. 15.6b), and a finger with a nail embedded like a human nail (Fig. 15.6c) were prepared. In addition, the finger presented in Fig. 15.6c with the tactile nail chip (Fig. 15.6d) was prepared. Acrylic bars and commercial nail tips

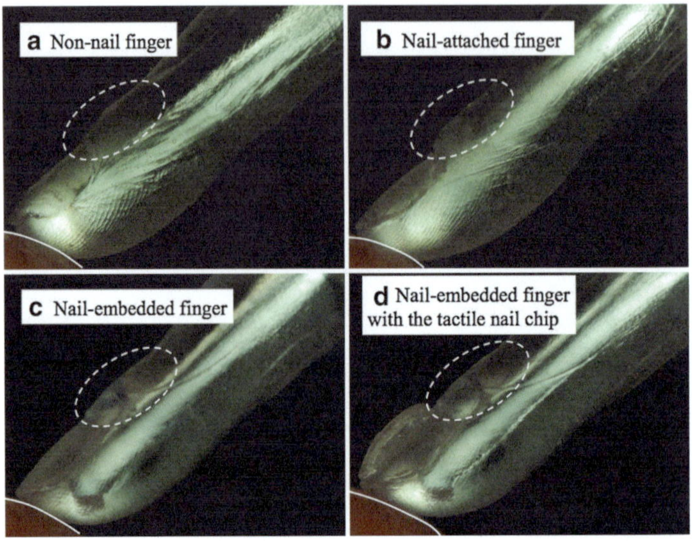

Fig. 15.6 Experimental results of the photoelasticity of artifactual human fingers. The area where stress is applied is bright. (**a**) Non-nail finger. (**b**) Nail-attached finger. (**c**) Nail-embedded finger. (**d**) Nail-embedded finger with the tactile nail chip. The *dotted circle* and *solid line* indicate the area of the nail root and the boundary with a rigid plate

were used as the bone and nail, respectively. The finger pad of each artifactual finger was pressed against a rigid plate. The stress distribution was observed employing circular polarization. Results in Fig. 15.6c, d show two specific stress distributions. One is the stress concentration observed on the nail root in Fig. 15.6c, d. This distribution shows that a load applied on the finger pad generates stress on the nail root. Here, Merkel cells were found to distribute in the matrix and proximal nail fold [19]. The stress distribution on the nail root is consistent with the mechanoreceptor distribution. Thus, this finding indicates that the stress transferred on the nail root is used for tactile perception. Mechanical properties of the nail such as thickness and shape affect the transfer of stress from the finger pad to the nail root. The second distribution is enhancement of the stress concentration in the finger pad in Fig. 15.6d. Stress in the finger pad appears larger in Fig. 15.6d than in Fig. 15.6c. This result suggests the possibility that bending the nail inward increases the stiffness of the finger. The obtained results account for the change in sensitivity of the fingertip when bending the nail. Although individual differences in the mechanical properties of the nail and the effect of bending the nail require further investigation, the tactile nail chip has the potential to control the sensitivity of the fingertip. The control of tactile sensitivity might allow the reduction of individual differences in tactile sensations and the enhancing or diminishing of tactile sensations.

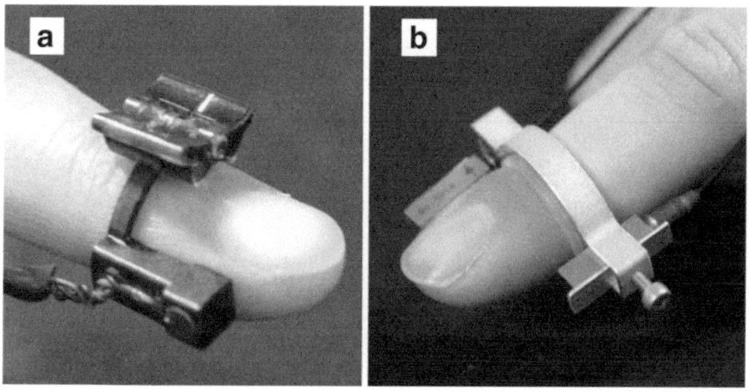

Fig. 15.7 Skin-propagated vibration sensors. (**a**) Sensor with two microphones. One microphone is in contact with the user's skin and the other is set on the back of the contact microphone facing away from the skin to measure the ambient sound. (**b**) Sensor with two accelerometers. One accelerometer measures skin-propagated vibration and finger vibration, and the other on the ring measures the finger vibration

15.3.5 Tactile Sensing with a Human Finger

Following the tactile sensor shown in Fig. 15.2, wearable tactile sensors having the properties of self-reference and bidirectionality of haptic perception were proposed by Tanaka et al. [20, 21]. Here, human fingers are employed as a sensing probe. The sensors allow the user to touch an object with their bare fingers and skin deformations and vibrations are measured. As related works, wearable contact force sensors that use the nail [22] and the side of the finger pad [23], measurement of skin-propagated vibration for the basic study of haptic perception [24, 25], and motion detection by measuring vibration propagating to the nail [26] have been proposed. Figure 15.7 shows finger-mounted tactile sensors that detect sound and vibration associated with mechanical interaction between the skin and the object using microphones [20] and that detect vibration propagating to the side skin from the contact finger using accelerometers [21].

Figure 15.8a shows an example of the sensor output intensity for sheets of sandpaper with different roughness attached on a flat or curved ($R = 42\,\mathrm{mm}$) surface for one participant. Although the sensor output depends on the user owing to different exploratory movements and different mechanical properties of the skin, there were good relations between the sensor output and roughness for most users. Moreover, it was observed that the sensor outputs of the same sample were similar for the flat and curved surfaces. It appears that participants rubbed their fingers thoroughly on the surface to evaluate the roughness. This result shows that a very simple sensor is available for evaluating roughness employing human haptic characteristics. Figure 15.8b shows examples of the sensor output for the same sandpaper when two participants employed a constant exerted force and scanning

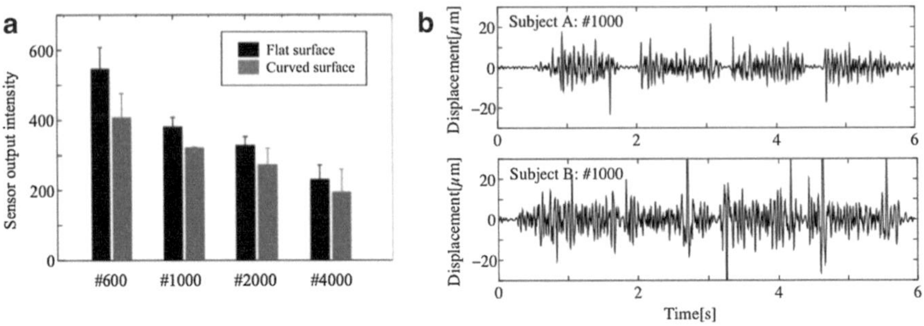

Fig. 15.8 Examples of outputs from skin-propagated vibration sensors. (**a**) Output for sheets of sandpaper with different roughness attached on a flat or curved ($R = 42$ mm) surface for one participant and the sensor using microphones. (**b**) Output for the same sandpaper using a constant exerted force and scanning velocity for two participants and the sensor using accelerometers

velocity. The sensor output differed between the participants even when the object and exploratory movement were the same. Since the sensor output reflects each user's exploratory movement and mechanical properties, the sensor could be useful in evaluating the tactile sensation and studying the haptic skill of individuals.

15.4 Palpation During Minimally Invasive Surgery

15.4.1 Requirements of Application

Medical devices must meet specific safety standards to protect against infection and the leakage of current. They must be able to be physically or chemically sterilized using, for example, steam or ethylene oxide gas. Devices without electronic elements are best in terms of preventing current leakage. Devices that do not generate electronic and magnetic noise are compatible for use with a radio knife and magnetic resonance imaging. Medical tactile sensors providing such safety have been proposed for minimally invasive surgery; e.g., a non-contact stiffness sensor using an air jet [27], a stiffness sensor using an air cushion [28], a force sensor using a Moire fringe [29], an optical force sensor [30], and a catheter force sensor that employs the bending of a wire in a Y-connector [31]. The measurement of stiffness is useful in localizing a tumor because a tumor is generally known to be stiffer than normal tissue. In addition, Tanaka et al. developed a tactile sensor that uses an expanding balloon to measure stiffness and surface conditions such as friction [32]. Its sensing principle is based on the sensing mechanism by which a human's finger can detect friction only through touch [33]. Here, the palpation skill of surgeons is still not sufficiently clear. Although many tactile devices have been tried, the palpation of surgeons is still not well substituted by any device.

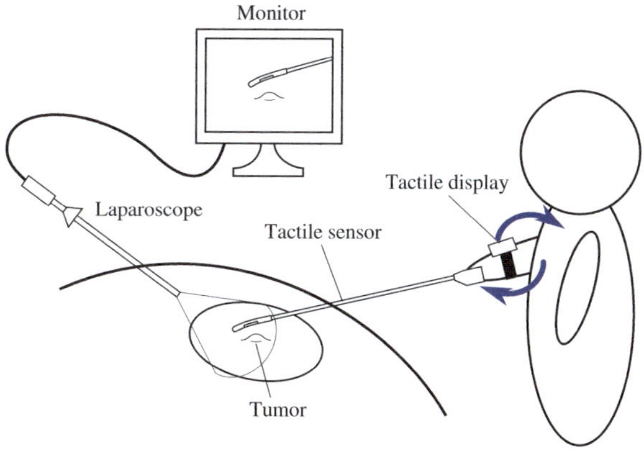

Fig. 15.9 Schematic illustration of the palpation system using a surgeon's haptic bidirectionality. The surgeon recognizes the sensor output through feedback from the tactile display

15.4.2 Sensor System with Sensory Feedback

Many medical tactile sensors have sensor-centered characteristics such as high sensitivity, high safety, and high flexibility. However, as a different approach, Tanaka et al. proposed a tactile sensor system that includes haptic bidirectionality for laparoscopic surgery [34]. Figure 15.9 shows the basic idea of the system. The sensor system comprises forceps having a tactile sensing function and a tactile display giving feedback of the tactile sensor output to the surgeon who manipulates the forceps. Thus, the surgeon perceives the tactile sensor output as a tactile sensation and adjusts the exploratory parameters (e.g., contact force, scanning velocity, and posture) of the forceps according to the perceived tactile sensation. When the surgeon perceives a clear tactile sensation from the tactile display by the sensor output for tumor, the sensor output is also clear. By including haptic bidirectionally in the system, the sensing condition can be optimized by the surgeon and the surgeon can identify the sensor output for a tumor. Such a system utilizes the haptic ability of the surgeon. Here, the tactile display is not required to give a realistic tactile sensation of the tumor. It is important that surgeons can discriminate the difference in the sensor output depending on the contact conditions and objects' properties. The visual and audial senses are also available for sensory feedback, but visual feedback might interrupt the surgeon watching an endoscopic monitor and audial feedback might interrupt communication with staff. The tactile sense has the same modality and is reasonable for practical applications.

A tactile sensor that employs acoustic reflection was assembled for the forceps having a tactile sensing function in the above-mentioned system [35]. Figure 15.10 presents the overview and geometry of the sensor, the sensor output characteristic, and an example of the sensor output for a tumor phantom. The structure is very simple. The sensor comprises a handle with a microphone and a speaker, an aluminum tube, and a sensor tip with an elastic deformable cavity as the sensing area. The shape and dimensions of the sensor are similar to those of actual general

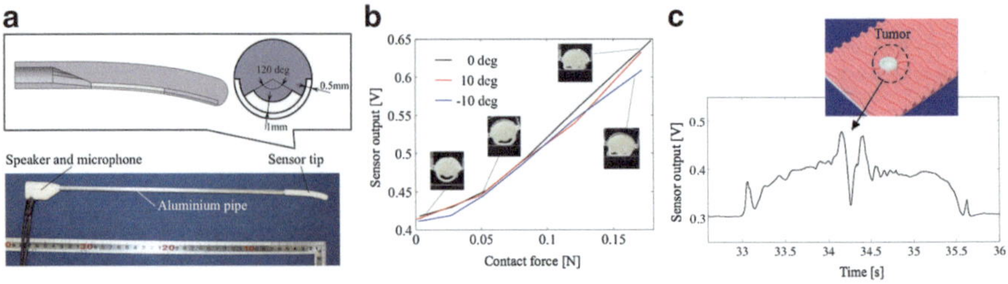

Fig. 15.10 Forceps-type tactile sensor using acoustic reflection. (**a**) Overview and geometry of the sensor. The sensor comprises a handle having a microphone and a speaker, an aluminum tube, and a sensor tip having an elastic deformable cavity as the sensing area. The contact force is measured from the acoustic wave generated in the tube. (**b**) Characteristic of the sensor for a normal contact force. (**c**) Typical sensor output obtained by scanning a phantom stomach wall with a tumor. The outer smooth surface was palpated in consideration of practical use

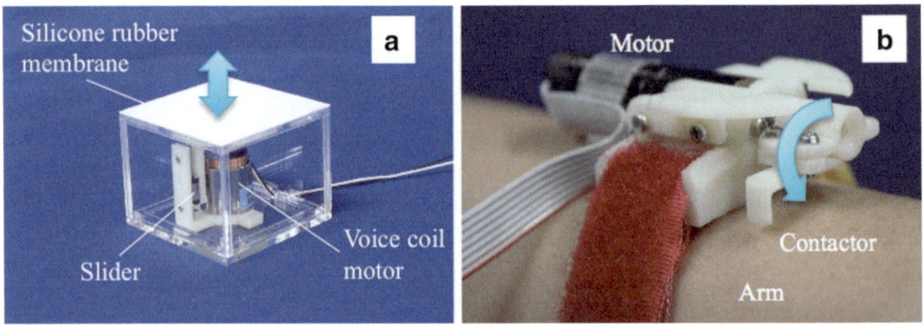

Fig. 15.11 Tactile display for the palpation system. Indentation stimulation is applied to the skin according to the sensor output. (**a**) Ground-based tactile display for the other hand using a voice coil motor. (**b**) Arm-mounted tactile display using a servomotor

forceps. The acoustic wave in the tube, which is detected by the microphone, is the superposition of the input wave from the speaker and two waves reflected at the closed edge of the sensor tip and the projection of the elastic deformable cavity generated by deformation yielded by contact with an object. Therefore, the contact force can be estimated from the acoustic wave in the tube. The sensor has no electrical elements in the tissue contact area, can be sterilized, and can be disposed because of its low cost. Figure 15.10c shows that the sensor well responds to a phantom of a common early-stage gastric cancer on a stomach wall.

An arm-mounted tactile display using a servomotor [36] and a ground-based tactile display for the other hand using a voice coil motor [34] provide the tactile display, as shown in Fig. 15.11. Both tactile displays give an indentation stimulation on the skin according to the sensor output. The tactile feedback is not realistic but the surgeon can recognize the sensor output sufficiently well. Fundamental experiments showed that both tactile displays are available to the palpation system. The surgeon can adjust and optimize the manipulation of the sensor-mounted forceps for tumor detection on the basis of the perceived sensor output.

15.5 Tactile Design

15.5.1 Potential of Tactile Design

There are many tactile and haptic illusions [37]. Such illusions help to explain human tactile and haptic perceptions. In addition, illusions have potential use in product design based on tactile or haptic perception. The velvet hand illusion is a tactile illusion that hard-wire derives a smooth and soft feeling like that of velvet. This illusion indicates that soft materials are not always required to generate a soft feeling and there is the possibility that even hard materials generate a soft feeling. Direct methods have been applied in the general tactile design of products; e.g., leathers are used on a product surface to give a soft feeling. Recently, soft-feel paint including elastic materials such as urethane resins, has been developed to give a soft feeling on a product surface. However, human tactile feelings are based on mechanical interactions, as mentioned in the self-reference. Thus, the latitude of the tactile design of products might be greater when human tactile and haptic perceptions are applied to product design.

15.5.2 Soft Feeling of Geometric Dimples on a Hard Surface

Automobile interior parts are often made of plastic materials and have patterned indented surfaces. Such patterned indented surfaces often mimic the grain of leather visually. In general, the texture feelings of plastic interior parts are evaluated as being poor. However, the feelings depend on the patterned indented surfaces, and one patterned indented surface has provided a good feeling. On the basis of that pattern, prototype plates with geometric patterns of dimples were fabricated from acrylonitrile butadiene styrene plastic, which is a popular material for automobile interior parts. Figure 15.12 presents drawings of the prototype plates. The

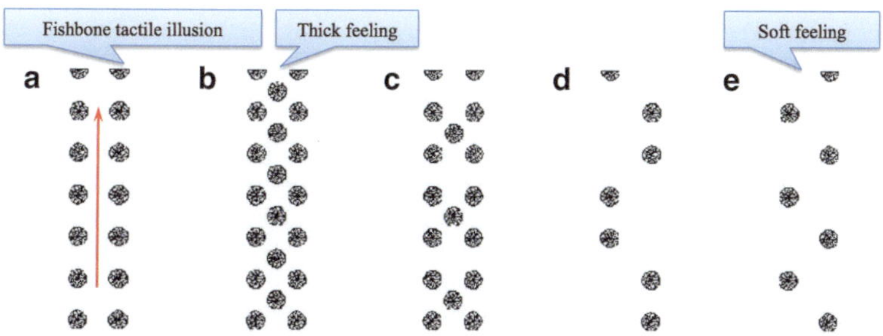

Fig. 15.12 Drawings of test samples that have geometric patterns of dimples. Test samples were fabricated from acrylonitrile butadiene styrene. Each sample has a different feeling when its surface is rubbed with fingers. In particular, plate (**e**) has a soft feel like that of leather

dimensions of a dimple were a diameter of about 2.4 mm and depth of about 0.4 mm for all plates. Participants scanned the center between the dimpled areas in a back-and-forward motion with their fingertips, as shown by the arrow in Fig. 15.12. The plate shown in Fig. 15.12a provided the fish bone tactile illusion [38]. Participants perceived an indentation line in the contact area and bank lines along the two rows of discrete dimples. The plate shown in Fig. 15.12b felt thicker than the other plates. By expanding the distance between dimples, the plate shown in Fig. 15.12e provided a soft-feeling texture that felt like that of leather. There is basically dimple perception, but a feeling of surface undulation is derived by alternately arranging dimples. Here, the perceived depth of the dimple is less than the actual depth. When the geometric pattern shown in Fig. 15.12e was applied to a large area, the surface had a soft and painted feeling. The perception mechanism of the leather-like soft feel of the geometrically patterned dimples has not yet been revealed but the after-effect [39] might be an important factor.

The described technology has great potential to combine a hard material and soft feel. Tactile feelings do not completely correspond to simple physical parameters. The technology shows that tactile design based on human perception expands the possibility of product design.

15.6 Conclusion

This chapter presented tactile technologies that are based on human tactile and haptic characteristics. Human tactile and haptic characteristics promote tactile devices that are simple and effective. Simple and effective tactile devices can be applied to meet the present requirements of manufacturing and medical applications. Thus, human tactile and haptic characteristics are important to the development of tactile technologies, and associated issues and ideas can be found from present needs and conditions of manufacturing and medical fields. Both basic and applied research promote the development of haptics and effective technologies.

References

1. Lederman, S.J., Klatzky, R.L.: The hand as a perceptual system. In: Connolly, K.J. (ed.) The Psychobiology of the Hand, pp. 16–35. Cambridge University Press, New York (1998)
2. Smith, A.M., Gosselin, G., Houde, B.: Deployment of fingertip forces in tactile exploration. Exp. Brain Res. **147**, 209–218 (2002)
3. Drewing, K., Lezkan, A., Ludwig, S.: Texture discrimination in active touch: effects of the extension of the exploration and their exploitation. In: Proceedings of the IEEE World Haptics Conference 2011, pp. 215–220. Istanbul (2011)
4. Kaim, L., Drewing, K.: Exploratory strategies in haptic softness discrimination are tuned to achieve high levels of task performance. IEEE Trans. Haptics **4**(4), 242–252 (2011)

5. Tanaka, Y., Bergmann Tiest, W.M., Kappers, A.M.L., Sano A.: Contact force and scanning velocity during active roughness perception. PLOS ONE **9**(3), e93363 (2014)
6. Kikuuwe, R., Sano, A., Mochiyama, H., Takesue, N., Fujimoto, H.: A tactile sensor capable of mechanical adaptation and its use as a surface deflection detector. In: Proceedings of the 3rd IEEE Conference on Sensors, pp. 256–259. Vienna (2004)
7. Tanaka, Y., Sato, H., Fujimoto, H.: Development of a finger-mounted tactile sensor for surface irregularity detection. In: Proceedings of the IEEE/RSJ International Conference on Intelligent Robots and Systems, pp. 690–696. Nice (2008)
8. Maeno, T., Kobayashi, K., Yamazaki, N.: Relationship between the structure of human finger tissue and the location of tactile receptors. JSME Int. J. Ser. C **41**, 94–100 (1998)
9. Scheibert, J., Leurent, S., Prevost, A., Debregeas, G.: The role of fingerprints in the coding of tactile information probed with a biomimetic sensor. Science **323**, 1503–1506 (2009)
10. Cauna, N.: Nature and functions of the papillary ridges of the digital skin. Anat. Rec. **119**, 449–468 (1954)
11. Tanaka, Y., Ito, T., Hashimoto, M., Fukasawa, M., Usuda, N., Sano, A.: Collagen fibers induce expansion of receptive field of Pacinian corpuscles. Adv. Robot. (2015). doi:10.1080/01691864.2014.1003194
12. Sano, T., Kikuuwe, R., Tanaka, Y., Fujimoto, H.: Enhancing tactile sensor's capability using elastic medium. In: Proceedings of the 24th Annual Conference of the RSJ, 1C22. Okayama (2006)
13. Tanaka, Y., Kinoshita, D., Fujimoto, H.: Tiny protrusion detecting sensor with force sensing resistor film. In: Proceedings of the 25th Annual Conference of the RSJ, 3O16. Chiba (2007)
14. Kikuuwe, R., Sano, A., Mochiyama, H., Takesue, N., Fujimoto, H.: Enhancing haptic detection of surface undulation. ACM Trans. Appl. Percept. **2**(11), 46–67 (2005)
15. Hayward, V., Cruz-Hern'andez, J.M.: Tactile display device using distributed lateral skin stretch. In: Proceedings of 8th Symposium on Haptic Interfaces for Virtual Environment and Teleoperator Systems, pp. 1309–1314. Orlando (2000)
16. Tanaka, Y., Sano, A., Ito, M., Fujimoto, H.: A novel tactile device considering nail function for changing capability of tactile perception. In: Proceedings of EuroHaptics 2008, pp. 543–548. Madrid (2008)
17. Jeong, H., Higashimori, M., Kaneko, M.: Improvement of touch sensitivity by pressing. In: Proceedings of International Conference of the IEEE Engineering in Medicine and Biology Society, pp. 2409–2414. Vancouver (2008)
18. Tanaka, Y., Sano, A., Fujimoto, H.: Effect of tactile nail chip for tactile sensitivity. In: Proceedings of 1st International Conference on Applied Bionics and Biomechanics. Venice (2010)
19. Moll, I., Moll, R.: Merkel cells in ontogenisis human nails. Arch. Dermatol. Res. **285**, 366–371 (1993)
20. Tanaka, Y., Horita, Y., Sano, A., Fujimoto, H.: Tactile sensing utilizing human tactile perception. In: Proceedings of the IEEE World Haptics Conference 2011, pp. 621–626. Istanbul (2011)
21. Tanaka, Y., Horita, Y., Sano, A.: Finger-mounted skin vibration sensor for active touch. In: Isokoski, P., Springare, J. (eds.) Haptics: Perception, Devices, Mobility, and Communication. Lecture Notes in Computer Science, vol. 7283, pp. 169–174. Springer, Berlin/Heidelberg (2012)
22. Mascaro, S.A., Asada, H.H.: Photoplethysmograph fingernail sensors for measuring finger forces without haptic obstruction. IEEE Trans. Robot. Autom. **17**(5), 698–708 (2001)
23. Nakatani, M., Shiojima, K., Kinoshita, S., Kawasoe, T., Koketsu, K., Wada, J.: Wearable contact force sensor system based on fingerpad deformation. In: Proceedings of the IEEE World Haptics Conference 2011, pp. 323–328. Istanbul (2011)
24. Bensmaia, S., Hollins, M.: Pacinian representations of fine surface texture. Attent. Percept. Psychophys. **67**(5), 842–854 (2005)
25. Delhaye, B., Hayward, V., Lefevre, P., Thonnard, J.L.: Texture-induced vibrations in the forearm during tactile exploration. Front. Behav. Neurosci. **6**(37), (2012)

26. Makino, Y., Murao, T., Maeno, T.: Life log system based on tactile sound. In: Proceedings of EuroHaptics 2010, Part I. Lecture Notes in Computer Science, vol. 6191, pp. 292–297. Amsterdam (2010)
27. Kaneko, M, Kawahara, T.: Co-axis type non-contact impedance sensor. In: Proceedings of the IEEE International Conference on Robotics and Automation, pp. 709–714. Barcelona (2004)
28. Zbyszewski, D., Althoefer, K., Seneviratne, L., Bhaumik, A.: Tactile sensing using a novel air cushion sensor–a feasibility study. In: Proceedings of the IEEE/RSJ International Conference on Intelligent Robots and Systems, pp. 41–46. Nice (2008)
29. Takaki, T., Omasa, Y., Ishii, I., Kawahara, T., Okajima, M.: Force visualization mechanism using a moire fringe applied to endoscopic surgical instruments. In: Proceedings of the IEEE International Conference on Robotics and Automation, pp. 3648–3653. Anchorage (2010)
30. Peirs, J., Clijnen, J., Reynaerts, D., Brussel, H.V., Herijgers, P., Corteville, B., Boone, S.: A micro optical force sensor for force feedback during minimally invasive robotic surgery. Sens. Actuators A: Phys. **115**(2–3), 447–455 (2004)
31. Nagano, Y., Sano, A., Sakaguchi, M., Fujimoto, H.: Development of force sensor for extra-fine and long objects: application in endovascular coil embolization. Trans. Soc. Instrum. Control Eng. **44**(3), 278–284 (2008)
32. Tanaka, Y., Yu, Q., Doumoto, K., Sano, A., Hayashi, Y., Fujii, M., Kajita, Y., Mizuno, M., Wakabayashi, T., Fujimoto, H.: Development of a real-time tactile sensing system for brain tumor diagnosis. Int. J. Comput. Assist. Radiol. Surg. **5**(4), 359–367 (2010)
33. Maeno, T., Kawai, T., Kobayashi, K.: Friction estimation by pressing an elastic finger-shaped sensor against a surface. IEEE Trans. Robot. Autom. **20**(2), 222–228 (2004)
34. Tanaka, Y., Nagai, T., Sakaguchi, M., Fujiwara, M., Sano, A.: Tactile sensing system including bidirectionally and enhancement of haptic perception by tactile feedback to distant part. In: Proceedings of the IEEE World Haptics Conference 2013, pp. 145–150. Daejeon (2013)
35. Tanaka, Y., Fukuda, T., Fujiwara, M., Sano, A.: Tactile sensor using acoustic reflection for lump detection in laparoscopic surgery. Int. J. Comput. Assist. Radiol. Surg. **10**(2), 183–193 (2015)
36. Tanaka, Y., Aragaki, S., Fukuda, T., Fujiwara, M., Sano, A.: A study on tactile display for haptic sensing system with sensory feedback for laparoscopic surgery. In: Proceedings of the 25th International Symposium on Micro-NanoMechatronics and Human Science. Nagoya (2014)
37. Lederman, S.J., Jones, L.A.: Tactile and haptic illusions. IEEE Trans. Haptics **4**(4), 273–294 (2011)
38. Nakatani, M., Howe, R.D., Tachi, S.: Surface texture can bias tactile form perception. Exp. Brain Res. **208**(1), 151–156 (2011)
39. Vogels, I.M.L.C., Kappers, A.M.L., Koenderink, J.J.: Haptic after-effect of curved surfaces. Perception **25**, 109–119 (1996)

Part IV
Search for New Fields

Chapter 16
Toward the Haptic Interaction in Daily Life

Takuya Nojima

Abstract It is common that usual people rarely have chance to experience emerging technologies. It is also true to the haptic research area. In spite of its several decades of haptic research history, usual people is not familiar with haptic display systems. To make things worse, they have little motivation to use artificial and active haptic signals in their daily lives. In this chapter, three different research systems will be described to broaden the applicability of haptic display systems. Those systems are designed to encourage people using "haptic displays". The first is named SmartTool that intended to enable augmented reality of haptics. The system is capable of changing the haptic feeling of actual objects in a real world, to support manual task and to entertain people. The second system is Tactile Flight Display, to support pilot's task by reducing visual workload. By using this system, the pilot could perceive flight related information through their haptic feeling. While using this system, they can keep the aircraft within a certain altitude based on flight information provided through their haptic feeling. At the same time, the pilot could watch outside for surveillance, not visual instruments inside. The last one is stravigation, a vibrotactile navigation system to entertain usual people. By reducing visual information and providing navigational information through haptic feeling, people could free from watching their navigation device when they walk around unknown place. That contributes to improving the quality of experience of sightseeing. In addition, this method also succeeded in users to feel some sort of sense of delight while they are walking with this navigation system.

Keywords Augmented reality • Vibrotactile • Mobile • Navigation • Pilot • Airplane

---

T. Nojima (✉)
University of Electro-Communications, Chofu, Japan
e-mail: tnojima@nojilab.org

H. Kajimoto et al. (eds.), *Pervasive Haptics*, DOI 10.1007/978-4-431-55772-2_16

16.1 Introduction

Haptic research in engineering has several decades of history, yet there is still much research needed to popularize haptic display technology. In fact, general consumers have little motivation to use artificial and active haptic signals in their daily lives.

To deal with this issue, searching for a new field is important to increase the chances for general consumers to experience the benefits of "using haptic signals/sensations", in our daily life. In this chapter, three studies motivating general consumers to use haptic display technology are described.

The first trial was "Augmented reality of haptics", which intended to enhance actual haptic sensation by merging artificial haptic signal, especially for a certain kinds of manual work with tools. Haptic signal is an important clue, especially for a certain kinds of manual works, e.g., milling a block of wood, cutting a paper. It should be able to reduce the working load if the haptic signal could convey much more information to the worker.

The second was "tactile flight display", more task-oriented way to use the haptic signal. When flying an aircraft, pilots' visual workload often becomes high. To improve the flight safety, reducing the visual workload is required. To comply with this issue, many kinds of visual display and GUI design are proposed. However, pilots still making use of haptic sensation when they fly aircrafts, even though it is equipped with autopilot system. When considering this fact, providing primarily required information through haptic sensation will contribute reducing pilot workload.

The last one was "vibrotactile navigation display", to support explorative sightseeing. The vibrotactile signal is often used for blind people to display character based information, navigational information, etc. Vibrotactile signal is quite important for blind people, but not limited to them. If that method is provided with an appropriate application, a sighted person willing to use such kind of display. This system is one of the trials to be used such kind of method for both blind person and sighted person.

16.2 Augmented Reality of Haptics

Augmented reality (AR) is often thought to be a research/application field concerned with the visual aspects. This is, in fact, not true. When considering its definition [1], we find the concept of AR spans an extremely large field, including not only visual effects, but also smell, taste, hearing, and haptics.

"AR of haptics" in this section means "merging virtual haptic sensation to real haptic sensation", which is derived from the original definition of AR. This definition is definitely different from "haptic (display) technology to be used in AR (visual) applications", which is commonly considered. This technology has the potential to transform haptic sensation of real world objects, which could be effective in supporting tasks in the real world and developing entertainment applications.

Research on the "AR of haptics" began in the early 2000s and is now a popular research field. In this section, a basic system for the "AR of haptics", called the SmartTool, is described to understand this concept.

16.2.1 Overview of the SmartTool

The SmartTool [2] is one of the earliest prototypes realizing the concept of the AR of haptics. It consists of sensor(s) to measure the real environment, a conventional tool to work with, and a haptic display for displaying sensory information through the user's force sensation. The characteristics of the SmartTool allow support of actual tasks in the real world. For example, consider the case of cutting a complex object with a restricted area that should not be cut, and an unrestricted area that can be cut freely. In this case, the SmartTool uses a knife as the conventional tool and sensor(s) to identify the different areas. The sensor on the SmartTool monitors what the tool (the knife in this case) touches. The haptic display on the SmartTool generates a repulsive force when the tool is about to penetrate and damage the restricted area. The generated force has three meanings. The first is to "change the object's haptic feeling". When cutting the unrestricted area, the user perceives the actual haptic sensation only. However, when the user is about to cut the restricted area, the user perceives an augmented haptic feeling, which is a combination of the actual haptic sensation and a virtual haptic feeling induced by the generated repulsive force. This ensures that users feel that the object's haptic feeling has changed. The second meaning is a simple message to the user that the tool is about to damage the restricted area. The final meaning is a "supporting force". The repulsive force can in fact, prevent the user from damaging the area by accident. As a result, the SmartTool assists the user with the actual task by changing the object's haptic feeling.

The SmartTool allows users to use various sensors that are not related to haptic sensation, such as optical, ultrasonic, electrical, and bio-chemical sensors, amongst others, if they suit the purpose of the system. The system displays information from these sensors through human haptic sensation. The system can transform the modality of any kind of sensation into a haptic sensation. Therefore, we can "feel" many different kinds of information (such as optical, ultrasonic, and electrical) through haptic sensations. In other words, the SmartTool is also a system that can "haptize" sensory information to enhance human haptic sensation.

16.2.2 Basic Structure of the SmartTool

16.2.2.1 Haptic Display Component

Figures 16.1 and 16.2 show an overview of the SmartTool and the schematics of the system, respectively. This system has six degrees of freedom (DOF), but only

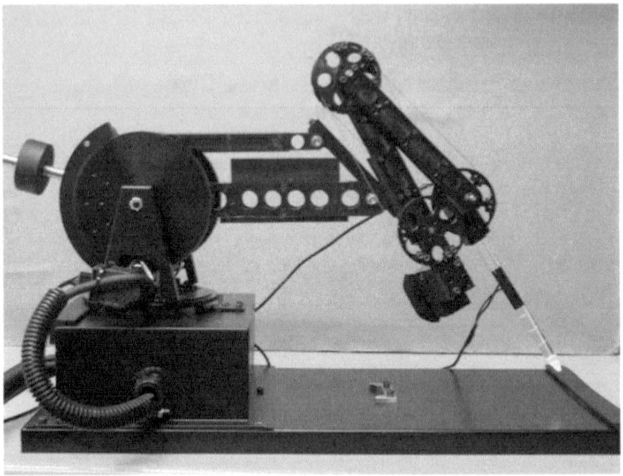

Fig. 16.1 Overview of the SmartTool

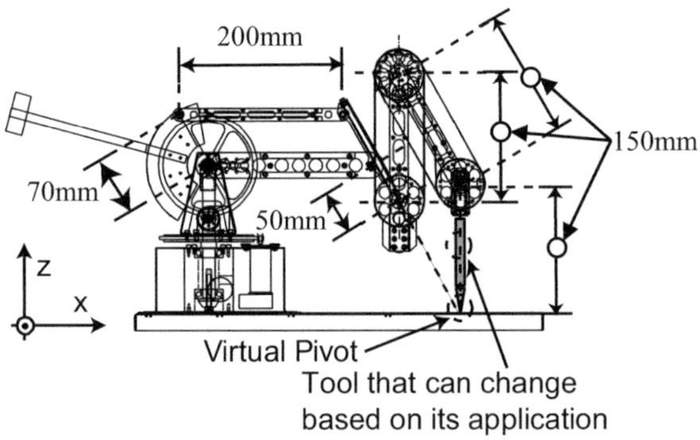

Fig. 16.2 Schematics of the SmartTool

three of these are active (no torque). Which means, the SmartTool is a kind of 3-DOF haptic display. To drive each active axis, a 10 W DC motor (Maxon Japan Corporation, RE25-118746) with 1/20 reduction gears is used.

The SmartTool have a specially designed wire based parallel link mechanism to install a working tool (Fig. 16.3). This parallel link mechanism is designed to compose a virtual pivot at exactly the same point of the tip of the tool, to create the necessary free space around the working point of the tool in use. The tool tip could move within the $280 \times 400 \times 400$ mm (x, y, z, as defined in Fig. 16.3) area, which means the working space. Additionally, the tool component has a socket on the end, which enables the user to use various kinds of conventional tools. The weight of the tool component is about 500 g with a counter-weight mounted on the base component. To measure the position of the tool tip, encoders (Maxon Japan Corp., HDES-5540 110511) are installed on each motor. The resolution of each encoder is 500 ppc and we use these multiplied by 4.

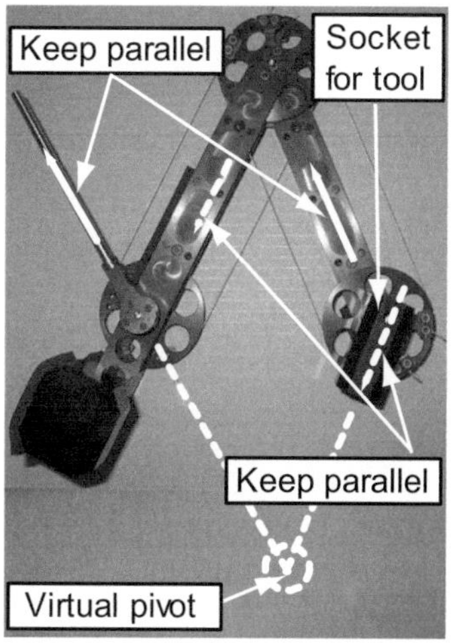

Fig. 16.3 A wire based parallel link mechanism to install a working tool

16.2.2.2 Sensors and Tools Component

The SmartTool needs sensor(s) to measure the environment to detect changes in the real environment in real-time depending on the task. Unfortunately, the exact area that needs to be measured is often occluded by the object and tool. For example, when inserting a needle into a certain object, the tip of the needle cannot be seen directly. However, the user typically needs to measure the area around the tip. Additionally, a higher sampling rate with low latency is preferable for the sensor(s) to achieve stable display of haptic sensation. Considering these conditions, small simple sensor(s) are installed on the tool tip according to its object to be used. The aim of the SmartTool is to support human tasks. Because various tools such as knives and pens may be needed depending on the task, the SmartTool is capable of changing its working tools.

16.2.3 Application Examples

16.2.3.1 Cutting the White of a Hard-Boiled Egg

A typical application of the SmartTool is to cut an object. For the first prototype, a hard-boiled egg was selected as the object to cut. In this application, the egg is assumed to represent a human body; the yolk of the egg denotes vital tissue, while the egg white indicates tissue that can be safely incised. A scalpel was used as the conventional tool on the SmartTool. Under these conditions, a restricted area was generated around the yolk. If the scalpel moves close to this area, a repulsive force is generated from the SmartTool to prevent damage to the yolk. The user holds

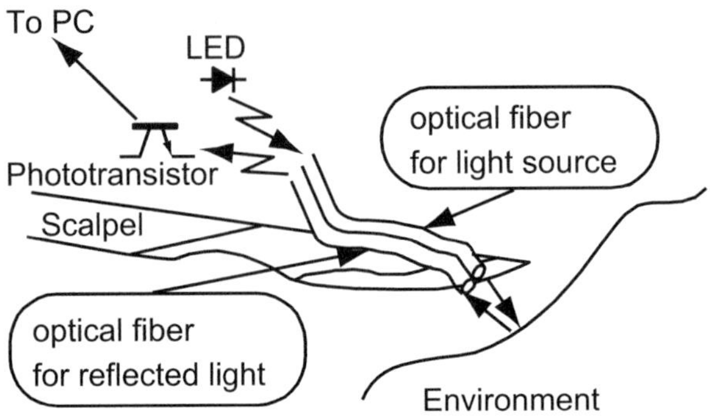

Fig. 16.4 Layout of the sensor used when cutting an egg

the scalpel and can move it freely if the sensor on the scalpel does not detect any restricted area. If the sensor detects the yolk, the user feels a sensation of hitting a solid virtual wall. By changing the yolk's soft haptic feeling to hard one, the SmartTool succeeded in sending material related information through the user's haptic sensation.

Regarding sensors, an optical sensor was used to detect the egg yolk (see Fig. 16.4). The sensor consists of two optical fibers; one is used to lead blue reference light (wavelength 470 nm) from the light source and the other is used to lead the reflected light from the egg to a phototransistor for the light receiver. These two fibers were installed on the tip of the scalpel. If blue light is scattered on a white object, the intensity of the reflected light is high. However, if blue light is scattered on a yellow object, the intensity of the reflected light decreases. The difference in reflectance can be used to detect a change in the material the scalpel is touching. Image (a) in Fig. 16.5 shows the light from the sensor on the tip, while images (b), (c), and (d) in the figure show the scalpel stopping at the yolk.

16.2.3.2 Touching the Interface Between Oil and Water

The second application involved touching the interface between oil and water. Generally, a human cannot touch the interface between two liquids. In this application, the SmartTool generates an artificial force at the interface between the oil and water layer, allowing a user of the system to feel an augmented haptic sensation between the layers. This application also shows that the system works in a changing environment. If a wave occurs in a water tank, the sensor on the tool tip detects the movement of the interface between the two layers in real-time, and the system sends the information to the user by means of haptic sensation. As a result, the user not only can touch the interface, but can also feel the movement of the wave in the tank.

For the tool of the SmartTool, an acrylic stylus was used to "touch" the interface. For the sensor, we used an electric conductivity sensor to detect the difference

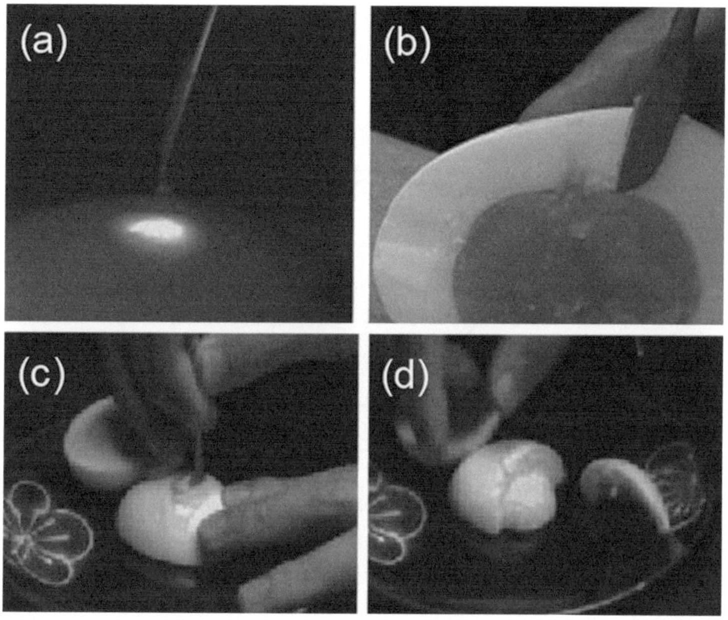

Fig. 16.5 (**a**) Using an optical sensor; (**b**) the scalpel stops at the yolk; (**c**, **d**) cutting the egg with the SmartTool

Fig. 16.6 Touching the interface between oil and water: (*left*) a normal tool penetrates the interface; (*right*) the tool of the SmartTool stops at the interface

between oil and water. The user holds the stylus, and if the tool moves from the oil layer to the water, the user feels a sensation similar to hitting a solid interface between the two liquids (Fig. 16.6).

The sensor, which consists of two metallic wires, measures the electrical conductivity between them. This sensor is small enough to install on the tool tip and works in real-time (Fig. 16.7).

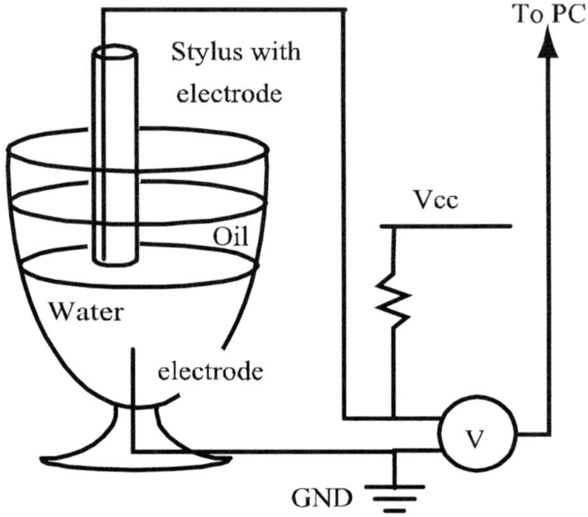

Fig. 16.7 Structure of a sensor for detecting water

16.2.4 Summary

In this section, we described a novel system for the AR of haptics, namely, the SmartTool. The SmartTool transmits information from its sensors to a human through haptic sensation, enabling the user to feel the sensory information. We also discussed the prototype system developed, and two experiments using this prototype.

16.3 Haptic Display for Airplane Pilots

To fly aircraft safely and smoothly, pilots must continually check the aircraft's status using instruments located inside the cockpit while monitoring outside. Based on the information obtained, pilots can control the aircraft by moving the control stick. The control stick of an airplane is not only a control device, but also a haptic display. Most of this haptic display is designed to indicate "force from air", which is essential for airplanes to fly. If the pilot pushes/pulls the control stick from its neutral position, the pilot feels a reactive force toward the neutral position. The strength of the reactive force changes according to the amount of movement of the control stick and airspeed of the airplane. Additionally, in the case of a stall (a situation resulting from loss of lifting power), the control stick vibrates to warn the pilot of the situation which is known as "stick shaker".

Allied to the increasing complexity of aircraft and flight situations, the amount of information that must be dealt with by pilots during a flight has also increased, resulting in increased pilot workload. To address this issue, using tactile sensation as the information transmission channel has been proposed. In this section, details of a tactile flight display are given.

16.3.1 Tactile Flight Display (TFD)

Pilots of airplanes and helicopters required to monitor the outside of the aircraft while at the same time monitor flight instruments inside. Looking outside, pilots scan for the presence of other aircraft and search for runway and its clearance when they are in landing phase. Simultaneously, they comprehend the flight state by monitoring flight instruments such as the altimeter, vertical speed indicator, attitude indicator, and airspeed indicator, and navigate using navigation instruments. Internal and external visual information are equally important for the safety of the flight, despite pilots' visual information channel being almost saturated. Therefore, various problems such as disorder in an aircraft, obstacles in the area, or bad weather can easily prevent pilots from adequately attending to visual information acquisition, particularly in single-pilot aircraft. In this section, we describe a tactile pilot support system aimed particularly at small airplanes and helicopters. The system uses tactile sensations to convey primary flight information to the pilot. This system should enable pilots to control an aircraft based on sensations from the tactile display without increasing visual workload.

16.3.2 TFD Display Method

Tactile sense has begun to be considered as an effective channel for displaying flight information [3], with research being carried out on tactile displays [4–6]. In this system, path angle error is displayed through the tactile display to maintain an appropriate vertical trajectory of the aircraft. The path angle error is defined as the error angle between the current flight path and an interception path to the target trajectory (Fig. 16.8).

If there is an error between these two paths, a tactile signal is generated. To display such information, three pin-array-type tactile display units (SC-9, KGS Inc.) are used. An overview of the TFD is shown in Fig. 16.9. As shown in the figure, the pin-array-type tactile displays are mounted on the joystick, which is used to control the flight simulator shown in the figure. The left-hand side of Fig. 16.10 shows a detailed configuration of the TFD. The pin stroke length is 0.7 mm and the

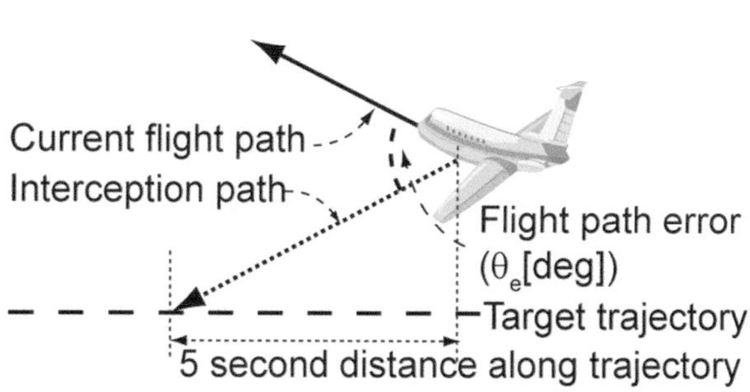

Fig. 16.8 Error between current and interception paths

Fig. 16.9 Overview of the TFD

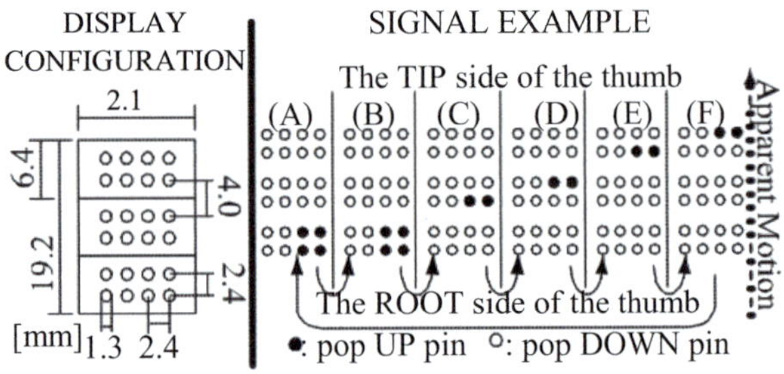

Fig. 16.10 Display configuration and signal

pins are pushed up with a force of about 18 gf. The right-hand side of Fig. 16.10 shows an example pin pattern for displaying flight information. In this figure, the pin pattern cycles through patterns (A) to (F) to generate apparent motion on the user's thumb. In this figure, this apparent motion indicates a pitch down command (push the control stick forward command) through the corresponding motion of the pins. Patterns (A) and (B) mean tapping twice to emphasize the starting position of the apparent motion for clearer understanding of the direction of the apparent motion.

Three different levels of the speed of the apparent motion are included to indicate the magnitude of the flight path angle error. The speed of the apparent motion is modified by changing the display time for each pattern shown in Fig. 16.10 from 80 to 150 ms. Between each pattern, all pins return to their starting positions for 50–150 ms.

16.3.3 Experiment

A series of flight simulation experiments were conducted to investigate the effectiveness of the proposed display device by comparing various combinations of tactile and visual information.

The flight simulator modeled a Dornier 228-202 aircraft. For ease of evaluation, only the pitch of the aircraft was manually controlled. The roll, yaw, and engine power of the aircraft were automatically controlled to maintain a lateral course and

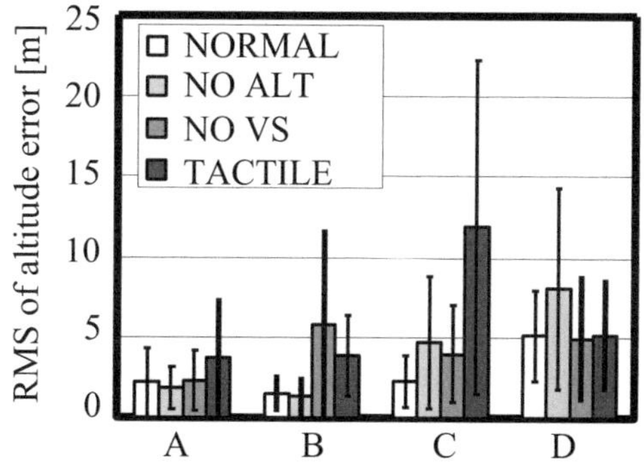

Fig. 16.11 Experimental results

airspeed. The wind was a direct headwind of 5 m/s, with a $\sigma = 0.8$ m/s random gust component. Subjects were required to track an 8 km-long straight level flight path at the assigned altitude.

Four display configurations of NORMAL, NO ALT, NO VS, and TACTILE were compared. For the NORMAL case, the pilot could use the visual altitude and vertical speed meters. For the NO ALT and NO VS cases, the tactile display was used with only the vertical speed indicator or the altimeter, respectively. For the TACTILE case, only the tactile display was used. The root-mean-square (RMS) values of altitude error together with standard deviation were compared for all cases. The experiment was conducted using four experienced pilots, subjects A, B, C, and D, with the results shown in Fig. 16.11.

Based on the results, except for subject C in the TACTILE case, no subject exceeded a 15 m altitude deviation in any of the cases. The tactile display successfully compensated for the lack of visual display information.

16.3.4 Summary

In this section, we presented details of the TFD, which is a support system for pilots using a tactile display. The preliminary evaluation showed that this system can assist a pilot in maintaining an assigned altitude with only partial reference to altitude and/or vertical speed instruments.

16.4 Haptics and Sightseeing

Vibration signals are often used for navigation purposes. Unfortunately, this is often thought to be a navigation method only for the blind. For sighted people, visual maps or navigation systems are commonly used to aid navigation. Such visual maps and navigation systems are capable of displaying a wealth of information, but at the same time, they can often distract users. In this section, a novel vibrotactile navigation method for sightseeing is described. This navigation method is intended to be used by both blind and sighted tourists.

16.4.1 Stavigation: A Vibrotactile Mobile Navigation for Exploration-Like Sightseeing

Wandering around an unfamiliar place, which is similar to exploration, is one of the ways of enjoying sightseeing. For instance, Venice is famous for its complex network of alleyways and its many bridges across canals. However, many tourists have found themselves lost while exploring the islands, and feel irritation when unable to reach their destination easily. However, while wandering around the city, tourists often discover something new. Wandering and exploring unknown places often lead tourists to novel experiences that are not mentioned in guidebooks. Such unique experiences highlight the enjoyment of sightseeing. In this research, we call this mode of sightseeing "exploration-like sightseeing". Although wandering can sometimes be enjoyable for tourists, they do not want to find themselves in situations where they are unable to return, for example, to their hotel. Current navigation devices are able to display appropriate and precise route information to guide tourists to their destination; these devices are a source of comfort, allowing tourists to enjoy their walk. However, such devices are incompatible with exploration-like sightseeing. Displaying a precise route implicitly forces tourists to stay on a specific track, requiring that they keep focusing on the displayed route. This diminishes the enjoyment of sightseeing.

Using visual displays to provide navigation-related information disrupts the enjoyment of wandering because it requires the user to focus on the display. To solve the problem, a new vibrotactile mobile navigation called "stravigation" is proposed [7]. The basic concept of stravigation involves two aspects. The first is reducing the amount of information; too much information can annoy tourists and impede their visiting the surrounding sights. Thus, stravigation provides only the distance and direction to a specific place, which is the minimum amount of information needed to prevent tourists from getting lost. The second aspect is making use of vibrotactile sensation. Stravigation provides distance and direction information by using vibrotactile signals, thus freeing tourists from having to focus on their navigation devices visually.

Vibrotactile navigation research falls into two categories: turn-by-turn and relative position-based methods. The turn-by-turn method indicates the correct direction each time the user reaches a specific waypoint or intersection [8, 9]. These systems are able to provide precise route information. However, vibrotactile signals at every intersection could annoy the tourists. The relative position-based method only indicates spatial relationships, such as direction and distance, between the current location and the destination [10–12]. This method does not provide precise route information. However, users can understand the spatial relationships, allowing them to choose freely between available routes to a destination, and to engage in what we term exploration-like sightseeing.

Most systems based on this method are able to provide a direction to the destination by having the user scan the surroundings with the device. A vibrotactile signal is generated only when the device points towards the destination. Stravigation is based on the same concept as this relative position-based method. However, previous systems do not have sufficient capability for navigation, and some are unable to provide distance information. This is not only essential for estimating time of arrival, but it also directly affects the reliability of such systems. The transition of such information, based on the tourists' wandering, forms the basis of them knowing whether the system is working. Additionally, the accuracy of directional information of previous systems is ±30°, which is far from satisfactory. Stravigation is capable of providing distance information as well as directional information with an accuracy of ±5°.

16.4.2 Overview of Stavigation System

When providing directional information, stravigation indicates the angular deviation between the device's axial direction and the direction to the destination. This means that a specific vibration occurs depending on the angular deviation (Fig. 16.12 left). If the angular deviation is greater than ±30° [13], the device does not provide any vibrational feedback. If it is between ±15° and ±30°, the device vibrates with a long period, while an intermediate vibration period is generated when the angular deviation angle is between ±5° and ±15°. If the angular deviation is less than ±5°, a short period vibration is generated. A similar principle is used to provide distance information with specific vibrations generated according to the distance.

Distance and directional information is provided through the same vibrotactile signal. Some sort of practical means is required for users to identify easily what information is provided. To solve this problem, we make use of the orientation of

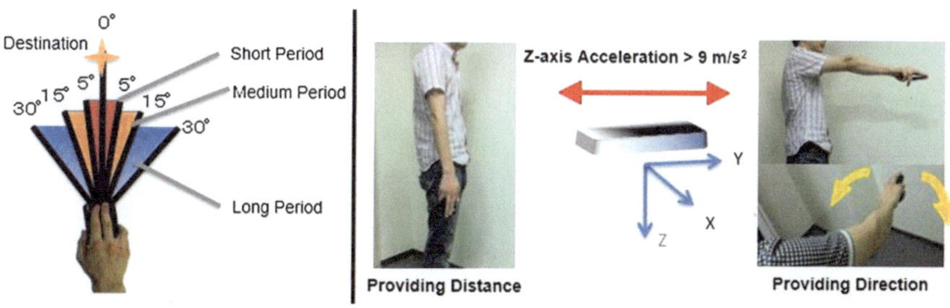

Fig. 16.12 (*Left*) Angular deviation and vibration period; (*right*) switching information by using device orientation

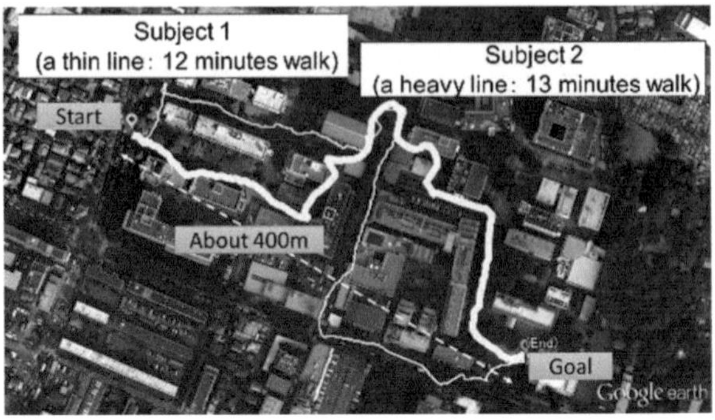

Fig. 16.13 Time taken and tracked route walked by subjects

the device. By using a three-dimensional accelerometer, we can determine whether the surface of the device is parallel to the ground (Fig. 16.12 right). Users are able to switch information modes through gestures. If its surface is parallel to the ground, the device switches to the direction-providing mode in which the user swings the device to the left and right to find the direction; otherwise the device switches to the distance-providing mode.

16.4.3 Evaluation Experiments

16.4.3.1 Navigational Capability

An experiment was performed to confirm the practicability of stravigation. Two subjects were asked to walk towards a specific place, which was 400 m away from their starting position, using stravigation. They were not told where the place was. The tracked routes and times taken are shown on the left of Fig. 16.13. The results show that stravigation has sufficient capability to guide users to a specific place.

16.4.3.2 User Test

User tests were performed to confirm that stravigation is able to put users in an exploratory-like mood while walking. We compared stravigation with a conventional map-based visual navigation system (Google Maps). Four participants were asked to answer a questionnaire after using these two navigation methods. All participants used both stravigation and Google Maps twice each. The results are shown in Fig. 16.14. The result for Q2 indicates that our system achieved better scores than navigation using maps. This means that stravigation does not hinder tourists from seeing the surrounding sights. At the same time, the result for Q5 indicates that using stravigation is more exciting than using conventional systems. As a result, tourists should be able to enjoy wandering without losing their way.

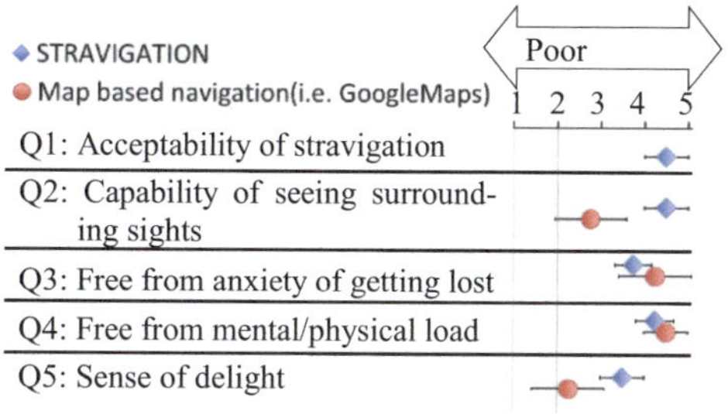

Fig. 16.14 Questionnaire responses

16.4.4 Summary

In this section, we discussed the details of a new navigation method called stravigation. Stravigation is a kind of vibrotactile navigation method intended to be used by tourists, enabling them to enjoy exploration-like sightseeing. The experimental results show that stravigation has higher accuracy in providing directional information. Moreover, stravigation had sufficient capability as a navigation aid and reliability to enable tourists to enjoy exploration-like sightseeing.

16.5 Conclusion

In this chapter, three different research systems were described. The first one was named SmartTool that intended to enable augmented reality of haptics. The research shows the actual construction of the proposed system and several applications. The second system was TFD, a system that could flight related information to pilots. The simulation test shows that pilots could keep their aircraft within a certain altitude range by using this display. The last one was stravigation, a vibrotactile navigation system. The experimental result shows that the system has enough capability as a navigational display. Even though that all participants were sighted people, but they could feel some sort of sense of delight while they are walking with this navigation system.

Overall, these systems are intended to increase the chances for general consumers to experience the benefits of "using haptic signals" and to broaden the applicable field of haptic displays. Unfortunately, further struggle is still required to make it reality. It is expected that this research will stimulate both engineers and researchers and contribute to its realization.

References

1. Milgram, P., Kishino, F.: A taxonomy of mixed reality visual displays. IEICE Trans. Inf. Syst. **77**, 1321–1329 (1994)

2. Nojima, T., Sekiguchi, D., Inami, M., Tachi, S.: The SmartTool: a system for augmented reality of haptics. IEEE Virtual Reality. IEEE. Comput. Soc. pp. 67–72 (2002)
3. Nojima, T., Funabiki, K.: Cockpit display using tactile sensation. In: Proceedings of the World Haptics Conference (WHC). IEEE, pp. 501–502 (2005)
4. Van Veen, H.A.H.C., van Erp, J.B.F.: Tactile Information Presentation in the Cockpit. In: Proceedings of the International Workshop on Haptic Human-Computer Interaction. pp. 174–181 (2000)
5. Van Erp, J.B.F., Veltman, H. (J.) A., van Veen, H.A.H.C.: A tactile cockpit instrument to support to altitude control. In: Proceedings of the Human Factors and Ergonomics Society, pp. 114–118 (2003)
6. Raj, A.K., Kass, S.J., Perry, J.F.: Vibrotactile displays for improving spatial awareness. In: Proceedings of the IEA/HFES Congress, pp. 181–184 (1998).
7. Kawaguchi, H., Nojima, T.: STRAVIGATION: A vibrotactile mobile navigation for exploration-like sightseeing. In: Proceedings of the Advances in Computer Entertainment, pp. 517–520 (2012)
8. Pielot, M., Poppinga, B., Boll, S.: PocketNavigator: vibro-tactile waypoint navigation for everyday mobile devices. In: International Conference on Human-Computer Interaction with Mobile Devices and Services (MobileHCI), pp. 423–426 (2010)
9. Rümelin, S., Rukzio, E., Hardy, R.: NaviRadar: A novel tactile information display for pedestrian navigation. In: 24th Symposium on User Interface Software and Technology, pp. 293–302. ACM Press, New York (2011)
10. Williamson, J., Robinson, S., Stewart, C., Murray-Smith, R., Jones, M., Brewster, S.: Social gravity: a virtual elastic tether for casual, privacy-preserving pedestrian rendezvous. In: 28th International Conference on Human Factors in Computing Systems, pp. 1485–1494 (2010)
11. Robinson, S., Jones, M., Eslambolchilar, P., Murray-Smith, R., Lindborg, M.: "I did it my way": moving away from the tyranny of turn-by-turn pedestrian navigation. In: 12th International Conference on Human-Computer Interaction with Mobile Devices and Services, pp. 341–344 (2010)
12. Szymczak, D., Magnusson, C., Rassmus-Gröhn, K.: Guiding tourists through haptic interaction: vibration feedback in the lund time machine. In: Proceedings of the Eurohaptics Part II, pp. 157–162 (2012)
13. Strachan, S., Murray-Smith, R.: Bearing-based selection in mobile spatial interaction. J. Pers. Ubiquit. Comput. **13**(4), 265–280 (2009)

Chapter 17
Haptic Interfaces That Induce Motion and Emotion

Hiroyuki Kajimoto

Abstract Recent attempts to control motion and emotion using a haptic interface are introduced. Tactile stimulation can induce motion, such as in the cases that the wire hanger attached to head induces involuntary head rotation by applying pressure, the tangential deformation of the skin induces forearm motion, and the pulling of an earlobe controls walking. On the other hand, the tactile presentation of a false physiological response can modulate emotion, such as in facilitating affection by presenting a false heartbeat and amplifying a feeling of surprise through artificial control of piloerection on the forearm. There is thus the possibility of broadening the application area of haptic interfaces, where users do not directly understand the presented information but involuntarily react and understand the information only through their own reactions.

Keywords Emotion • Hanger reflex • Motion • Navigation • Piloerection • Vection field

17.1 Introduction

Most haptic interfaces are designed to present information. They are haptic "displays" that stimulate haptic mechanoreceptors under the skin, in muscle, or in joints. The presented information is grasped and understood by the user, who responds accordingly. The reaction is sometimes emotional, such as a laugh in response to a ticklish stimulus, but the information or the intention of the information is still well understood by the user, and the user feels that the reaction is undertaken by him or herself.

The present chapter focuses on another possible type of haptic interface where users do not directly understand the information presented but the information induces an involuntary reaction from the user. In this case, the user understands information only through his or her own reaction. The tickle-induced laughter mentioned above is not classified in this category since we directly understand

H. Kajimoto (✉)
The University of Electro-Communications, Tokyo, Japan
e-mail: kajimoto@kaji-lab.jp

H. Kajimoto et al. (eds.), *Pervasive Haptics*, DOI 10.1007/978-4-431-55772-2_17

from the ticklish sensation that its intention is to induce laughter. The study focuses instead on a kind of reflex, having a broad meaning, where the users do not understand the intention of the stimulus before the reaction is automatically generated.

This chapter reviews works on two types of such haptic interface. One type induces motion and the other type induces emotion. The interfaces show the possibility of broadening the application area of haptic interfaces, from information displays to human controllers.

17.2 Haptic Interface That Induces Motion

Several studies have focused on presenting illusionary forces mainly to achieve force display with compact space. Amemiya et al. [1] employed asymmetric acceleration for this purpose and Rekimoto et al. [2] showed that asymmetric vibration of relatively high frequency induces similar effects. Minamizawa et al. [3] and Inaba et al. [4] proposed devices that squeezed the skin of a finger to produce a pseudo-force sensation. Additionally, several studies have presented information by tangential skin deformation, especially at the forearm. Shull et al. [5] produced tangential skin deformation on the arm with the primary purpose of emphasizing the direction of arm joint rotation. Yem et al. [6] presented a directed force through skin deformation for surgeon training. Kuniyasu et al. [7] demonstrated that a pseudo-force sensation is induced by tangentially deforming skin on the forearm using a simple haptic device (Fig. 17.1). Furthermore, studies have found that tangential deformation of the skin is an important cue for posture [8–10].

Overall, the primary purpose of the above works has been to present an illusory force, or illusory posture. In other words, they intend to use the illusionary force as a compact haptic (sensation) “display”. This section shows that with a similar setup, it is possible to develop a haptic interface that induces motion or a haptic interface that functions as a human controller.

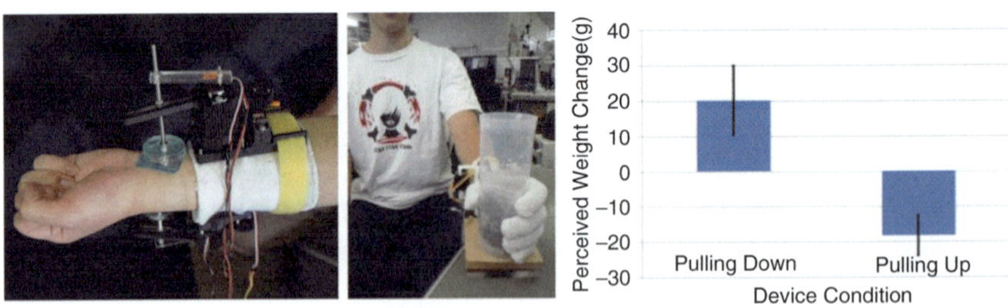

Fig. 17.1 Induction of forearm motion by tangential deformation of the skin. (*Left*) System structure. (*Middle*) Experiment setup. (*Right*) Result of the experiment [7]

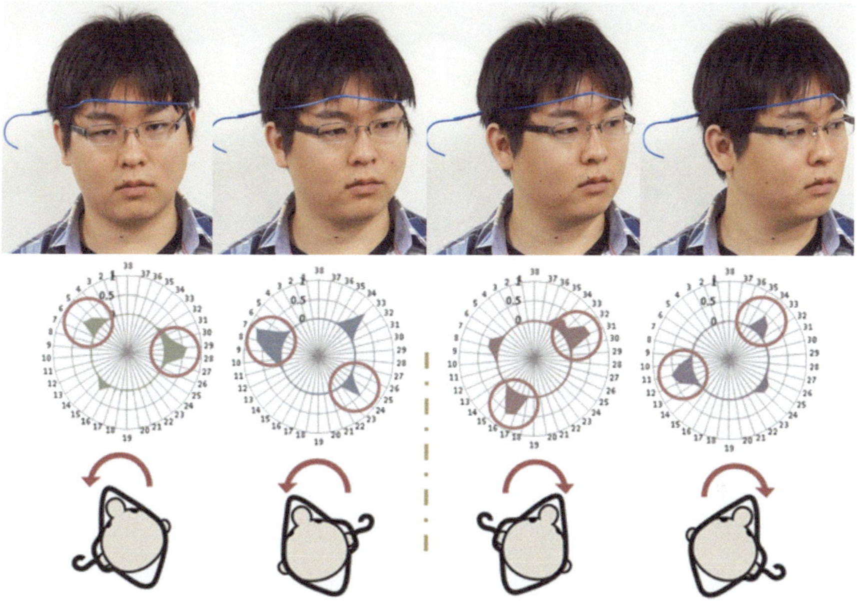

Fig. 17.2 (*Top*) Hanger reflex, which is an involuntary head rotation by wire hanger. (*Bottom*) Relationship between the pressure distribution and direction of rotation [11]

17.2.1 Hanger Reflex

The hanger reflex is a phenomenon wherein the head rotates involuntarily when a wire hanger is wrapped around its temporal regions (Fig. 17.2). The rotational force is felt by the user as an external force, but the force is actually exerted by the user him or herself. People tend to say that their head "rotated unexpectedly" when they did not resist the force, while they tend to say they feel an "external force" when they resist the movement.

Sato et al. [11] measured pressure distribution when the hanger reflex occurs, and discovered that the pressure on the side of the forehead or on the contralateral side of the forehead plays a major role. It is highly possible that skin stretch around the head induces an illusory force sensation, and this force sensation induces involuntary head motion.

The hanger reflex is characterized by a strong involuntary motion, which surprises most users. Therefore, it can be used for not only a haptic display for virtual-reality application but also a medical or training device that induces appropriate motion.

Asahi et al. [12] applied the hanger reflex as a simple, nonelectrical and mechanical treatment of cervical dystonia. Cervical dystonia is a disease related to neck posture, and general treatment involves the injection of botulinus toxin

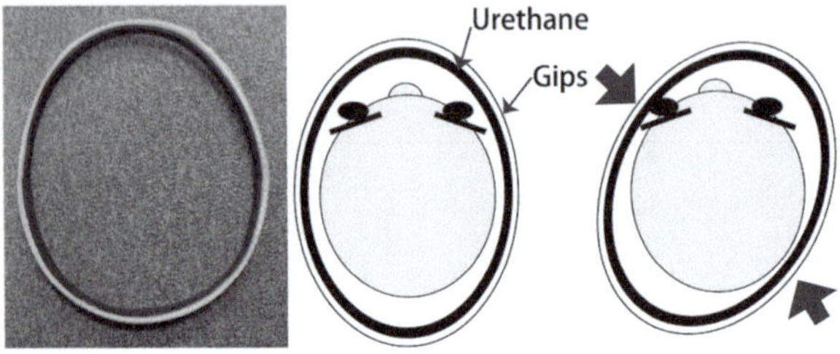

Fig. 17.3 Wearable device for cervical dystonia [12]

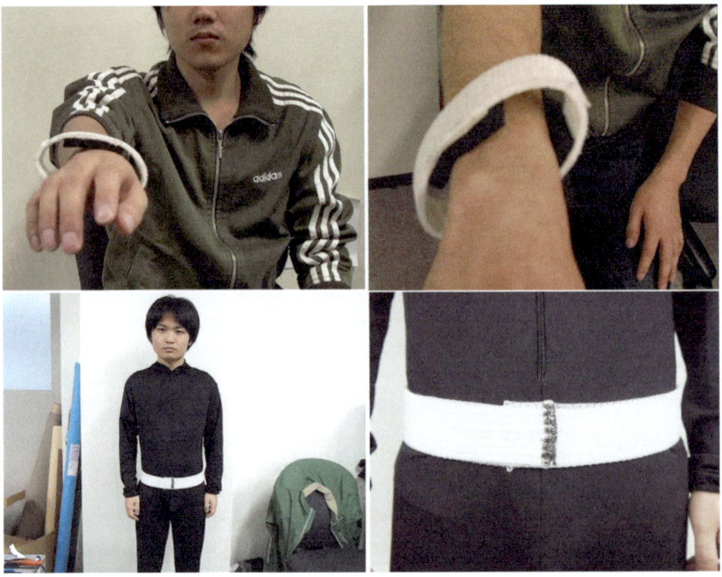

Fig. 17.4 Application of the hanger reflex to the wrist and waist [13, 14]

to paralyze the neck muscle, or even deep brain stimulation that requires brain surgery. Treatment with the hanger reflex seems to counter the disease, at least temporarily, and is a low-cost treatment both economically and mentally (Fig. 17.3).

Nakamura et al. [13, 14] applied the hanger reflex phenomenon at the wrist and waist (Fig. 17.4). As both these body parts have an elliptical cross-section similar to that of the head, similar skin deformation using a simple mechanical device was expected. The experiment showed that rotational motion was generated at both the wrist and waist.

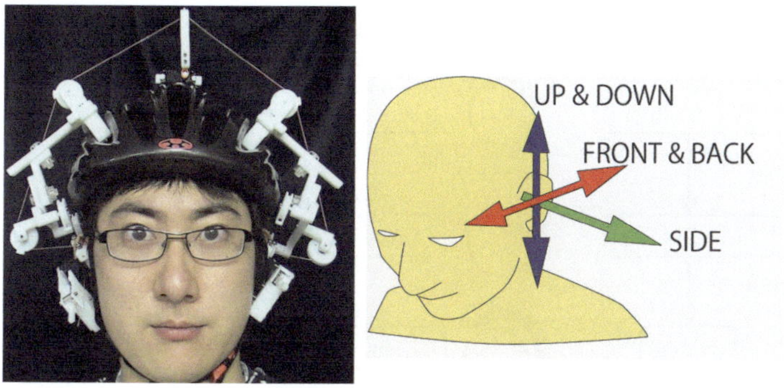

Fig. 17.5 Control of walking by pulling the earlobe [15]

17.2.2 Pulling Ears

The pulling of body parts has a similar effect. Kojima et al. [15] developed a navigation interface that pulls the user's ears. The interface, named Pull-Navi, comprises two clips attached to the earlobes, six DC motors that pull each ear in three directions, a microprocessor that controls the motors and a helmet on which the motors are mounted. The user wears the helmet and attaches it to his or her earlobes via the clips (Fig. 17.5).

Kojima et al. [15] found that the user was inevitably tempted to move left/right when their left/right ear was pulled to the left/right. When both ears were pulled forward/backward, the user was tempted to walk more quickly/slowly. Interestingly, when both ears were pulled up/down, the user was tempted to walk up/down stairs if there were stairs in front of him or her. The experiment conducted on blindfolded users showed that the walking path of the user curved in the direction of the pulling of the ears even though the user did not notice this effect during the experiment.

Previous studies on navigation for walking have mainly used visual or acoustic sensations, which are not intuitive and can even be dangerous because they may block visual and auditory information from the surrounding environment. Other studies have used tactile stimulation, which is more intuitive and less annoying, on the hand or arm to generate a pseudo-pulling force. However, the devices tend to be large and heavy. In contrast, the system reviewed here controls the user's behavior, while the user does not notice they are being controlled. The concept is similar to that in the work by Maeda et al. [16], where walking navigation was realized through galvanic vestibular stimulation [17], but has fewer clinical challenges, since galvanic vestibular stimulation requires electrical stimulation of head.

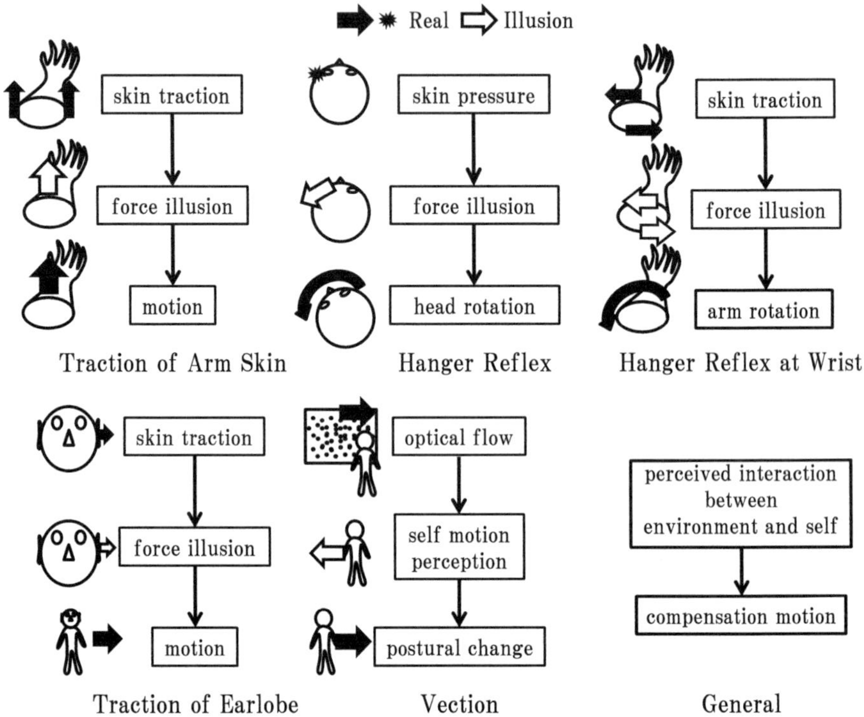

Fig. 17.6 Possible processes of the traction of arm skin, hanger reflex, hanger reflex at the wrist, and traction of the earlobe, and comparison with vection

17.2.3 Discussion

Three types of haptic interface that induce motion were presented above: the hanger reflex acting on the head, the hanger reflex acting on the wrist and waist, and the traction of the earlobe (Fig. 17.6). In fact, the arm skin traction introduced in Fig. 17.1 is frequently accompanied with tiny movement. In all these cases, an illusory force is generated, and involuntary motion is probably induced by this force. A natural question arises; Why the motion is induced and why it is toward the direction of the force, and not in the opposite direction? If there is a mechanism that stabilizes our body, a motion "against" a perceived force is also conceivable.

A comparison with other illusions that induce motion might be important in solving the above question. When we perceive a large image moving sideways (vection), our body also moves. If the motion of the image is rightward, the body inclines to the right [18], which is a phenomenon that can be used for a walking navigation interface [19] (Fig. 17.7). A general explanation of vection-induced motion is that we perceive our body to be moving in the opposite direction, and the body tries to correct this illusory motion, leading to an inclination in the direction of vection.

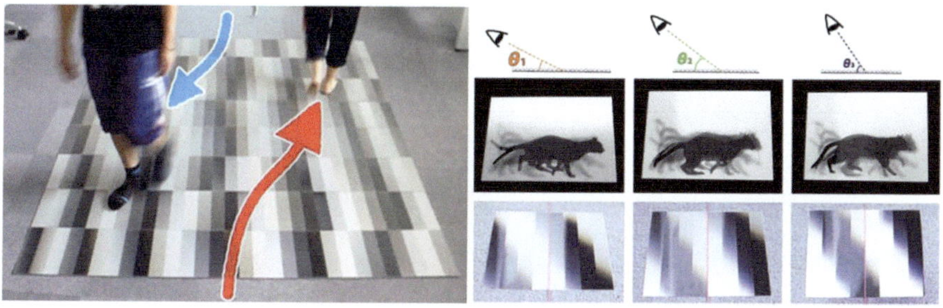

Fig. 17.7 Vection field for pedestrian traffic control [19]

In haptics, we observe that motion is induced in the direction of an illusory external force, whereas in vision, motion is induced against the direction of illusory self-motion. These might be commonly understood as a compensation function of perceived interaction between environment and self. In haptics case, an illusory external force can be regarded as a result of relative motion, i.e. the environmental motion to the direction of the force, or the self-motion against the direction of the force, and the real motion is induced to make the relative motion smaller. In vision case, the real motion is also induced to make the relative motion between the environment and oneself smaller. In both cases, the motions are induced to follow the relative motion of the environment.

17.3 Haptic Interface That Induces Emotion

We saw in the previous section that haptic stimuli can affect motion. In this section, we will see that haptic stimuli can also affect emotion, a motion of mind. This is simply understood by the examples in which we laugh when tickled or feel embarrassed when touched.

Several works have used haptic stimuli to enhance audio–visual contents. EmotiChair [20] is a chair-shaped tactile display that creates an immersive environment by turning audio information to vibration. Tactile Jacket [21] is a jacket-shaped tactile display that creates an immersive experience by rendering emotional states of a movie onto the viewer's body as vibration. These technologies improve the immersive feeling or realism by embodying various cues in the contents on a body. In other words, they are haptic versions of background music.

While these haptic stimuli enhance emotional experience, the intention of the stimuli are quite obvious to the users. By contrast, another approach that we discuss here is to induce emotion with haptics, in a less obvious way, so that the intention of the stimuli is not noticed.

One approach to achieve this goal is to use haptic sensation to present the state of mind of the user. According to James–Lange theory [22], emotion is experienced

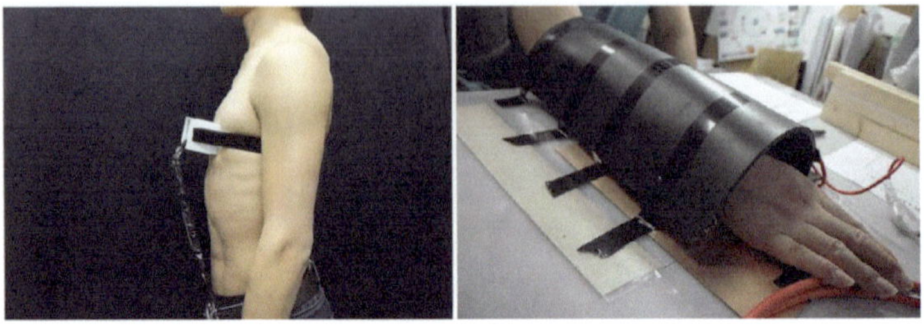

Fig. 17.8 (*Left*) Facilitation of affection by tactile feedback of a false heartbeat [25]. (*Right*) Facilitation of a surprised feeling by artificial control of piloerection on the forearm [26]

as a result of a physiological change of the body induced by the autonomic nervous system. Therefore, there is the possibility that the presentation of a pseudo reaction of the body induces an emotional experience.

The first reaction is the heartbeat, which is one of the most easily implemented and easily understood cues of our mind state. Valins [23] reported that the preference for semi-nude photographs is affected by modulating the frequency of a false heartbeat, which is presented aurally and synchronized with the presentation of the photographs. Nakamura et al. [24] showed that displaying a false heart rate on a visual monitor affects the real heartbeat. As these are audial or visual methods of presenting a pseudo reaction, the intention of the stimuli are easy to understand.

In contrast, Nishimura et al. [25] made a tactile device composed of an audio speaker as a vibrator. The vibrator was attached on the left side of the participant's chest and a pseudo heartbeat vibration was presented (Fig. 17.8 (Left)). The vibration can be felt as the user's own heartbeat and the intention of the stimuli can be hidden. Furthermore, the stimulus can be controlled for each user and does not interfere with the main audio-visual content.

The second reaction is piloerection, which is involuntarily induced when a person is cold or experiences emotion such as fear and surprise. Fukushima et al. [26] conducted an experiment in which the standing of forearm hairs was induced using an electrostatic force generated by a polarized acrylic plate set above the forearm (Fig. 17.8 (Right)). The acrylic plate was set 2 cm above the forearm with a copper electrode ($38 \times 100 \times 0.08$ mm) on the top of the acrylic plate and crude rubber on top of the electrode. The electrode was connected to a high-voltage source, and the forearm was connected to ground. The acrylic plate is polarized by applying a high voltage (0–20 kV) to the electrode. The forearm hair is then attracted to the acrylic plate by the electrostatic force. Although the piloerection itself was felt as a soft breeze, once it was presented in synchrony with content, it affected the emotional state.

17.3.1 Discussion

In the previous section, we discussed the mechanism underlying the haptics-induced motion, and showed that it can be understood as a mechanism to compensate relative motion between the environment and oneself, i.e. the motions are induced to follow the relative motion of the environment. What I would like to mention here is that haptics-induced emotions might also be understood similarly. The heartbeat change or the piloerection indicates the change of the environment, and our mind follows that change.

17.4 Conclusion

This chapter introduced recent attempts to control motion and emotion employing haptic interfaces. In this category of haptic interfaces, the user does not directly understand the presented information, but an involuntary reaction of the user is induced, and the user understands the information only through his or her reaction. It was shown that skin deformation induces motion and that imitation of an autonomous nerve reaction induces emotion.

The control of motion and emotion mostly involves the use of visual or auditory cues; e.g., vection (visual motion) induces involuntary motion and audial presentation of the heartbeat affects the emotional state. Haptic presentation is generally superior when the intention of the presented stimulus is hidden, the stimulation is controlled for each user, and the stimulation is applied to a specific part of the body.

References

1. Amemiya, T., Maeda, T.: Asymmetric oscillation distorts the perceived heaviness of handheld objects. IEEE Trans. Haptics **1**(1), 9–18 (2008)
2. Rekimoto, J.: Traxion: a tactile interaction device with virtual force sensation. In: Proceedings of the 26th ACM Symposium on User Interface Software and Technology, pp. 427–432. ACM Press (2013)
3. Minamizawa, K., Prattichizzo, D., Tachi, S.: Simplified design of haptic display by extending one-point kinesthetic feedback to multipoint tactile feedback. In: Proceedings of IEEE Haptics Symposium, pp. 257–260 (2010)
4. Inaba, G., Fujita, K.: 2006. A pseudo-force-feedback device by fingertip tightening for multi-finger object manipulation. In: Proceedings of the 2006 Eurohaptics Conference, pp. 75–278 (2006)
5. Shull, P., Bark, K., Cutosky, M.: Skin nonlinearities and their effect on user perception for rotational skin stretch. In: Proceedings of the 2010 IEEE Haptics Symposium, pp. 77–82 (2010)
6. Yem, V., Kuzuoka, H., Yamashita, N., Ohta, S., Takeuchi, Y.: Hand-skill learning using outer-covering haptic display. In: Proceedings of EuroHaptics '14, Lecture Notes in Computer Science, pp. 201–207 (2014)

7. Kuniyasu, Y., Sato, M., Fukushima, S., Kajimoto, H.: Transmission of forearm motion by tangential deformation of the skin. In: Proceedings of the 3rd Augmented Human International Conference (2012)
8. Edin, B.B., Johansson, N.: Skin strain patterns provide kinaesthetic information to the human central nervous system. J. Physiol. **487**(1), 243–251 (1995)
9. Collins, D.F., Prochazka, A.: Movement illusions evoked by ensemble cutaneous input from the dorsum of the human hand. J. Physiol. **496**(3), 857–871 (1996)
10. Ebied, A.M., Kemp, G.J., Frostick, S.P.: The role of cutaneous sensation in the motor function of the hand. J. Orthop. Res. **22**(4), 862–866 (2004)
11. Sato, M., Matsue, R., Hashimoto, Y., Kajimoto, H.: Development of a head rotation interface by using hanger reflex. In: IEEE International Symposium on Robot and Human Interactive Communication, pp. 534–538 (2009)
12. Asahi, T., Sato, M., Kajimoto, H., Taira, T., Hayashi, A., Oyama, G., Fujii, M., Takashima, S., Kuroda, S., Endo, S.: Clinical application of the hanger reflex to the treatment of cervical dystonia—multicenter trial. In: 27th Japan Neurosurgery English Forum (JNEF) (2012)
13. Nakamura, T., Nishimura, N., Sato, M., Kajimoto, H.: Application of hanger reflex to wrist and waist. In: IEEE VR 2014 (2014)
14. Nakamura, T., Nishimura, N., Sato, M., Kajimoto, H.: Development of a wrist-twisting haptic display using the hanger reflex. In: Proceedings of the 11th Advances in Computer Entertainment Technology Conference (2014)
15. Kojima, Y., Hashimoto, Y., Kajimoto, H.: Pull-Navi. Emerging Technologies Session, ACM SIGGRAPH (2009)
16. Maeda, T., Ando, H., Amemiya, T., Inami, M., Nagaya, N., Sugimoto, M.: Shaking the world: galvanic vestibular stimulation as a novel sensation interface. ACM SIGGRAPH Emerging Technologies. Los Angeles (2005)
17. Fitzpatrick, R.C., Fitzpatrick, R.C., Wardman, D.L., Taylor, J.L.: Effects of galvanic vestibular stimulation during human walking. J. Physiol. **517**(3), 931–939 (1999)
18. Bronstein, A.M., Buckwell, D.: Automatic control of postural sway by visual motion parallax. Exp. Brain Res. **113**(2), 243–248 (1997)
19. Furukawa, M., Yoshikawa, H., Hachisu, T., Fukushima, S., Kajimoto, H.: "Vection field" for pedestrian traffic control. In: Proceedings of the 2nd Augmented Human International Conference (2011)
20. Baijal, A., Kim, J., Branje, C., Russo, F., Fels, D.I.: Composing vibrotactile music: A multi-sensory experience with the emoti-chair. In: Proceedings of IEEE Haptics Symposium, pp. 509–515 (2012)
21. Lemmens, P., Crompvoets, F., Brokken, D., van den Eerenbeemd, J., de Vries, G.-J.: A body-conforming tactile jacket to enrich movie viewing. In: Proceedings of World Haptics, pp. 7–12 (2009)
22. Cannon, W.: The James–Lange theory of emotions: a critical examination and an alternative theory. Am. J. Psychol. **39**, 106–124 (1927)
23. Valins, S.: Cognitive effects of false heart-rate feedback. J. Pers. Soc. Psychol. **4**(4), 400–408 (1966)
24. Nakamura, K., Katayama, T., Terada, T., Tsukamoto, M.: Evaluation on effect of presenting false information for biological information visualization systems. IPSJ SIG Technical Report vol. 2011-UBI-30 No. 1 (2011)
25. Nishimura, N., Ishi, A., Sato, M., Fukushima, S., Kajimoto, H.: Facilitation of affection by tactile feedback of false heartbeat. In: Extended Abstracts on Human Factors in Computing Systems, pp. 2321–2326 (2012)
26. Fukushima, S., Kajimoto, H.: Facilitating a surprised feeling by artificial control of piloerection on the forearm. In: Proceedings of the 3rd Augmented Human International Conference (2012)

Chapter 18
Bidirectionality of Haptics

Satoshi Saga

Abstract In the chapters of "Component Design", we introduced sensing technologies and displaying technologies treating both independently. However, these technologies are related. In this chapter we consider the links that can be made. Gibson coined the term, "active touch." This term implis haptic interfaces also require mutual interactivity. We discuss about necessity of bidirectionality between input/output and the way how to design haptic devices. Furthermore, we consider attractive content of haptic media and their several bidirectionalities.

Keywords Bidirectionality • Inter-world interface • Haptic media • Content design

18.1 Introduction

With the popularity of touchscreen interfaces and the rise in smartphones and tablets, demand for tactile/haptic technologies is expected to increase. Touchscreens and other positioning sensors offer a means to access digital contents such as augmented-reality contents or computer graphics games in a more intuitive way. In the previous chapters, we introduced sensing technologies and displaying technologies treating both independently. However, these technologies are related. Here we consider the links that can be made. The links between them indicate designing method of ideal haptic devices. Furthermore, to popularize the haptic technology, we have to consider not only about hardware but also about attractive content design for haptic media.

S. Saga (✉)
University of Tsukuba, 1-1-1, Tennodai, Tsukuba, Japan
e-mail: saga@saga-lab.org

H. Kajimoto et al. (eds.), *Pervasive Haptics*, DOI 10.1007/978-4-431-55772-2_18

18.2 Bidirectionality of Haptic Devices

In this section we discuss about a bidirectionality between input/output and the way how to design haptic devices.

18.2.1 Haptics as Mutual Interaction

Gibson coined the term, "active touch [1]." This emphasizes the difference between visual or audial sensation and haptic sensation. With visual or audial sensation, information comes to receptors naturally. However, with haptic sensations, haptic information does not come to the receptors naturally. Receptors cannot receive any signals without some form of touching action. This is evidence that haptic sensations are special for which input and output coincide with each other. Indeed, this is also true of haptic interfaces; users cannot feel any haptic sensation without any touching action. This implies that haptic interfaces require mutual interactivity.

Here we look back at recent haptic displays and sensors. Many studies have employed high-frequency mechanical vibration for textural displaying, such as "texture on tablets" [2], Verro Touch [3], and Techtile Toolkit [4]. In contrast, touch sensors, such as Skinput [5], Touché [6], or Touch & Activate [7], employ machine-learning-based classification obtained from some kind of touch signal.

The importance of these devices is that they employ "mutual interaction" as implicit key technology. In textural display, such as textures on tablets, the devices record and display mechanical vibrations between finger and textiles. In impacting displays, such as badminton demos in Techtile Toolkit, the devices record and display vibrations between racket and shuttle. The signals used in the displays are recorded during several active movements of hands or tools by sensors placed on them. That is, the active movement information itself is very important for displaying phase. While in the touch sensing, such as Touché, the device encodes the interaction between touched object and hand. If either the touched object or the subject touching the object is different, the sensing signal will be different. Hence, the sensor cannot distinguish true information. Therefore, the interaction between subject and object in the contact phase is very important for haptic devices.

18.2.2 Design of Interactive Haptic Device

In the previous section, we considered the importance of interaction. In designing haptic devices, to what should we pay attention? Because the interaction itself is important, one input and one output should exist on each boundary. Hence, at the boundary between a human and a device, there should be an input and an output. For example, for a touchscreen, it senses the touch of a human fingertip as input, and creates visual information as output. This is an interactive mode; however, the output is not haptic information. With this view, the device is a partially interactive haptic device. In contrast, haptic device, such as PHANToM, senses the position of the fingertip and creates force feedback. This also seems interactive, though, the input and output forms a different modality.

Many general haptic device systems employ so called impedance control method. The method measure position of the fingertip and according to the amount of penetration toward virtual object and calcurate the reactive force. Thus the position and force information is linked by the method. Similar method, admittance control, links the force and position information, vice versa. However the impedance method cannot display internal force against virtual hard object, which is hardly deformed with human operation. The internal force with little deformation is also very important information for object manipulation. Similarly, admittance method is difficult to display soft material because of its less backdrivability. Better interactive haptic device should measure and display force and position information simultaneously.

Furthermore, if we want to use an interactive haptic device as an interface between human and outer world (Fig. 18.1), we have to consider the boundary not only as between human and device, but also between device and the outer world. Thus the inter-world interface should have two boundaries, and each boundary should have a set of input and output.

18.2.3 Inter-worlds Interface

Many conventional interface devices, not only haptic interface but also smartphones, are only connected to virtual information world. However, in our opinion, many interface devices should have dense connection to real world or augmented world in the near future. With such devices, human can receives and generates new valuable lives. In this section, we discuss about such inter-worlds interfaces.

Fig. 18.1 Communication between human and world: If human communicate with outer world, we have to consider the boundary (*white line*). The boundary is not only as between human and device, but also between device and the outer world

First, we introduce some existing inter-world interfaces. The first is SmartTool [8]. It responds to the real environment using a real time sensor and a force feedback device. The sensor on the SmartTool measures the real environment then sends us that information through haptic sensation. The second is the Haptic Teaching System [9]. The system records the haptic operations of a trainer in a real world, and plays the inverted force to a trainee. The trainee can then learn how to exert force applied during the operation. Finally, there is the stiffness modulator [10]. The study proposes an augmented haptic system by modifying the stiffness of the touched object via a haptic interface. The last two devices employ force sensors on a force feedback device to acquire interaction between device and the outer world.

Here we have to pay attention to the number of boundaries and sets of input/output. These systems have two boundaries, and they should have two sets of input/output. However, they have only one output, a force feedback device, and one input, a photo detector or a force sensor. The peculiarity is the reason for the first inter-worlds interfaces (Fig. 18.2). Specifically, the tip of the haptic interface is a rigid body, and by operating the outer-world via the tip, input forces from the outer-world and the device itself are combined as a resultant force. The user feels the resultant force from both sides. Consequently, through the characteristic of the rigid body, the boundaries are virtually integrated as one boundary, and one set of input/output can be omitted.

In contrast, general haptic interfaces should have one set of input/output per each boundary. Furthermore, to recognize one interaction as an output of the users input, the input and output position should be precisely the same (Fig. 18.3). How can we design such haptic devices? Ultimately, haptic devices should measure all information on the boundary and display all responses as distribution information. To realize such an interface, sensing and displaying should take place at the same boundary, like a mirror of a haptic device (Fig. 18.4). While there are no such devices existing in the world, the idea points the way to giving greater value to haptic devices and open new worlds of novel interfaces.

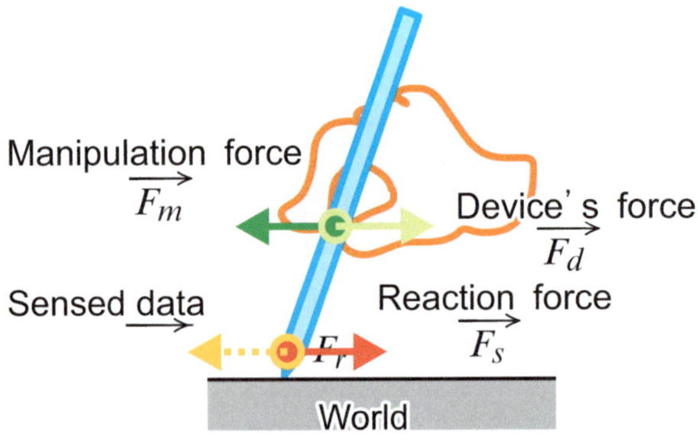

Fig. 18.2 Input and output of a haptic device: The stylus is a tip of haptic device, and each force is integrated on the stylus

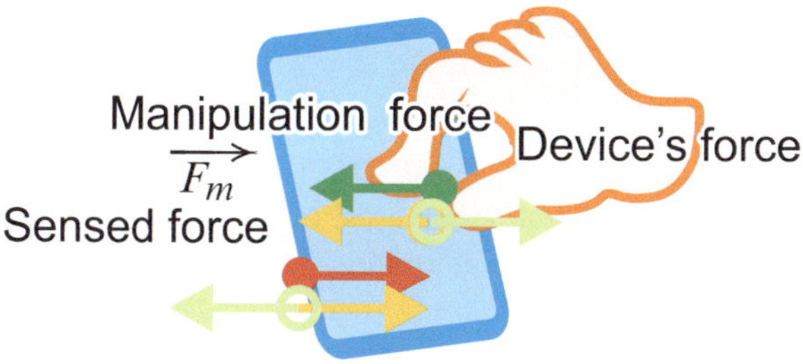

Fig. 18.3 Input and output of a touchscreen device: The screen is a touchscreen device. The difference from haptic device is the pairs of broken line vectors. Touchscreen device should have each pair of input and output toward device for each boundaries

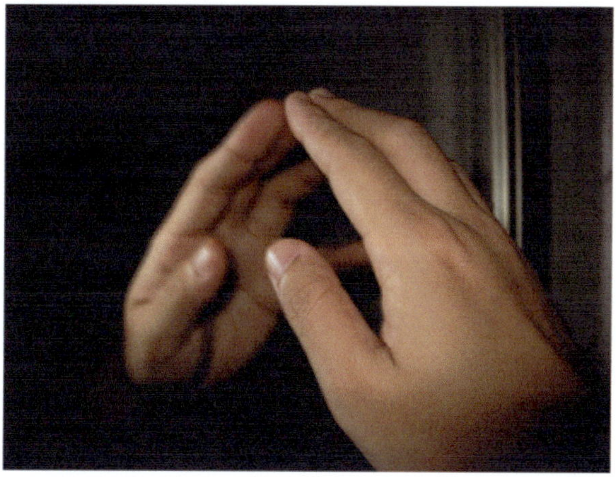

Fig. 18.4 Ideal I/O should be a mirror: Haptic devices should measure all information on the boundary and display all responses as distribution information. To realize such an interface, sensing and displaying should take place at the same boundary, like a mirror of a haptic device

18.3 Bidirectionality of Content Design for Haptic Media

So far, we have discussed the bidirectionality of haptic devices and their design. Furthermore, to popularize haptic technology, we have to consider not only about hardware but also about attractive content design for haptic media. So what is important point for the attractive content of haptic media? Here we discuss several "bidirectionalities" of the content design. In our opinion, there are three points that would make haptic content to be attractive;

1. Displaying impactive haptic stimuli
    - Contextual information
    - Responsiveness
2. Ease in content creation
3. Creating new value in communication

18.3.1 Displaying Impactive Haptic Stimulus

Here we take a glance at several haptic devices that have attracted the attention of many people. TouchSense [11] is a haptic technology which enables various stimulations by employing vibration patterns. E-Sense [12], TeslaTouch [13], and Electro-Tactile System [14] are electrostatic-force-based texture displays. New Feelings Haptic Response Touch Panel [15] generates click feelings. Conventional eccentric motors generate symbolic haptic stimulation only; however, recent devices display realistic textures and click feelings. Furthermore, in the research field, more and more realistic haptic displays are being developed. For example, Techtile Toolkit [4], "Texture on tablets" [2], VerroTouch [3], Texture on floor [16], Funbrella [17], or Eternal Sharpener [18]. In this manner, realistic displays will draw people's considerable attention.

18.3.1.1 Contextual Information

In the previous section, we introduced realistic displays. In recalling these devices, all of the displays employ a context-based displaying method. That is, an LCD displays button graphics and if the button is pressed, a haptic feedback will be generated and the animation will be displayed at the same time. On other devices, graphics of the texture are displayed and the user feels the texture by rubbing on the graphics. In these cases, the impact of the content is enhanced by the additional context information, that is, the recognition of what kind of action is done on what kind of situation raises the impact. This is a bidirectionality between input and output in a long period.

18.3.1.2 Responsiveness

Furthermore, we have to consider the high responsiveness of haptics. When human touch objects, the sensations from the touch are transferred immediately. The same thing should be true of haptic devices. If the user touches some object via a haptic device, the response transfer should be immediately. Hence the refresh rate of the input should be as fast as possible. Because of this, with common sense, the refresh rate of the haptic device should be controlled faster than 1 kHz. This is a bidirectionality between input and output in a very short period.

However, touchscreens are currently unable to provide such refresh rates. The sensing rate of a normal touchscreen is around 60 Hz, and as Albert et al. [19] pointed out, the speed is too slow for good interactivity. Thus, if normal touchscreens are used for haptic device, and rubbing movement are detected, the device has to generate interpolated multiple information per one input information for displaying fine haptic information like textures. For example, in our touchscreen system, it is not an input position but input velocity and a precise timer that are used as key input for displaying interpolated multiple texture information [20]. The bidirectionality in a very short period requires several contrivance to realize.

18.3.2 Ease in Content Creation

Here we consider the content itself. Currently, the ease in creating content with the smartphone is spreading the use of digital media, such as photos or videos. In other words, the simplicity of content creation makes everyone creators. Thus, for the spreading haptic technology, creation of haptic content should be simple as well. This means that the bidirectionality between creators and receivers is also important.

In that sense, the record and playing method of haptic information may be a good way to spread content. For example, our system realizes record and playing method of several haptic skills, like calligraphy or sculptures [9]. Minamizawa et al., Romano et al. and others developed vibration-based record and playing systems [2, 4, 20]. Another simple creation method is, for example, shape-information-based display. Nowadays, Kinect or other spacial sensing device realizes quick sampling of shape information. By employing such information, several researchers display realistic shape information in real-time [21, 22].

18.3.3 Creating New Value in Communication

Ultimately, we should not forget that haptic media will create new value in our daily lives. If novel communication technology use not only a verbal but also haptic channel to transmit social media, a brand new world will certainly open in the near future. For example, haptic interaction are largely connected to emotional information. Several researchers employ haptic communication for medical therapy [23, 24]. Thus haptic media will generate emotional channel on telecommunication. If the communication is augmented by the haptic technology, further new value will be generated, like a relation between SNS and other media technologies. There are another bidirectionality between communication and other media technologies, including haptic one. Thus such new value will generate further brand new value in chain.

18.4 Conclusion

In this chapter, we considered the links between sensor and display, and discussed about bidirectionality from the viewpoints of mutual interaction, inter-world interface, and design of haptic content. Compared with vision and auditory devices, haptic devices are difficult to implement, because of this bidirectionality. Furthermore, the bidirectionality is a very special characteristic of haptic media compared with others. Thus, taking this special feature into consideration, the realization of good sensorial transfer must be the mission of haptic media.

Acknowledgements This work was partially supported by KAKENHI (26540095), Grant-in-Aid for challenging Exploratory Research.

References

1. Gibson, J.J.: Observations on active touch. Psychol. Rev. **69**, 477–491 (1962)
2. Romano, J.M., Kuchenbecker, K.J.: Creating realistic virtual textures from contact acceleration data. IEEE Trans. Haptics **5**(2), 109–119 (2012)
3. Kuchenbecker, K.J., Gewirtz, J., McMahan, W., Standish, D., Martin, P., Bohren, J., Mendoza, P.J., Lee, D.I.: Verrotouch: high-frequency acceleration feedback for telerobotic surgery. Haptics: Gener. Perceiving Tang. Sensat. **6191**, 189–196 (2010)
4. Minamizawa, K., Kakehi, Y., Nakatani, M., Mihara, S., Tachi, S.: TECHTILE toolkit – a prototyping tool for design and education of haptic media. In: Proceedings of Virtual Reality International Conference (Laval Virtual 2012), Laval, p. 22 (2012)
5. Harrison, C., Tan, D., Morris, D.: Skinput: appropriating the body as an input surface. In: Proceedings of the 28th International Conference on Human Factors in Computing Systems, Atlanta, pp. 453–462. ACM (2010)
6. Sato, M., Poupyrev, I., Harrison, C.: Touché: enhancing touch interaction on humans, screens, liquids, and everyday objects. In: Proceedings of the 2012 ACM Annual Conference on Human Factors in Computing Systems, Austin, pp. 483–492. ACM (2012)
7. Ono, M., Shizuki, B., Tanaka, J.: Touch & activate: adding interactivity to existing objects using active acoustic sensing. In: Proceedings of the 26th Annual ACM Symposium on User Interface Software and Technology, St Andrews, pp. 31–40. ACM (2013)
8. Nojima, T., Sekiguchi, D., Inami, M., Tachi, S.: The smarttool: a system for augmented reality of haptics. In: Proceedings of IEEE Virtual Reality, Orlando (2002)
9. Saga, S., Vlack, K., Kajimoto, H., Tachi, S.: Haptic video. In: ACM SIGGRAPH 2005 Emerging Technologies, Los Angeles, p. 7. ACM (2005)
10. Jeon, S., Choi, S.: Stiffness modulation for haptic augmented reality: extension to 3D interaction. In: Haptics Symposium, Waltham, 2010 IEEE, pp. 273–280. IEEE (2010)
11. Immersion Corporation: Touchsense (2010). http://www.immersion.com/products/touchsense-tactile-feedback/
12. Toshiba Corporation: E-sense (2010). 13th Embedded Systems Expo
13. Bau, O., Poupyrev, I., Israr, A., Harrison, C.: TeslaTouch: electrovibration for touch surfaces. In: Proceedings of the 23nd Annual ACM Symposium on User Interface Software and Technology, New York, pp. 283–292. ACM (2010)
14. Radivojevic, Z., Beecher, P., Bower, C., Cotton, D., Haque, S., Andrew, P., Henson, B., Howard, I., Ingram, J., Wall, S., Wolpert, D.: Programmable electrostatic surface for tactile perceptions. In: Proceedings of 2012 Society for Information Display International Symposium, Boston (2012)
15. Kyocera Corporation: New feelings haptic response touch panel (2011). CEATEC
16. Visell, Y., Law, A., Cooperstock, J.R.: Touch is everywhere: floor surfaces as ambient haptic interfaces. IEEE Trans. Haptics **2**, 148–159 (2009)
17. Fujita, K., Itoh, Y., Yoshida, A., Ozaki, M., Kikukawa, T., Fukazawa, R., Takashima, K., Kitamura, Y., Kishino, F.: Funbrella: recording and replaying vibrations through an umbrella axis. In: Proceedings of the International Conference on Advances in Computer Entertainment Technology, Athens, pp. 66–71. ACM (2009)

18. Kojima, Y., Hashimoto, Y., Kajimoto, H.: Eternal sharpener – a rotational haptic display that records and replays the sensation of sharpening a pencil. In: The 18th IEEE International Symposium on Robot and Human Interactive Communication (RO-MAN 2009), Toyama, pp. 18–21. IEEE (2009)
19. Ng, A., Dietz, P.H.: 39.3: the need for speed in touch systems. In: SID Symposium Digest of Technical Papers, Vancouver, vol. 44, pp. 547–550. Wiley Online Library (2013)
20. Saga, S., Raskar, R.: Simultaneous geometry and texture display based on lateral force for touchscreen. In: Proceedings of IEEE World Haptics 2013, Daejeon, pp. 437–442 (2013)
21. Yano, H., Miyamoto, Y., Iwata, H.: Haptic interface for perceiving remote object using a laser range finder. In: Proceedings of IEEE World Haptics 2009, Salt Lake City. IEEE (2009)
22. Saga, S., Deguchi, K.: Lateral-force-based 2.5-dimensional tactile display for touch screen. In: Proceedings of IEEE Haptics Symposium 2012, Vancouver, pp. 15–22 (2012)
23. Marino, L., Lilienfeld, S.O.: Dolphin-assisted therapy: More flawed data and more flawed conclusions. Anthrozoos: Multidiscip. J. Interact. People Anim. **20**(3), 239–249 (2007)
24. Wada, K., Ikeda, Y., Inoue, K., Uehara, R.: Development and preliminary evaluation of a caregiver's manual for robot therapy using the therapeutic seal robot Paro. In: The 19th IEEE International Symposium on Robot and Human Interactive Communication (RO-MAN 2010), Viareggio, pp. 533–538 (2010)

Chapter 19
Remote Transmission of Multiple Tactile Properties

Masashi Konyo

Abstract To transmit realistic tactile textures remotely, we need to represent multiple tactile properties and to compensate communication delays. This chapter introduces a real-time remote transmission system that can deliver multiple tactile properties using a master-slave robot system. The author discusses required tactile properties to represent texture feelings and potential ways to transmit them through a robotic master-slave system in real-time. Three tactile properties are introduced: roughness, friction, and softness, which are associated with physical properties such as wavelengths, friction coefficients, and spring constants. Vibrotactile stimulation methods proposed by the authors can transmit the multiple tactile properties. Synchronization between hand exploration and tactile stimulation is also necessary to transmit natural texture feelings. The presented methods compensate time delays between the master and the slave by generating vibrotactile stimuli with local models on the master side. Tactile stimulations are synthesized in real time with the tactile properties transmitted from the slave side. Discrimination experiments of materials showed the feasibility of the proposed tactile transmission system.

Keywords Tactile display • Tactile sensor • Tactile transmission • Remote operation • Master-slave system • Tactile properties

19.1 Introduction

Transmission of tactile sensation is necessary for robotic teleoperation systems such as a robotic surgery system and a dexterous telemanipulation system to transmit detailed contact information in the remote site. Such a tactile transmission technology could be useful for a feature communication systems. This chapter introduces a real-time remote texture transmission system with a robotic master-slave system developed by the author's group.

M. Konyo (✉)
Graduate School of Information Sciences, Tohoku University, 6-6-01 Aramaki Aza Aoba Aoba-ku, Sendai 980-8579, Japan
e-mail: konyo@rm.is.tohoku.ac.jp

H. Kajimoto et al. (eds.), *Pervasive Haptics*, DOI 10.1007/978-4-431-55772-2_19

One of the difficulties of the texture transmission is that tactile information requires multiple tactile properties related to multiple physical properties. For example, tactile information can be expressed in multiple adjective words such as rough/smooth, soft/hard, sticky/slippy, and warm/cool. Such variety of feelings is a typical feature of our tactile sensing capability. Another issue is synchronization between hand exploration and tactile feedback under a limited communication environment with a time delay. Human tactile perception is closely related to the active hand exploration. A decoupled tactile feedback with the hand exploration could provide insufficient tactile information. For example, the authors confirmed that a delay of about 40 ms in tactile feedback in response to hand movement caused changes in tactile sensation [1].

Several studies reported remote tactile transmission systems that deliver tactile information detected by a remote tactile sensor to the operator. For example, tactile transmission systems for a medical gripper or glove used in minimally invasive surgery or palpation were reported [2, 3]. These studies aimed to transmit pressure distribution and a total load at the contact area between the probe and body tissue. A tactile transmission system for a prosthetic leg, which transferred the position of the center of gravity and the total load in a sole was developed [4]. A prosthetic hand that transferred pressure in hand to the user was also developed [5]. Several master-slave type robot systems aimed to transfer tactile information such as pressure [6] and high-frequency vibrations [7, 8] from the slave hand to the master side to support telemanipulation. Yamamoto et al. developed a master-slave system that transfers vibration detected by a tactile sensor running on material surfaces for the transmission of texture information [9]. Studies mentioned above basically focused on a special use and a particular tactile property. On the other hand, Caldwell et al. [10] reported a transmission system for multiple tactile properties. Their system delivered roughness on a textured surface, vibration delivered from stick-slip friction, total load, and temperature change to the operator. However, a part of tactile properties is not intended to deliver natural sensation for the operator because the total load was represented by stepwise vibration intensity in a symbolic manner. Also, few studies dealt with the compensation for communication delay to synchronize tactile stimulation with hand exploration.

The authors have been proposed tactile display methods using vibratory stimulation for representing roughness, softness [11], and friction [12]. These methods were based on frequency response characteristics of human mechanoreceptors. Superposition of these vibratory stimulations is expected to represent multiple tactile properties simultaneously. We have partly confirmed that such superposition could represents several tactile feelings like cloth materials [11], in which the tactile display methods for each tactile properties were not well connected to physical properties of the materials. In addition, the authors also have been developed a tactile sensor that can estimate multiple physical factors including wavelength of textured surfaces, friction coefficients, and Young's modulus, which are related to the tactile properties [13].

The author's group has been studied tactile display methods using vibrotactile stimulation for representing roughness, softness [11], and friction [12]. These methods were based on frequency response characteristics of human mechanoreceptors. Superposition of these vibrotactile stimuli could represent multiple tactile properties simultaneously. We confirmed that such superposition could represent several tactile feelings like cloth materials [11]. Also, the authors also have been developed a tactile sensor that can estimate multiple physical factors including wavelengths of textured surfaces, friction coefficients, and Young's modulus [13].

This chapter reports the author's trial to integrate the tactile sensing and display systems mentioned above. The authors have proposed a remote transmission for multiple tactile properties through a master-slave robot system in real-time [14, 15]. The proposed systems investigated the connection between texture properties and physical properties of the materials. A key technology for the integration is how to represent texture information in real-time by connecting the measured signals to the displayed signals.

First, the author discusses what kind of tactile properties should be transmitted and how to connect them in real-time. Real-time synchronization between the master and slave sides is a critical issue for tactile transmission as mentioned above. We apply a time-delay compensation method proposed by the authors [14]. In this method, tactile stimulation can be generated in synchronization with hand exploration at the master side by using a local tactile generation model. Then, the author introduces a master-slave system, tactile sensing methods, and tactile display methods. Finally, the verification results are shown to confirm the feasibility of the total system. The demonstration in a remote environment is also introduced.

19.2 System Design

19.2.1 Selection of Multiple Tactile Properties

Haptic cues for object recognition are categorized into two types of properties: material properties and geometric properties [16]. Geometric properties involve volume or shape of the object. A force feedback device is commonly used for presenting such properties. This chapter does not focus on force feedback because many studies have been proposed for a master-slave system including time-delay communication [16, 17]. On the other hand, material properties are highly related to tactile sensations, which are perceived through skin deformations.

A general classification of tactile properties is not standardized yet. Several studies reported different classifications of tactile properties because the classifications depend on the target material. The detailed material properties are described in Chapter 1. As shown in the chapter, apparent categories found by many studies are roughness, softness, and temperature properties. Friction property is another candidate for material properties. Note that friction coefficients are not attributed to an independent tactile property because it affects on multiple tactile properties [18].

In our study, we selected three tactile properties: roughness, softness, and friction, which have a reasonable agreement with the conventional classifications. Furthermore, these properties have been targeted in our tactile display and sensing methods. We tried to combine the both tactile and sensing methods in real-time transmission and verified the tactile transmission quality [15].

19.2.2 Synchronization of Hand Exploration and Tactile Feedbacks

To deliver natural tactile sensation through a master-slave robotic system, we need to compensate temporal gap between operator's manipulation and sensory feedback. Conventional force-feedback-type master-slave systems often use a local model to represent remote environments, such as a geometric model [15].

The same approaches could be applied for the tactile feedback if local models of the material properties for producing texture sensation are available. For example, we proposed a tactile roughness transmission system based on a local model involving the surface wavelengths of textures [14]. This chapter introduces an extended trial for friction and softness transmission in addition to the roughness [15]. The local models for representing texture sensations, involving roughness, friction, and softness, require physical properties related to the hand exploration. We assumed that the physical parameters of them were wavelengths, friction coefficients, and spring constants, respectively.

Figure 19.1 shows the block diagram of the system developed. This system is based on a standard force-reflecting type force feedback for kinetic information. Besides, the tactile sensor on the slave system estimates physical parameters of the target material in real-time. The estimated physical parameters are transferred to the local models on the master system. The local models always use updated physical parameters transmitted from the slave system. On the master system, the local models generate tactile stimuli using the updated local hand movements of the operator. In this case, there is no temporal gap between the tactile feedback and operators' movements.

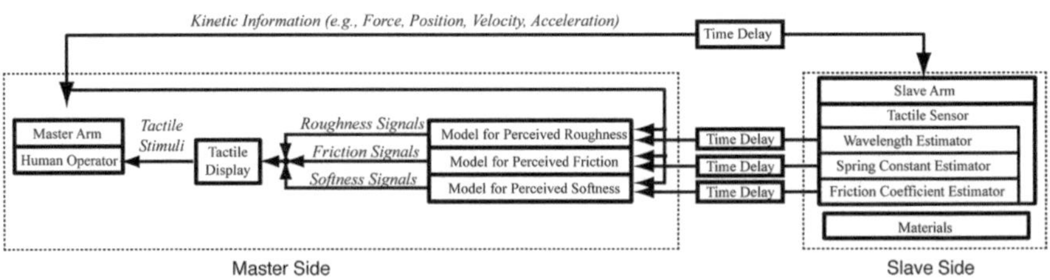

Fig. 19.1 Block diagram of the master-slave-typed tactile transmission system for multiple tactile properties

Note that the proposed method does not compensate the actual delay between the master and slave systems, which consists of the estimation time for the physical parameters and communication delays. If the remote environment changes rapidly during hand exploration, the local models cannot follow the changes of the material properties. In general, the speed of the hand exploration is slow enough, even at more than several Hz, the change of physical properties could not be so serious. As described later, the update rates of physical parameters are high enough comparing with hand movement; at 250 Hz for wavelengths and friction coefficients, and at 125 Hz for spring constants.

19.2.3 *Local Models to Generate Tactile Stimuli from Transmitted Signals*

A key feature of our proposal is the local models to generate tactile stimuli from the transmitted signals of the sensor at the slave system. As mentioned in Sect. 2.2, the physical parameters of the texture sensation should be extracted from the measured signals. Here, the relationship between the measured and the displayed signals are introduced for the individual tactile properties.

For the roughness property, we have confirmed that vibrotactile stimulation based on surface wavelengths and hand's speed can represent roughness sensation [19]. The detailed estimation and display methods are described in Sects. 4.2 and 5.1.

For the friction property, human perception mechanism is not well understood. We have suggested that high-frequency vibrations generated by stick-slip phenomena on the finger can be a cue for displaying friction sensation [12]. Based on this assumption, we proposed a friction display method that reproduces approximate stick-slip transition based on a model of the 1-DOF vibration system with Coulomb friction. However, it is very difficult to measure the stick-slip phenomena in the quite same manner as a human skin because the tactile sensor has different friction characteristics. In our study, we selected simple kinetic friction coefficients calculated by a 3-axis force sensor as a physical parameter for representing the friction property (Sect. 4.3). As for the display side, the stick-slip transition model, which was determined by observations of contact between a human skin and an acrylic plate [12], is used (Sect. 5.2). In this sense, the local model of the friction property in this study does not reflect the real material property strictly. However, we have confirmed that the proposed friction display method can control the magnitude of perceived friction sensation by changing the friction coefficients in the model [12]. Therefore, the measured friction coefficients can reflect on the magnitude of the friction sensation.

For the softness property, human perception mechanism of softness is also not well understood. In this study, we assume that a spring constant could be a physical parameter for the stiffness, which is related to the softness property. Note that Young's Modulus, which was used in [13], does not reflect the stiffness attributed to the geometric shape such as thickness. The spring constant would be better as the parameter of stiffness though we have to approximate the structure as a simple spring. In this study, we propose a real-time estimation method for the spring constant at the slave side (Sect. 4.4). For the display, we assume that the amount of the activity of SA I type mechanoreceptor is related to the softness property. For instance, this assumption could be supported by a study which reported that the perceived softness can be controlled by changing contact area against pressing force [20], because human detection of contact area is highly related to the activity of SA I. To control the activity of SA I, we use the pressure display method proposed by the author's group [11]. This method produces static pressure sensation using vibrotactile stimuli at a very low frequency (<5 Hz), where SA I is most sensitive than the other mechanoreceptors. Thus, we can control the activity of SA I by changing the amplitude of the low-frequency vibration. However, this method cannot be applied to the tactile transmission system immediately, because the relationship between the estimated spring constant and the vibration amplitude is not clear. In this study, we assume a simple inversely proportional relationship between the spring constants and the displayed amplitudes (Sect. 5.3).

19.3 Master-Slave System Developed

The overview of the master-slave system is described for explaining the whole system in this chapter. Figure 19.2 shows the master-slave system developed. A force feedback device, PHANToM Premiums (SensAble Technologies), is adopted for the robot arms. The force feedback method is a standard force-reflecting type based on the force sensor attached to the slave arm. A human mimetic tactile sensor, which is described in IV-A, is mounted on the tip of the slave arm. A tactile display on the master side requires to generate 1-DOF vibrations against the finger with enough force. In this study, a voice coil (AURA, SOUND/NSW1-205-8A) is used for the vibrator. The corn of the voice coil is covered with a 0.5 mm thick acrylic plate. The voice coil is attached to the robot arm with a gimbal mechanism, which allows operators move their hands smoothly.

19.4 Tactile Sensor and Sensing Methods

19.4.1 Tactile Sensor

A tactile sensor for the tactile transmission system required for estimating surface wavelengths, friction coefficients, and spring constants as described in Sect. 2.3. The author's group had proposed a human mimetic tactile sensor that satisfies above

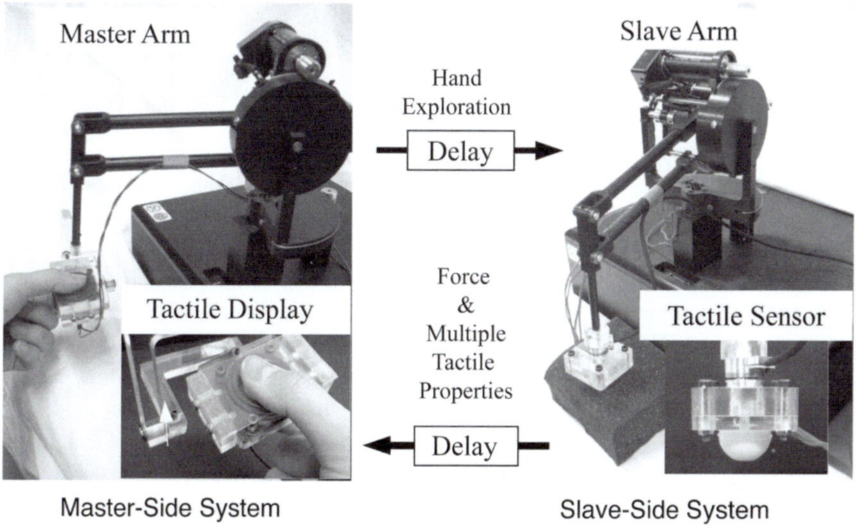

Fig. 19.2 Schematic view of the master-slave system

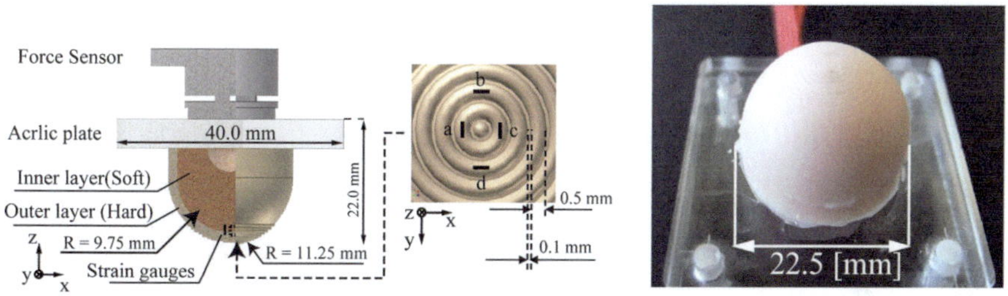

Fig. 19.3 Schematic and overview of the tactile sensor [21]

requirements [13]. The tactile sensor adopted in this study is a hemispherical version [21] of this tactile sensor to scan the texture in the planar direction.

Figure 19.3 shows the tactile sensor used. The sensor consists of a hemispherical silicone rubber and a torque sensor (NITTA, EFS-18M20A25-M10). The hemispherical rubber forms a hard outer layer and soft inner layer. The outer layer is covered with distal ridges whose width is 0.5 mm and close to that of epidermal ridges of the human finger. As transducers, four strain gauges are embedded at the border of two layers just beneath the distal ridges. The disposition of the gauges was designed so that the gauges efficiently respond to deformations of the ridges [13]. The strain gauges are connected to the control computer with a strain amplifier. The strain signals are sampled at 5 kHz.

19.4.2 Estimation of Surface Wavelengths

The estimation method for the surface wavelength of textures is the same as the FFT-based method adopted by the authors [14]. In scanning textures, because of the surface roughness of the textures, the sensor vibrates. The strain gauges respond to this vibration. By computing FFT of the output voltages of strain gauges, vibratory frequency of the sensor is acquired. The surface wavelength of the texture is determined by $\lambda = |v(t)/f(t)|$, where λ, $v(t)$, and $f(t)$ are the surface wavelength, scanning velocity of the sensor, and vibratory frequency of the sensor, respectively.

In order to estimate the vibratory frequencies of the sensor, the computed power spectrum is used. N peaks are extracted from the power spectrum. The frequencies corresponding to each peak are named $f_1, f_2, \ldots, f_N$. Also, the corresponding powers are $A_1, A_2, \ldots, A_N$, where $A_1 > A_2 > \cdots > A_N$, which means the power peaks are extracted in descending order. Ai is used for determining the amplitude of roughness stimuli in Sect. 5.1. We have confirmed that this sensor could detect at least 0.2 mm wavelength when the velocity was 20 mm/s and the normal force was 1.0 N. Note that the estimation accuracy becomes worse when the normal force is small and the hand movement is slow. Thus, the friction coefficients are updated only when the normal force was more than 0.2 N and the hand velocity was more than 50 mm/s. The update rate was 250 Hz.

19.4.3 Estimation of Friction Coefficients

The stick-slip model for the friction display requires both kinetic and static friction coefficients as described in Sect. 5.2. However, it is difficult to estimate the static friction coefficients because the tactile sensor developed has a short sticking phase during strokes with materials. In this study, transmitting the friction coefficients is focused rather than the reproduction of accurate stick-slip phenomena. Thus, we assumed that the perceived friction is related to the kinetic friction coefficient, and the static friction coefficients are simply proportional to the kinetic coefficients. Friction coefficients between the tactile sensor and materials were estimated using the force sensor. The friction coefficients were determined by

$$\mu_k = \frac{F_t}{F_n} \tag{19.1}$$

$$\mu_s = 2\mu_k \tag{19.2}$$

where F_t and F_n were the lateral force and the normal force exerted to the sensor. The force data was sampled at 1 kHz, and the estimated values were determined by averaging the forces for 20 ms. The estimation could become unstable when the normal force was small, and the hand exploration was in turn-around points. Thus, the friction coefficients are updated only when the normal force was more than 0.2 N and the hand velocity was more than 40 mm/s. The update rate was 250 Hz.

19.4.4 Estimation of Spring Constants

When the tactile sensor is thrust to elastic objects, the deformation of the sensor depends on the elasticity of the objects and the normal force applied if the elasticity of the tactile sensor is known. The amount of the deformation reflects on the outputs of the strain gauges embedded in the sensor. The authors have proposed the method to estimate the elasticity of target objects from the variance of the strain gauges' outputs [13]. However, the variance was susceptible to manufacturing error in the arrangement of the strain gauges.

This study adopts a more robust method using a mapping table determined by a prior experiment to prepare the mapping table. In the experiment, the sensor was thrust into three silicone rubber samples that has different spring constants (4.71, 11.76, 22.29 [kN/m]) by changing thrust forces from 0.2 N to 2.0 N by 0.2 N. Figure 19.4 shows an example of the observed relationships between spring constants of the samples, the normal forces, and the outputs from a strain gauge. Note that the spring constants are corresponding values assumed that the sample is a cylinder in 30 mm thick and 3.14×10^{-4} m^2 in cross section. Since the tactile sensor has four strain gauges, the mapping table is constructed on each strain gauge. The estimated spring constant is interpolated with the table linearly as shown in Fig. 19.4. For example, when ν_i and F_n are measured by the tactile sensor, three neighborhood points are found on the plane. The final value of the estimated spring constant is determined by averaging values from the four mapping tables.

To validate this estimation method, an experiment was performed. In the experiment, the sensor moved in the normal direction and thrust into the softness specimen made from silicone rubber. No lateral movement was applied to the

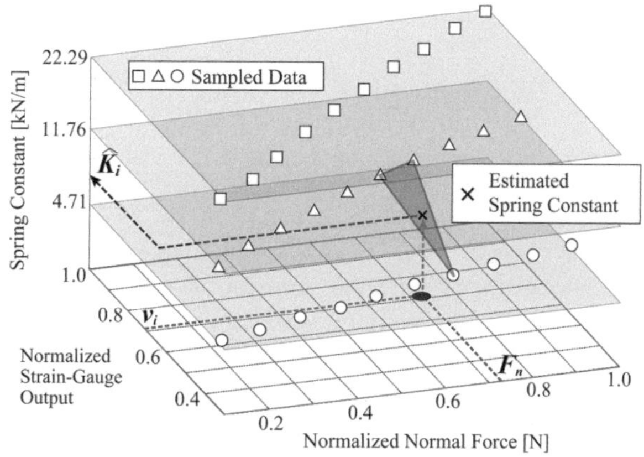

Fig. 19.4 Mapping model for estimating spring constants

Fig. 19.5 Validation of estimation method for spring constants

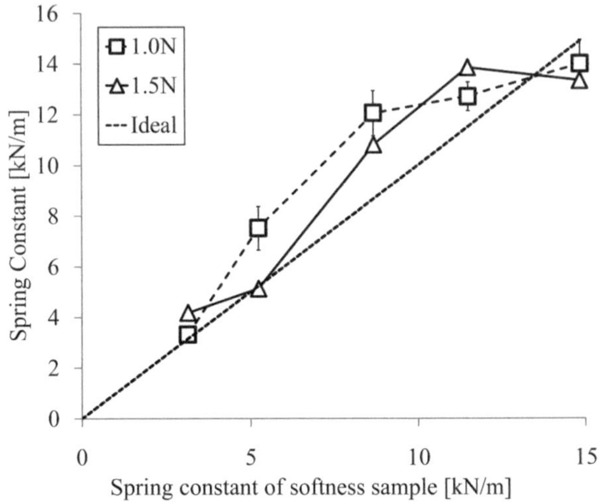

sensor. Figure 19.5 shows the result of the estimation. The figure shows the average and standard deviation of estimated spring constant when $F_n = 1.0$, and 1.5 N. The maximum error of the estimation was 44 %. This error seems to come from the limited number of the rubber samples for the mapping model. More gradual sampling in spring constants will improve the accuracy of the estimation.

For the remote transmission, when the lateral force is applied to the tactile sensor, this estimation method is not available because the estimation table was constructed from the data acquired without lateral forces. Therefore, the estimation of spring constant was performed only when $|F_t| < 0.3$ N. While the estimation is out of work, the adjacent estimated value is used as a current spring constant. In this condition, the slave arm usually estimates the spring constant only at the initial contact with target objects. This can be a limitation of our method, but the estimation is done in enough short time and robustly. The update rate of the estimation was 150 Hz.

19.5 Tactile Display Methods

19.5.1 Roughness Property

As described in Sect. 2.2, the tactile display method and the local model for the roughness property are the same as the method of the previous study [14]. This method had successfully transmitted the roughness sensations of the grating scales whose surface wavelengths varied from 0.8 mm to 2.0 mm by 0.2 mm. In this study, we extended it to multiple wavelengths of texture surface. When the surface wavelengths consist of $\lambda_1, \lambda_2, \ldots, \lambda_N$, the vibration at the finger in stroking the surface is approximated as

$$y(t) = \sum_{i=1}^{N} A_i \sin\left(2\pi \frac{x(t)}{\lambda_i}\right), \tag{19.3}$$

where x(t), A_i, N are the position of the finger on the surface, amplitude of skin deformation caused by the component of λ_i, and the number of wavelengths, respectively. λ_i and A_i are determined from the FFT from the outputs of the tactile sensor as described in Sect. 4.2. N affects the quality of roughness transmitted to the operator. In this paper, N was set to 3 because the preliminary evaluation showed little difference in quality of roughness when N is larger than 3. The electric voltage applied to the tactile stimulator is given by

$$y_r(t) = \alpha y(t), \tag{19.4}$$

where α is a scaling constant to transform the roughness stimuli to the voltage.

19.5.2 Friction Property

As described in Sect. 2.3, the local model and the tactile display method for the friction property are based on our proposed method [12]. In this study, we made several improvements on the waveforms. Figure 19.6 shows the concept of the display method. The method represents the friction sensation based on the stick-slip transition by controlling FA II activities using high-frequency vibration, which is sensitive to FA II.

As shown in Fig. 19.6, at the transition from the sticking phase to the slipping phase, vibrotactile stimuli is generated with a burst-like-waveform at the frequency of 400 Hz, at which FA II has more sensitive than the other mechanoreceptors. In the previous study [12], the envelope of vibration raised up in a step-wise manner. This

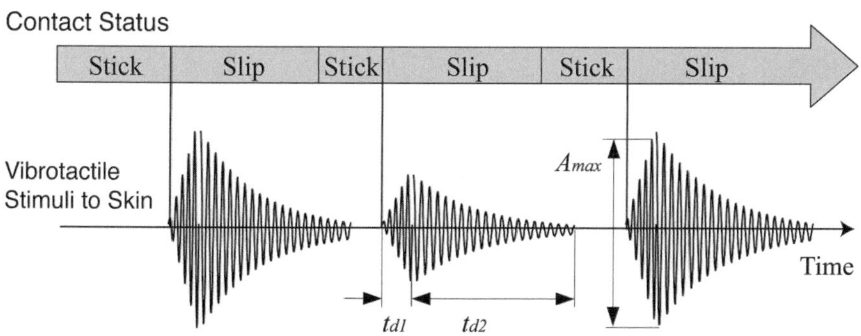

Fig. 19.6 Concept of the displaying method for friction sensation (Modified from [12])

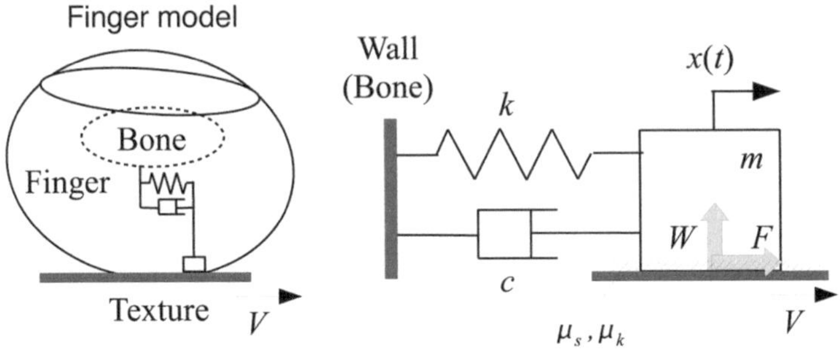

Fig. 19.7 Single DOF vibration model for simulating stick-slip transitions

may cause in unnatural feeling. In this study, we have the envelope of the vibration raised up gradually (t_{d1} period). After the peak, the envelope decay gradually on the assumption of the viscosity of the skin (t_{d2} period). The peak amplitude $A_{\max}$ depends on the elastic energy that is stored during the sticking phase.

Figure 19.7 illustrates 1-DOF vibration model with a coulomb friction to generate the approximated stick-slip transition at the finger skin as described in Sect. 2.3. The mass in the model is coupled with the wall (Bone). This model assumes the texture moves and the wall is connected to the ground. $x(t)$, V, and W are the displacement between the mass and wall, the moving velocity of the texture, and normal force applied to the skin, respectively. μ_s and μ_k are the static and kinetic friction coefficients, respectively. F is the lateral force applied to the skin, which depends on the contact status of the skin. When the skin is in sticking phase ($\dot{x}(t) = V$), F is equal to the static friction force. When the force exerted to the mass exceeds the maximum static friction force, the contact status changes from the sticking to slipping phase. This condition is described as

$$m\ddot{x} + \dot{x} + k\,(x - x_0) > \mu_s W, \tag{19.5}$$

where x_0 is the equilibrium position of the spring.

After the mass starts slipping, it vibrates if the system is not over damping. In the slipping phase, F is equal to kinetic friction force. Therefore, the equation of motion is given by

$$m\ddot{x} + \dot{x} + k\,(x - x_0) + \mathrm{sgn}\,(\dot{x} - V)\,\mu_k W = 0. \tag{19.6}$$

When the velocity $\dot{x}(t)$ becomes equal to V, the contact status changes into the sticking phase again. $\dot{x}(t)$ in the slipping phase is analytically given by solving (19.6).

The physical parameters in the model were determined by the observation of stick-slip phenomena in contact between the human finger skin and an acrylic plate. m, c, and k were 1×10^{-6} kg, 1×10^{-5} Ns/m, and 2 N/m, respectively.

The waveform applied to the tactile stimulator is

$$y_f(t) = A_f \sin(2\pi f_1 t), \tag{19.7}$$

where A_f is the amplitude of vibration and f_1 is the frequency in 400 Hz.

A_f is determined by

$$A_f = \begin{cases} \frac{A_{\max}}{t_{d1}}(t - t_s) & \text{if } (t - t_s) < t_{d1} \\ \frac{A_{\max}}{t_{d2}^2}(t - t_s - t_{d2})^2 & \text{if } t_{d1} < (t - t_s) < t_{d2}, \end{cases} \tag{19.8}$$

where A_{max} and t_s are the maximum amplitude of vibration, and the time at which the stick-to-slip transition occurred, respectively. t_{d1} and t_{d2} are constant values to determine the decay-period of the stimuli. They were set to 5 and 30 ms determined by the preliminary adjustments. A_{max} was defined to be proportional to the restoring force of the spring. A_{max} was given by

$$A_{\max} = \beta k (x - x_0), \tag{19.9}$$

where β is the scaling constant to transform the vibratory stimuli to the voltage applied to the stimulator.

19.5.3 Softness Property

As described in Sect. 2.3, the tactile display method for the softness property is based on stimulation of SA I because SA I is related to the perception of contact area that corresponds to the human perceived softness [19]. The proposed method controls the activity of SA I by changing amplitude of vibration with a low-frequency in less than 5 Hz. When the spring constant of the object becomes small, the contact area at the finger increases and the population of activated SA I is also expected to rise. In this study, we simply assume that the relationship between the spring constant and amount of SA I stimulation is inversely proportional. The actual relationship should be determined in the future. The waveform of the vibration for the softness sensation ys(t) is determined by,

$$y_s(t) = \frac{\gamma}{K} \sin(2\pi f_2 t), \tag{19.10}$$

where K is the spring constant of the object. f_2 is the vibration frequency ($f_2 = 5$ Hz). γ [10^4 V N/m] is a scaling constant to transform the vibrotactile stimuli to the voltage applied to the stimulating actuator.

19.5.4 Integration of All Properties

For integrating of all the tactile properties, the superposition of Eqs. 19.4, 19.7, and 19.10 was applied for the electric voltage $V(t)$ to the voice coil as follows,

$$V(t) = y_r(t) + y_f(t) + y_s(t), \tag{19.11}$$

where $y_r(t)$, $y_f(t)$, and $y_s(t)$, are the displayed signals for roughness, friction, and soft properties, respectively.

The magnitude of each perceived property is adjusted by the scaling constants, which are α, β, and γ in Eqs. 19.4, 19.7, and 19.10. To present natural feeling to the operator, we need to select the optimal scaling constants. However, we found that such adjustments depends on individual preference in the preliminary experiments with several participants. In this study, the scaling factors were determined by an expert, who has good tactile sensitivity. As a result, we determined that $\alpha = 0.4$, $\beta = 3.5$, and $\gamma = 2.0$.

19.6 Experiment: Material Discrimination

To evaluate the performance of the tactile transmission system, we conducted material discrimination experiments with the developed master-slave system. We also compared with the discrimination performance only with the force feedback without tactile feedbacks as for a reference. We confirm that the tactile feedback improves the discrimination performance.

19.6.1 Experimental Procedure

The target of the experiment is to evaluate the discrimination performances to identify five kinds of materials under two different feedback conditions: (1) both tactile and force feedback, and (2) force feedback only.

The participants explored five kinds of materials through the developed master-slave system. Five materials were (1) an embossed coated paper (unevenly rough and hard), (2) an acrylic grating scale with the surface wavelength of 1.5 mm (uniformly rough and hard), (3) a copying paper (flat and hard), (4) a fleece fabric about 3 mm thick (smooth and soft), and (5) a boa fabric about 7 mm thick (rough and soft).

The participants were instructed to answer the most similar material from the five materials, for which the same materials were located in front of the participants to explore them directly with their bare fingers during the whole experiment.

Ten trials were performed on each material. In total, 50 trials were performed in a random manner for each participant. After the first 25 trials, the participants took a 5-min break. The time limit was not set for the trials. The participants were not allowed to see the slave side and heard a pink noise over the headphones. The participants practiced before the main trials for approximately 3 min. The number of participants was three. All the participants were students and had experiences to use haptic interfaces.

19.6.2 Experimental Results

Table 19.1 shows the averaged answer ratings and the standard deviation. The vertical column shows the displayed materials and the horizontal rows are the answer ratings. The italic bold values represent the correct answer ratings. Values in the parentheses in each cell are the ratings for the case of the force feedback only.

The correct answer ratings with the tactile feedback showed the good potential over 70 % except for the paper (57 %).

Table 19.1 Answer ratings of the material discrimination experiment

		Answer ratings				
		Embossed paper	Grating scale	Copying paper	Fleece fabric	Boa fabric
Displayed materials	Embossed paper (unevenly rough and hard)	***0.90 ± 0.10*** (***0.23 ± 0.06***)	0.10 ± 0.1 (0.10 ± 0.10)	0.00 ± 0.00 (0.53 ± 0.15)	0.00 ± 0.00 (0.10 ± 0.00)	0.00 ± 0.00 (0.03 ± 0.06)
	Grating scale (uniformly rough and hard)	0.13 ± 0.23 (0.23 ± 0.12)	***0.73 ± 0.15*** (***0.27 ± 0.15***)	0.07 ± 0.06 (0.17 ± 0.06)	0.03 ± 0.06 (0.27 ± 0.15)	0.03 ± 0.06 (0.07 ± 0.06)
	Copying paper (flat and hard)	0.00 ± 0.00 (0.21 ± 0.10)	0.00 ± 0.00 (0.03 ± 0.06)	***0.57 ± 0.21*** (***0.31 ± 0.12***)	0.27 ± 0.21 (0.41 ± 0.20)	0.17 ± 0.21 (0.03 ± 0.06)
	Fleece fabric (smooth and soft)	0.00 ± 0.00 (0.10 ± 0.00)	0.00 ± 0.00 (0.13 ± 0.06)	0.20 ± 0.10 (0.17 ± 0.21)	***0.70 ± 0.10*** (***0.43 ± 0.21***)	0.10 ± 0.10 (0.17 ± 0.06)
	Boa fabric (rough and soft)	0.00 ± 0.00 (0.17 ± 0.06)	0.03 ± 0.06 (0.03 ± 0.06)	0.07 ± 0.06 (0.07 ± 0.12)	0.20 ± 0.10 (0.17 ± 0.06)	***0.70 ± 0.17*** (***0.57 ± 0.15***)

Values in the *parentheses* are the answer ratings for the case of the force feedback only. *Italic bold* values represent the correct answer ratings.

It is also indicated, however, that the correct answer ratings depended on the materials. The correct answer of the tactile feedback tends to be higher than those of the force feedback only. Especially, the correct answer ratings for the embossed paper and grating scale were approximately at the chance level in the case of the force feedback only, while the correct answer ratings with the tactile feedback were much higher. A two-way ANOVA was applied to the correct answer ratings with two factors: the types of the feedback and kinds of the material. The test showed that there was a significant difference in the types of the feedback $\left(F\,(1,20) = 40.5, p = 3.29 \times 10^{-6}\right)$, without significant interaction between the two factors.

From the results, we confirmed the feasibility of the proposed system.

19.7 Demonstration in Remote Environments

We demonstrated the tactile transmission system developed in the actual remote environment at the 14th Annual Conference of Virtual Reality Society of Japan (September 9th–11th, 2009, Waseda University). Figure 19.8 shows an overview of the demonstration. We could deliver multiple tactile properties from Tohoku

Fig. 19.8 The world's fist demonstration of a remote tactile transmission between Tokyo and Sendai

University in Sendai to the operators at Waseda University in Tokyo; the direct distance is 300 km. As for communication delay, the round trip time of PING command was 20–25 ms, which were relatively stable.

Many participants could discriminate touch of different materials (Fleece, embossed papers, fake leather, etc.) located at the remote site. The frequent comments are "the feelings are similar in some specific features," and "Discrimination of the prepared materials is possible, but the quality of the feeling is not so convincing to identify the material." Several participants pointed that their experiences were not similar to the direct touch, but similar to the feelings when they touch the materials through a rubber glove attached on the hand. That may come from the effect of the silicone rubber used for the tactile sensor. It is important to make the tactile sensor have the same mechanical properties as human skin properties in terms of the visco-elasticity and tribology.

We think this demonstration is one of the world's first remote tactile transmissions involving multiple tactile properties.

19.8 Conclusions

This chapter reported a tactile transmission system that transmits multiple tactile properties composed of roughness, softness, and friction sensations from remote environments. The author introduced our trial to integrate the tactile sensing and display systems through a robotic master-slave system. The author discussed the system design in terms of the selection of multiple tactile properties, synchronization of hand exploration and tactile feedbacks, and transmission method between measured and displayed signals. For the compensation of communication time-delay for the three tactile properties, the local models and the display methods involving the physical parameters and exploring movements of operators were constructed. This chapter also reported the tactile sensing methods to achieve detailed information in real time. To confirm the feasibility of the proposed system, this study conducted the material discrimination experiment. The results showed the potential of the system. The demonstration in a remote environment is also introduced.

Acknowledgement This work was supported in part by Grant-in-Aid for Scientific Research B (18760179) and a grant from the Ministry of Internal Affairs and Communications SCOPE (082102006).

References

1. Okamoto, S., Konyo, M., Saga, S., Tadokoro, S.: Detectability and perceptual consequences of delayed feedback in a vibrotactile texture display. IEEE Trans. Haptics **2**(2), 73–84 (2009)
2. Howe, R.D., Peine, W.J., Kontarinis, D.A., Son, J.S.: Remote palpation technology. IEEE Eng. Med. Biol. **14**(3), 318–323 (1995)

3. Ottermo, M.V.: Virtual palpation gripper, Ph.D. Thesis of Norwegian University of Science & Technology (2006)
4. Fan, F.E., Culjat, M.O., King, C., Franco, M.L., Boryk, R., Bisley, J.W.: A haptic feedback system for lower-limb prostheses. IEEE Trans. Neural Sys. Rehabil. Eng. **16**(3), 270–277 (2008)
5. Warwick, K., Gasson, M., Hutt, B., Goodhew, I., Kyberd, P., Andrews, B., Teddy, P., Shad, A.: The application of implant technology for cybernetic systems. Arch. Neurol. **60**(10), 1369–1373 (2003)
6. Shimojo, M., Suzuki, T., Namiki, A., Saito, T., Kunimoto, M., Makino, R., Ogawa, H., Ishikawa, M.: Development of a system for experiencing tactile sensation from a robot hand by electrically stimulating sensory nerve fiber. In: Proceedings of the 2003 IEEE International Conferene on Robotics & Automation, pp. 1264–1270 (2003)
7. Kontarinis, D.A., Howe, R.D.: Tactile display of vibratory information in teleoperation and virtual environments. Presence **4**(4), 387–402 (1995)
8. Dennerlein, J.T., Millman, P.A., Howe, R.D.: Vibrotactile feedback for industrial telemanipulators. In: Proceedings of the Sixth Annual Symposium on Haptic Interfaces for Virtual Environment and Teleoperator Systems, ASME International Mechanical Engineering Congress & Exposition, pp. 189–195 (1997)
9. Yamamoto, A., Nagasawa, S., Yamamoto, H., Higuchi, T.: Electrostatic tactile display with thin film slider and its application to tactile telepresentation system. IEEE Trans. Vis. Comput. Graph. **12**(2), 168–177 (2006)
10. Caldwell, D.G., Gosney, C.: Enhanced tactile feedback (tele-taction) using a multi-functional sensory system. In: Proceedings of the 1993 IEEE International Conference on Robotics and Automation, pp. 955–960 (1993)
11. Konyo, M., Tadokoro, S., Yoshida, A., Saiwaki, N.: A tactile synthesis method using multiple frequency vibrations for representing virtual touch. In: Proceedings of the 2005 IEEE/RSJ International Conference on Intelligent Robots and Systems, pp. 3965–3971 (2005)
12. Konyo, M., Yamada, H., Okamoto, S., Tadokoro, S.: Alternative display of friction represented by tactile stimulation without tangential force, haptics: perception. In: Manuel, F. (ed.) Devices and Scenarios, Lecture Notes in Computer Science, vol. 5024, pp. 619–629. Springer, Heidelberg (2008)
13. Mukaibo, Y., Shirado, H., Konyo, M., Maeno, T.: Development of a texture sensor emulating the tissue structure and perceptual mechanism of human fingers. In: Proceedings of the 2005 IEEE International Conference on Robotics and Automation, pp. 2565–2570 (2005)
14. Okamoto, S., Konyo, M., Maeno, T., Tadokoro, S.: Remote tactile transmission with time delay for robotic master-slave systems. Adv. Robot. **25**(9-10), 1271–1294 (2011)
15. Yamauchi, T., Okamoto, S., Konyo, M., Hidaka, Y., Maeno, T., Tadokoro, S.: Real-time remote transmission of multiple tactile properties through master-slave robot system. In: Proceedings of the 2010 IEEE International Conference on Robotics and Automation, pp. 1753–1760, Anchorage (2010)
16. Bejczy, A.K., Kim, W.S., Venema, S.C.: The phantom robot: predictive displays for teleoperation with time delay. In: Proceedings of the 1990 IEEE International Conference on Robotics and Automation, pp. 546–551 (1990)
17. Kotoku, T.: A predictive display with force feedback and its application to remote manipulation system with transmission time delay. In: Proceedings of the 1992 IEEE/RSJ International Conference on Intelligent Robots and Systems, pp. 239–246 (1992)
18. Shirado, H., Nonomura, Y., Maeno, T.: Realization of human skin-like texture by emulating surface shape pattern and elastic structure. In: Proceedings of the 2006 IEEE Symposium on Haptic Interfaces for Virtual Environment and Teleoperator Systems, pp. 295–296 (2006)
19. Okamoto, S., Konyo, M., Maeno, T., Tadokoro, S.: Transmission of tactile roughness through master-slave systems. In: Proceedings of the 2009 IEEE International Conference on Robotics and Automation, pp. 1467–1472 (2009)

20. Bicci, A., Scilingo, E.P., De Rossi, D.: Haptic discrimination of softness in teleoperation: the role of the contact area spread rate. IEEE Trans. Robot. Autom. **16**(5), 496–504 (2000)
21. Hidaka, Y., Shiokawa, Y., Tashiro, K., Maeno, T., Konyo, M., Yamauchi, T.: Development of an elastic tactile sensor emulating human fingers for tele-presentation systems, In: Proceedings of the IEEE SENSORS 2009 Conference, pp. 1919–1922 (2009)

Index

A
Acoustic radiation pressure, 122–126, 137
Activating function, 80–84
Active sensing, 160–161
Aerial interface, 121, 122, 134
Airborne ultrasound tactile display (AUTD), 121–137
Airplane, 256–259
Assistive technology, 226
Augmented reality, 250–256, 263
AUTD. *See* Airborne ultrasound tactile display (AUTD)

B
Bidirectionality, 232–234, 236, 237, 239, 241, 275–282
Braille display, 73–74

C
Chronaxie, 87
Coldness, 5, 7, 11–12, 16
Content design, 275, 279–281

D
Diabetes mellitus, 74, 75

E
Electrocutaneous display, 79
Electro-tactile display, 79–95
Emotion, 265–273

F
fMRI. *See* Functional magnetic resonance imaging (fMRI)
Force perception, 39–52
Friction, 4, 5, 7, 10, 16
Friction control, 97, 98, 104, 105
Functional magnetic resonance imaging (fMRI), 26, 27

H
Hanger reflex, 267–268, 270
Haptic aids, 221–229
Haptic content, 185–199
Haptic experience, 187–198
Haptic illusion, 111, 112
Haptic media, 275, 279–282
Haptic sensing, 140, 146, 148
Haptic signal processing, 150–154
Hardness, 5, 7, 9, 16
Higher-level tactile perception, 57–59, 61, 75
Human interface, 168, 169, 172

I
Image processing, 62, 63
Inter-world interface, 277–279, 281
Intrinsic force sensing, 139–154

L
Lateral force, 111–119

H. Kajimoto et al. (eds.), *Pervasive Haptics*, DOI 10.1007/978-4-431-55772-2

M
Massage, 202–211, 218, 219
Master-slave system, 285–288, 290, 291, 298, 301
Mechanical principles, 232, 233, 235, 236, 238–240, 243, 244
Micro vibration, 57–76
Mobile navigation, 260–261
Motion, 266–271, 273
Musculoskeletal model, 40, 41, 43, 44, 52

N
Navigation, 250, 257, 259–263, 269, 270
Neuroimaging, 22, 24–27, 29, 31

O
Object recognition, 21–33
Optical lever, 159

P
Perceptual dimension, 3–16
Phased array, 122–125, 137
Piloerection, 272, 273
Pilot, 250, 256–259, 263
Pseudo-tactile sensation, 72–73

R
Remote operation, 285–301
Rheobase, 86, 87
Roughness, 3–10, 16

S
Self-reference, 232, 233, 236, 237, 239, 243
Sensorimotor function, 41, 48–52
Sensory language, 201–219
Shape memory alloy, 57–76
Softness, 4, 5, 7, 9, 16
Steering wheel operation, 41–48
Strength-duration curve, 84–88, 95

T
Tactile display, 97–109, 111–119, 121–137, 221, 223, 225–228, 286–288, 290, 294–298, 301
Tactile enhancement, 235, 236, 238
Tactile illusion, 232, 243, 244
Tactile information, 29
Tactile interaction, 57–76
Tactile properties, 285–301
Tactile recognition, 22
Tactile score, 201–219
Tactile sensing, 57–72, 74, 139–154, 203, 204, 206–213, 217–219
Tactile sensor, 157–165, 286–295, 301
Tactile transmission, 285–290, 298, 300, 301
TECHTILE, 185–199
Texture, 3–6, 12–14, 58, 62–70, 73
Thermal display, 167–180
Thermal sensor, 167–180
Total-internal-reflection, 158
Touch, 21, 22, 24, 25, 27–33
Touchscreen, 111, 112, 114, 119

U
Ultrasonic vibration, 98, 100, 103, 104, 108, 109

V
Vection field, 270, 271
Vibration in solid, 98
Vibratory stimuli, 194–196
Vibrotactile navigation, 250, 260–261, 263
Virtual reality, 167
Visually impaired, 221–229
Visuo-tactile interaction, 28–32

W
Warmness, 4, 5, 7, 11–12, 16
Wearable assistive device, 40, 41, 48–52

MIX
Papier aus verantwortungsvollen Quellen
Paper from responsible sources
FSC® C105338

If you have any concerns about our products,
you can contact us on
ProductSafety@springernature.com

In case Publisher is established outside the EU,
the EU authorized representative is:
Springer Nature Customer Service Center GmbH
Europaplatz 3, 69115 Heidelberg, Germany

Printed by Libri Plureos GmbH
in Hamburg, Germany